Intelligent Wireless Sensor Networks and the Internet of Things

The edited book Intelligent Wireless Sensor Networks and the Internet of Things: Algorithms, Methodologies and Applications is intended to discuss the progression of recent as well as future generation technologies for WSNs and IoTs applications through Artificial Intelligence (AI), Machine Learning (ML), and Deep Learning (DL). In general, computing time is obviously increased when the massive data is required from sensor nodes in WSN's.the novel technologies such as 5G and 6G provides enough bandwidth for large data transmissions, however, unbalanced links faces the novel constraints on the geographical topology of the sensor networks. Above and beyond, data transmission congestion and data queue still happen in the WSNs.

This book:

- Addresses the complete functional framework workflow in WSN and IoT domains using AI, ML, and DL models.
- Explores basic and high-level concepts of WSN security, and routing protocols, thus serving as a manual for those in the research field as the beginners to understand both basic and advanced aspects sensors, IoT with ML & DL applications in real-world related technology.
- Based on the latest technologies such as 5G, 6G and covering the major challenges, issues, and advances of protocols, and applications in wireless system.
- Explores intelligent route discovering, identification of research problems and its implications to the real world.
- Explains concepts of IoT communication protocols, intelligent sensors, statistics and exploratory data analytics, computational intelligence, machine learning, and Deep learning algorithms for betterment of the smarter humanity.
- Explores intelligent data processing, deep learning frameworks, and multi-agent systems in IoT-enabled WSN system.
- This book demonstrates and discovers the objectives, goals, challenges, and related solutions in advanced AI, ML, and DL approaches.

This book is for graduate students and academic researchers in the fields of electrical engineering, electronics and communication engineering, computer engineering, and information technology.

Wireless Communications and Networking Technologies: Classifications, Advancement and Applications
Series Editor: D.K. Lobiyal, R.S. Rao and Vishal Jain

The series addresses different algorithms, architecture, standards and protocols, tools and methodologies which could be beneficial in implementing next generation mobile network for the communication. Aimed at senior undergraduate students, graduate students, academic researchers and professionals, the proposed series will focus on the fundamentals and advances of wireless communication and networking, and their such as mobile ad-hoc network (MANET), wireless sensor network (WSN), wireless mess network (WMN), vehicular ad-hoc networks (VANET), vehicular cloud network (VCN), vehicular sensor network (VSN) reliable cooperative network (RCN), mobile opportunistic network (MON), delay tolerant networks (DTN), flying ad-hoc network (FANET) and wireless body sensor network (WBSN).

Wireless Communication: Advancements and Challenges
Prashant Ranjan, Ram Shringar Rao, Krishna Kumar and Pankaj Sharma

Wireless Communication with Artificial Intelligence: Emerging Trends and Applications
Anuj Singal, Sandeep Kumar, Sajjan Singh and Ashish Kr. Luhach

Computational Intelligent Security in Wireless Communications
Suhel Ahmad Khan, Rajeev Kumar, Omprakash Kaiwartya, Raees Ahmad Khan and Mohammad Faisal

Networking Technologies in Smart Healthcare: Innovations and Analytical Approaches
Pooja Singh, Omprakash Kaiwartya, Nidhi Sindhwani, Vishal Jain and Rohit Anand

Artificial Intelligence in Cyber Physical Systems: Principles and Applications
Anil Kumar Sagar, Parma Nand, Neetesh Kumar, Sanjoy Das and Subrata Sahana

Intelligent Wireless Sensor Networks and the Internet of Things: Algorithms, Methodologies, and Applications
Bhanu Chander, Anoop BN, Koppala Guravaiah and G. Kumaravelan

For more information about this series, please visit: https://www.routledge.com/Wireless%20Communications%20and%20Networking%20Technologies/book-series/WCANT

Intelligent Wireless Sensor Networks and the Internet of Things

Algorithms, Methodologies, and Applications

Edited by

Bhanu Chander

Anoop BN

Koppala Guravaiah

G. Kumaravelan

CRC Press

Taylor & Francis Group

Boca Raton London New York

CRC Press is an imprint of the
Taylor & Francis Group, an **informa** business

Designed cover image: shutterstock

First edition published 2024
by CRC Press
2385 NW Executive Center Drive, Suite 320, Boca Raton FL 33431

and by CRC Press
4 Park Square, Milton Park, Abingdon, Oxon, OX14 4RN

CRC Press is an imprint of Taylor & Francis Group, LLC

ISBN: 978-1-032-45951-6 (hbk)
ISBN: 978-1-032-76497-9 (pbk)
ISBN: 978-1-003-47452-4 (ebk)

DOI: 10.1201/9781003474524

Typeset in Sabon
by SPi Technologies India Pvt Ltd (Straive)

Contents

Preface

Recent developments in wireless communications and computer networks now have a new computation and communication technology as wireless sensor networks (WSNs) and the Internet of Things (IoTs). Design and development of intelligent, adopted, and automated models for the progress of WSNs and IoTs performances is one of the foremost challenging issues. Due to the continuous increase in IoT devices, the massive amount of network data needs to be stored securely and processed quickly for real-world appliances. These requirements attract the artificial intelligence (AI), machine learning (ML), and deep learning (DL) approaches to implement effective and efficient systems. Most real-time IoT–related schemes like drones, smart healthcare devices, automatic driving and traffic lights, and security robots extremely depend on ML– and DL–based technologies, which are boosted with the augmentation of hardware skills. However, both ML/DL algorithm-based appliances still have problems with accurate precision and time requirements.

Nowadays, with digital transmission, the data are ever increasing from different devices from various networks. This increase goes into different shapes like volume, variety, and veracity, and its challenging task is to extract the user perceptions and make an appropriate decision. The existing theory and practices of AI/ML are not so much help with high-volume datasets. But the recent advances in DL promise benefits for decision-making with multicriteria approaches. AI/ML/DL approaches with IoT and wearable sensor nodes in the healthcare sector have been active and have shown brilliant improvement in disease diagnosis and tracking of human activities. With the tremendous upcoming technologies in wearable embedded devices, more appliances are being explored and industrialised. As the technology moves forward, WSNs are altered in tremendous shades. Nevertheless, the implementation of quantum technology in WSN has not yet been systematically studied. In addition, cryptography and blockchain bring significant advancements in securing the sensed data; in particular, blockchain technology has uses in IoT due to its most straightforward and secure computation complexities. Adopting existing solutions is also required to develop and improve communication in 5G and 6G technologies.

Hence, this book aims to capture all the multifaceted natures of WSNs, IoT, AI, ML, and DL in one single place. From the table of contents, it is clear that each chapter provides dissimilar issues and challenges, and suggests possible solutions and future research directions for WSN and IoT in terms of AI–based approaches. Due to these reasons, the authors are interested in introducing different kinds of efficient and emerging concepts with the help of worldwide researchers. These concepts will be helpful to the audience in solving their real-time problems using WSN and IoT. Researchers, network managers, academicians, industry engineers in computer science and artificial intelligence, IT professionals, and cybersecurity experts benefit from the various applications included in the book. Mainly, this book helps as a front-runner for undergraduate and post-graduate students to graph all the advanced AI–based healthcare, security, and privacy approaches in the research of WSN and IoT scenarios.

Chapter 1: Discusses the sensor nodes energy consumption and optimal load balancing with a DL approach. Authors clearly describe different data reduction approaches in WSNs and the addressed various challenges of maintaining these networks.

Chapter 2: Authors discoursed about a wide range of WSN routing protocols which are used in body area sensor networks, underwater sensor networks, underground sensor networks, wireless multi-media networks, and unmanned aerial networks based on the AI–based techniques. In addition, the chapter debates different properties of networks that influence the routing, along with giving some research directions.

Chapter 3: This chapter provides information about MANET, especially regarding energy use, resource utilisation, and packet delivery issues. Authors presented the Crescent Zone–based Location Aided Routing (CZLAR) protocol that integrates the zone-routing technique with location-assisted node selection methods. In both low- and high-density networks, the CZLAR protocol exhibits efficiency by resulting in a greater packet delivery ratio. According to the simulation results, CZLAR significantly improves packet delivery ratio compared to the current protocol while significantly reducing energy consumption.

Chapter 4: This chapter thoroughly discuss the widespread deployment of IoT technologies and how they make the human lifestyle more intelligent by offering flexibility and facility in a range of day-to-day applications, namely smart city, smart meter, smart home, intelligent transport, smart healthcare, and many more real-time applications.

Chapter 5: Due to enormous bandwidth, frequency, and antennas used in 5G communications, these services now offer precise localisation, sensing, and communication. 6G technology promises to deploy cloud applications while providing more bandwidth and it will carry on this trend by creating even higher frequencies and antennas. The problems caused by wireless communication are discussed in this chapter and also effectively addressed with IOT technology, blockchain, intelligent systems, and algorithms.

Chapter 6: Conferred about in-network data aggregation of WSN's where ML aids in enhancing sensor network performance by monitoring a dynamic environment that changes quickly. A mechanism called the routing protocol is used to choose a suitable route for data to go from source to destination. When choosing the route, which depends on the network type, transmit traits, and operational metrics, the process flows into a number of problems. Declarative syntax access to information generated by numerous sensor nodes is made possible by distributed query processing in networks of wireless sensors.

Chapter 7: Presents industry 4.0 as a database-driven manufacturing structure that is profoundly altering how people live and work. The general public is nonetheless hopeful about the prospects it might bring for sustainability and the prospects of quality labour in the globalised digital economy. Industry 4.0 is being lauded as the answer to enhancing efficiency, fostering economic progress, and guaranteeing the survival of production and manufacturing companies. Additionally, it tries to increase the industrial systems' durability, tractability, and flexibility. Particularly, chapter authors discuss how Industry 4.0 deals with digital transformation with numerous technologies in each sector of business where the healthcare sector is not an exemption. Healthcare 4.0 is the result of Industry 4.0, and this term is also known as Hospital 4.0.

Chapter 8: The major objective of this chapter is showing how healthcare companies provide good service and keep track of their supplies. The supply chain of the pharmaceutical industry is broken up by drug shortages and poor inventory management, which costs money and worsens healthcare services. Authors explain the planning for comparable events must be prioritised and discuss how the pharmaceutical supply chain plays a crucial role in ensuring appropriate use of medications in the right circumstances and at the appropriate times.

Chapter 9: In the era of rapid technological advancements, the sharing and exchange of data has become increasingly prevalent. However, the challenges of effectively managing and sharing large volumes of data have increased. Existing methods for sharing files frequently rely on centralised servers, which can be subject to data loss, censorship, and hacking. In this chapter authors designed a decentralised file-sharing system based on interplanetary file system (IPFS) and blockchain technology that is safe, dependable, and scalable. To address this issue, in this work we present a potential solution in the form of a data sharing platform that incorporates blockchain and IPFS technologies to provide a secure, efficient, and distributed method for sharing files. This platform involves a blockchain alliance that utilises a consensus mechanism, encryption/decryption, and quick recovery to facilitate seamless data sharing.

Chapter 10: The chapter addresses an advanced intelligent machine learning algorithm incorporating promising signal analysis technique, namely, a flexible analytic wavelet transform approach to detect driver fatigue using neurophysiological brain electroencephalogram (EEG) signals. The performance of the proposed method is examined in terms of various parameters including accuracy, F-score, AUC and kappa. It achieves optimum individual accuracies in the range of 96–98% in classifying the FATIGUE and REST states with F-score of 97.50%.

Chapter 11: The chapter presents the development of an advanced learning algorithm employing discriminant correlation analysis–based fusion framework (DCABFF). The DCABFF aims to enable interaction of multiple sets of modality data in two different domains and to extract low-order features by combining them using standard statistical technique that could indicate the inherent characteristics of underlying phenomena. The main focus of the algorithm is to explore the efficacy of the low order combined features extracted from high-order input searched space.

Chapter 12: In this chapter, the authors chose one of the most important fields in life, electronic health care. The emergency system for the elderly and heart disease was implemented through the cooperation between WSN technology and IOT devices to make the emergency system use the heart sensor and IOT devices to send short message services connected to the Global Positioning System to locate the patient remotely by clicking on the link with the availability of Wi-Fi or 3G access to save the patient's life in the shortest time and in the most assured way. This technology is not only used in healthcare specialty but also in different disciplines such as sports, entertainment, military, and so on.

Chapter 13: The primary goal of this chapter is to explain use of WSNs in IoT systems. A WSN comprises of distributed gadgets with sensing elements that are utilised for supervising atmospheric and physical conditions.

Chapter 14: Finally, this chapter provides information related to smart appliances stored in cloud storage. Various security attacks and their identification with ML and DL approaches discussed.

We are sincerely thankful to the authors for their contributions. Our gratitude is also extended to many anonymous referees involved in the revision and acceptance process of the submitted manuscripts. It would not have been possible to reach this publication quality without the contributions of the referees. The editors are sincerely thankful to the series editors **Prof. Vishal Jain** and **Prof. Jyotir Chatterjee** for providing constructive input and allowing the opportunity to edit this important book. As the editors, we hope this book will stimulate further research in medical image processing, theories, and applications and utilise them in real-world wireless applications. Special thanks go to our publisher, CRC Press/Taylor &

Francis Group. We hope that this book will present promising ideas and outstanding research results.

Bhanu Chander
Dept of CSE, IIIT Kottayam, Kerala India
Guravaiah
Dept of CSE, IIIT Kottayam, Kerala India
Kumaravelan Gopalakrishnan
Dept of CSE, Pondicherry University, Pondicherry, India
Anoop BN
Post Doc University of Texas Health Science
Centre at San Antonio, TX, USA

About the editors

Dr. Bhanu Chander

Assistant Professor, Department of Computer Science and Engineering, Indian Institute of Information Technology (IIIT-K)

Dr. Bhanu Chander is Assistant Professor at Indian Institute of Information Technology (IIIT-K), Pala, Kerala, India. Dr. Chander graduated from Acharya Nagarjuna University, AP, in 2013 and earned a post-graduate degree from the Central University of Rajasthan, Rajasthan in 2016. He received a PhD on Machine Learning in Wireless Sensor Networks for sensor data classification from Pondicherry University in 2022. He has published six articles in peer-reviewed journals, five international conferences, and eight book chapters (Elsevier, Wiley, CRC and Springer). Presently his main interest areas include wireless sensor networks, IoT and healthcare, cryptography, machine learning, deep learning.

Dr. Anoop BN

Postdoctoral Research Fellow, University of Texas Health Science Centre.

Dr. Anoop is Postdoctoral Research Fellow at Glenn Biggs Institute for Alzheimer's & Neurodegenerative Diseases, University of Texas Health Science Centre, US. He received an MTech on Signal Processing from National Institute of Technology, Calicut, 2013 and a PhD on Development of Automated Methods for Retinal Optical Coherence Tomography Image Analysis from National Institute of Technology, Surathkal, India in 2021. He published six articles in peer-reviewed journals, and two book chapters. Presently his main areas of interest include medical image processing, deep learning, GANs, and auto-encoders.

Dr. Koppala Guravaiah

Assistant Professor, Department of Computer Science and Engineering, Indian Institute of Information Technology

Dr. Koppala Guravaiah is Assistant Professor at Indian Institute of Information Technology (IIIT-K), Pala, Kerala, India. Dr. Guravaiah completed his PhD on the topic of performance of routing protocols in wireless sensor networks using river formation dynamics from National Institute of Technology Tiruchirappalli, India. His research interests include IoT, software defined networks, wireless ad hoc, and sensor network routing protocols and their applications. He has published five journal articles, eight conferences, and two book chapters in the areas mentioned above by his name.

Dr. G. Kumaravelan

Assistant Professor and Head (i/c), Department of Computer Science, School of Engineering & Technology, Pondicherry University

Dr. G. Kumaravelan is Assistant Professor in the Department of Computer Science and Engineering at Pondicherry University, Karaikal Campus, Pondicherry, India. He received a PhDin Communications and Information Systems from Bharathidasan University in 2008. He has more than 12 years of teaching and research experience and has published 20 articles in peer-reviewed journals and international conferences. He is an active reviewer in various international conferences and peer-reviewed journals. His research interests include the internet of things, cloud computing, big data analytics, wireless communications, and networking.

List of contributors

T. Abirami
Associate Professor, Department of Information Technology
Kongu Engineering College
Perundurai, Erode, India

Nashwan Ghaleb Al-Thobhani
Computer Network Engineering, Technology Department
Sana'a Community College
Sana'a, Yemen

S. Anbukkarasi
Assistant Professor, Department of Computer Science and Engineering
SRM University
Tamil Nadu, India

G. Bhanu Chander
Assistant Professor, Department of Cyber Security
IIIT Kottayam
Kottayam, India

N. Bhavatharini
Department of Artificial Intelligence
Kongu Engineering College
Perundurai, Erode, Tamil Nadu, India

P. Bora
Department of Physics
Mayang Anchalik College
Marigaon, Assam, India

J. Deka
Department of Physics
Cotton University
Guwahati, Assam, India

M. N. Dharshini
Student, Department of Artificial Intelligence
Kongu Engineering College
Erode-638060, India

Koppala Guravaiah
Assistant Professor, Department of CSE
IIIT Kottayam
Kottayam, India

A. Hazarika
Department of Physics
Mayang Anchalik College
Marigaon, Assam, India

S. Hemalatha
Assistant Professor, Department of Computer Science and Design
Kongu Engineering College
Tamil Nadu, India

Swati Jain
Department of Commerce, School of Business and Commerce
Manipal University
Jaipur, India

P. Jayadharshini
Assistant Professor, Department of Artificial Intelligence
Kongu Engineering College
Erode-638060, India

A. Jayanthiladevi
Professor, Institute of Computer Science and Information Science
Srinivas University
Mangalore, Karnataka, India

V. N. Jinesh
School of CSE & IS
Presidency University
Bangalore, India

K. Jothimani
Assistant Professor, Department of Computer Science and Design
Kongu Engineering College
Tamil Nadu, India

G. K. Kamalam
Associate Professor, Kongu Engineering College
Tamil Nadu, India

G. V. Kamalam
Associate Professor, Department of Information Technology
Kongu Engineering College
Perundurai, Erode, Tamil Nadu, India

P. Karthikeyan
National Chung Cheng University
Chiayi, Taiwan

C. J. Kumar
Cotton University
Guwahati, India

Radhakrishnan Maivizhi
Department of Computational Intelligence
SRM Institute of Science and Technology
Kattankulathur, India

B. Patir
Department of Computer Science
Cotton University
Guwahati, Assam, India

Galiveeti Poornima
School of CSE & IS
Presidency University
Bangalore, India

G. Rithanya
Department of Artificial Intelligence
Kongu Engineering College
Perundurai, Erode, Tamil Nadu, India

Margani Rohith
Department of Computer Science and Engineering
Indian Institute of Information Technology
Kottayam-686635, India

J. Ruthranayaki
Student, Department of Artificial Intelligence
Kongu Engineering College
Erode-638060, India

Vidya Sagar
Research Scholar, Department of CSE
IIIT Kottayam
Kottayam, India

N. Sai Chandu
Department of Computer Science and Engineering
Indian Institute of Information Technology
Kottayam-686635, India

A. Saikia
Department of Physics
Cotton University
Guwahati, India

S. Santhiya
Assistant Professor, Department of Artificial Intelligence
Kongu Engineering College
Erode-638060, India

Jeevesh Sharma
Manipal University Jaipur
Jaipur, India

C. Sharmila
Assistant Professor, Department of CSE
Kongu Engineering College
Erode-638060, India

Radhika Sreedharan
Department of Computer Science and Engineering
Presidency University
Bangalore, India

S. Subashini
Assistant Professor, Kongu Engineering College
Tamilnadu, India

Jamil Sultan
Faculty of Engineering & IT, University of Modern Sciences
Sana'a, Yemen.

Mudavath Sai Teja
Department of Computer Science and Engineering
Indian Institute of Information Technology
Kottayam-686635, India

Mahesh Vanam
Department of Computer Science and Engineering
Indian Institute of Information Technology
Kottayam-686635, India

P. Vanitha
Assistant Professor, Kongu Engineering College
Tamil Nadu, India

J. Viji Gripsy
Post-Doctoral Research Fellow, Institute of Computer Science and Information Science
Srinivas University
Mangalore, India

J. Vinay
Cloud Operations, US

Palanichamy Yogesh
Department of Information Science and Technology
CEG, Anna University, Chennai, India

Key features

1. Addresses the complete functional framework workflow in WSN and IoT domains using AI, ML, and DL models.
2. Explores basic and high-level concepts of WSN security and routing protocols, thus serving as a manual for those in the research field as the beginners to understand both basic and advanced aspects sensors, IoT with ML and DL applications in real-world related technology.
3. Based on the latest technologies such as 5G and 6G and covers the major challenges, issues, and advances of protocols, and applications in wireless system.
4. Explores intelligent route discovering and identification of research problems and its implications to the real world.
5. Explains concepts of IoT communication protocols, intelligent sensors, statistics and exploratory data analytics, computational intelligence, machine learning, and deep learning algorithms for betterment of the smarter humanity.
6. Explores intelligent data processing, deep learning frameworks, and multi-agent systems in IoT–enabled WSN systems.
7. This book demonstrates and discovers the objectives, goals, challenges, and related solutions in advanced AI, ML, and DL approaches.
8. Readers improve and gain interest in the association between AI–based WSN and IoT with network complexities and the security difficulties.

About the book

The edited book *Intelligent Wireless Sensor Networks and the Internet of Things: Algorithms, Methodologies and Applications*, is intended to discuss the progression of recent as well as future generation technologies for WSNs and IoTs applications through artificial intelligence (AI), machine learning (ML), and deep learning (DL). Recent developments in the field of wireless communications and computer networks now have a new computation and communication technology with the help of wireless sensor networks (WSN) and Internet of Things (IoT). Design and development of intelligent, adopted, and automated models to the progress of WSNs and IoTs performance is measured as one of the foremost challenging issues. Due to the continuous increase in IoT devices, the massive amount of network data needs to be stored securely and processed quickly for real-world appliances. These requirements attract the ML, and DL approaches to implement effective and efficient systems. In addition, cryptography and blockchain bring significant advancements in securing the sensed data; in particular, blockchain technology has uses in IoT due to its most straightforward and secure computation complexities. Adopting existing solutions is also required to develop and improve communication in 5G and 6G technologies.

The new technologies in various domains and ultra-low latency for huge data transfer intensely expand the sensing application field. Mobile-based WSNs accurately monitoring the target fields in the combatant and non-combatant military area for security reasons. In particular, the sensor nodes sense whether the mistrustful target intrudes into the ground for target sensing and launch a communication connection with the sink node. Controlling data is issued for scheduling potential sensors to execute a sensing task on the sink node. Hence, the sink node is calculated as a controller with appropriate computing capacity for the sensor nodes. But still the optimal sensing time converts low owing to high latency reply in the outdated centralised manner. In general, computing time is obviously increased when the massive data is required from sensor nodes in WSNs. Novel technologies such as 5G and 6G provide enough bandwidth for large data transmissions, however, unbalanced links face the novel constraints on the geographical topology of the sensor networks. Above and beyond, data transmission congestion and data queue still happen in the WSNs. As a consequence, sensing time moves backwards in turn, missing the best sensing opportunity. Most agricultural research centres have started implementing IoT and WSNs to monitor a wide range of farm service stations in recent times. There are momentous prospects for agriculture supervision through the active employment of AI–based ML and DL approaches. Due to climate change and water crisis, farmers go through many troubles such as crop flattening, soil erosion, drought, and so on. These problems can be quickly suppressed by using AI, ML, and DL–based farming systems. IoT devices can collect health care data, including blood pressure, sugar levels, oxygen, and weight. Data is stored online and can be accessed anytime by a physician. IoT automates the workflow by allowing the provision of adequate health care services to the patients.

Due to these reasons, the editors are interested in introducing different kinds of efficient and emerging concepts with the help of worldwide researchers. These concepts will be helpful to the audience to solve their real-time problems using IoT. This proposed book will help researchers and practitioners to understand the design architecture of different problems of healthcare with AI, ML, and DL algorithms through IoT.

Chapter 1

Energy-efficient WSN with deep learning using dispersed data mining strategy-based LSTM

S. Santhiya, C. Sharmila, P. Jayadharshini, M. N. Dharshini, and J. Ruthranayaki
Kongu Engineering College, Erode, India

1.1 INTRODUCTION ON IMPROVING THE LIFETIME OF A WSN

The primary intent of this investigation is to present information on emerging functionalities of machine intelligence methods in connection with various WSN problems. We highlight the significant challenges associated with data collection (DC), aggregation, and dissemination in WSNs and offer a complete overview of recent research that employs various AI strategies to accomplish specific WSN goals between 2010 and 2021. Finally, in order to help the research community choose the ways most suitable to meet various WSN difficulties, we provide a broad review or rather contrast of the various AI methodologies utilized in WSNs. We also go over the benefits of using various AI techniques to address either of these WSN–related problems. The vast AI techniques now in use are examined throughout this study, along with how they are used in WSNs to handle the DC, aggregation, or even conveyance issues. Researchers concentrate on providing a brief synopsis of these WSN challenges and researching alternative AI techniques. Then, we show how applying AI approaches to these issues can enhance the performance of WSNs in regard to packet transfer rate, WSN lifetime, energy efficiency, and several variables. With this study, the author will have a good understanding of the challenging problems that WSN faces and the efficiency of AI techniques in addressing them. Others, though, focused on resolving some of the challenges WSN encountered. Researchers look at a number of AI methods that let us research fresh ideas for improving the efficacy of the WSN and solving present WSN problems. Researchers also discuss a number of optimization techniques that address various DC, aggregation, and dissemination-related problems in WSNs. The primary concern in our assessment is on WSN AI schemes, even if we explore various aspects of AI in resolving these WSN issues.

DOI: 10.1201/9781003474524-1

1.2 TRAINING LSTM-RNNS—THE HYBRID LEARNING APPROACH

When managing time mode information of traffic flow, the conventional forecasting models primarily have two shortcomings: (1) Conventional approaches, particularly traditional RNNs, are difficult to train if the traffic flow series has lengthy time lags, which means that if the time window size n in Eq. (1) is too big, traditional recurrent neural networks (RNNs) will perform poorly. (2) It is challenging to determine the ideal time frame size n since the association between traffic flows at various time periods is influenced by a variety of intricate elements, including weather, speed, and unforeseen incidents [1]. While preserving the CEC in LSTM memory block cells (Figure 1.1), two research methodologies are still used: real-time recurrent learning (RTRL) inspires system components before and including cells, and backpropagation through time (BPTT) to train systemic accessories after units. The latter units are compatible with RTRL because certain derivatives in part (linked to statehood for the cell) must always be figured out at each phase regardless of the likelihood a desired score is specified at that step or not. For the time being, researchers only shorten the flux to make room for the extra recurrent connections and let the gradient of the cell converse over time.

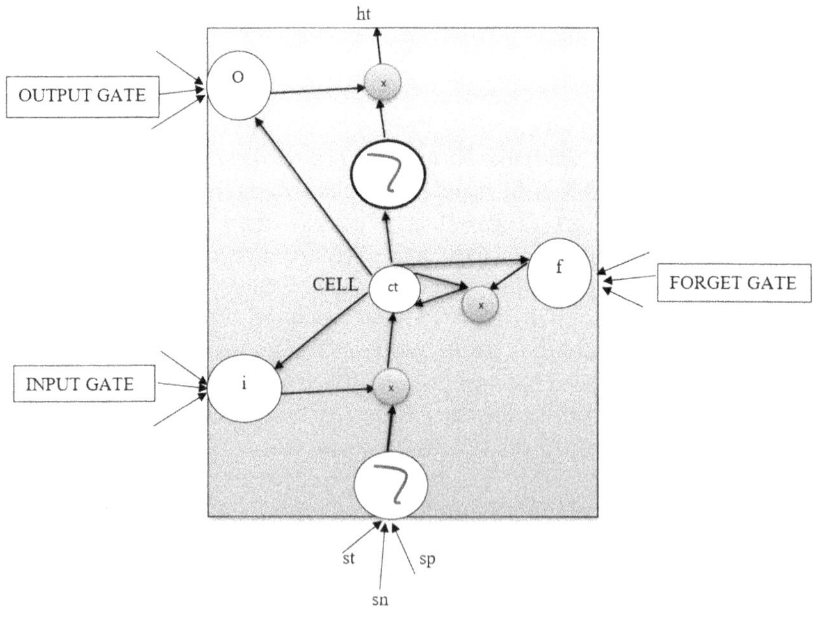

Figure 1.1 LSTM structure.

1.2.1 Using network optimization techniques to create an efficient sustainable network

The suggested technologies are designed to deliver a wireless body area network that is highly secure, sustainable, and energy efficient. By applying two protocols on the data and scheduling with the help of the smart grid and renewable energy systems, the goal is attained. A wireless body area network's (WBAN) general architecture is depicted in Figure 1.2. To ensure intra–WBAN communication via the smart grid and renewable energy systems, the sensors are connected to the patients and each can interact with the body coordinator. The personal device assistant (PDA), through which the coordinator communicates, sends the data gathered to the medical server for recording. A copy of the data is given to the patient's doctor. The doctor can access earlier records on the medical server and make decisions based on them. To change a prescription or provide other remote support, the doctor can contact the patient via PDA. When there is urgency, that doctor might receive a message instructing them to transport the patient in an ambulance. A WBAN network may extend its coverage by using wireless private area networks technologies as gateways. Via gateway devices, wearable technology can be linked to the internet [2]. Because medical personnel may access patient data online, the location of the patient is immaterial. Precisely how much power that network uses can be found by using a mathematical function, such as:

$$E = f(x)$$

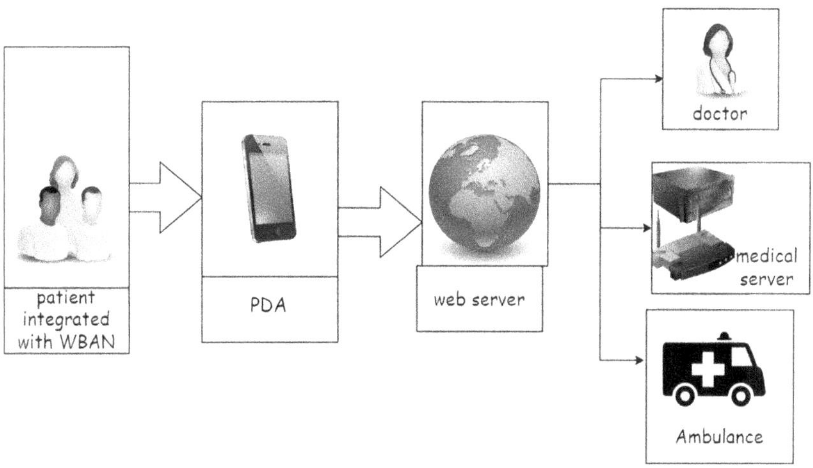

Figure 1.2 A wireless body area network's general architecture.

where E is precisely how much power that network uses, and x represents various network parameters that affect energy consumption, such as transmission power, routing protocol, and data aggregation techniques.

$$\text{Minimize}\, E = f(x)$$

subject to:

- Network connectivity is maintained
- Data is transmitted within a certain delay
- Data accuracy meets a certain threshold
- Resource constraints, such as node memory and processing power, are not violated

1.2.2 Energy- efficient secured routing protocol

A secured routing protocol for energy efficiency is being created to boost network energy consumption and enhancing network performance and security. The suggested SRPEE's flow diagram is shown in Figure 1.3. The protocol initializes by sending hello packets to the sensor nodes. The link between the nodes is formed and ready for data flow after the recipient node gets the hello packet [4]. The base station uses the smart grid and renewable energy systems to encrypt data using the Rivest–Shamir–Adleman encryption method and compress data using the Huffman coding compression technique. The particle swarm optimization routing algorithm chooses the best node. The coordinator performs the implementation of the suggested

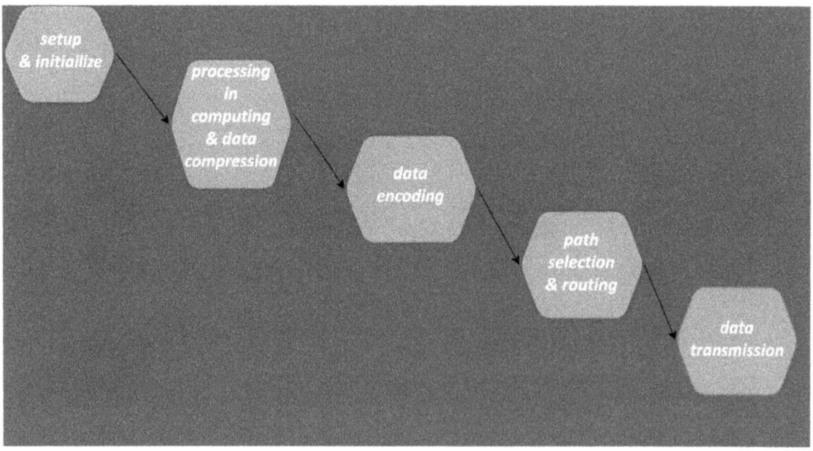

Figure 1.3 A secured routing protocol for energy efficiency flow diagram.

approach. The energy-efficient secured routing protocol (EESRP) is a protocol used in wireless sensor networks to ensure secure and energy-efficient communication between nodes. Here is an example of a mathematical formula that could be used in the EESRP:

Let's say that each node in the network has a certain amount of energy (E) that can be used for communication. The amount of energy used for communication (Ec) is the size of the packets being transferred on the distance throughout both nodes.

To calculate the amount of energy used for communication between nodes i and j, we could use the following formula:

$$\mathrm{Ec}_{ij} = k * d_{ij}^{\ 2} * P_{ij}$$

where, k is a constant that depends on the specific wireless channel being used

The gap connecting clusters i and j is measured by d_{ij}.
P_{ij} corresponds to size of the packets being transmitted between nodes i and j

It may compute the entire amount of energy utilized to interact in network over a specific time period using this formula. It can increase the network's longevity and lessen the frequency of battery changes by reducing the overall energy consumed.

In addition to energy efficiency, the EESRP also incorporates secure routing protocols to protect against attacks from malicious nodes. The mathematical formulas used to ensure secure communication between nodes depend on the specific security protocols being used but could include elements such as encryption keys and message authentication codes (MACs) to verify the integrity of the transmitted data.

1.3 RECURRENT NEURAL NETWORKS

The analysis of sequential data, comprising written material, audio recordings, and video recordings, has seen a significant increase in the application of RNNs. However, RNNs constructed from Tanh cells have trouble extracting relevant details from the source data when the input gap is wide. In order to successfully handle long-term dependency issues, LSTM–allocated networks are completed, including gate functionalities within the cell structure [3]. Many exciting findings in RNN–based research have been attributable to LSTM, which has grown to be the main focus of deep learning. The learning capability of the LSTM cell and its numerous iterations is examined in this article.

1.3.1 Basic architecture

RNNs indicate two straightforward RNNs and range in degree of connectivity from whole to partial. With the ones that follow an Elman network, analogous in any three-layer neural networks, so-called context cells are where the outcomes from the buried tier are stored and provide the original signal and a context cell's output to the buried neuron in a circle. As a result, each individual buried neuron has a unique context cell that plays a crucial role in information processing by receiving input from both context cells and input layer. Contextual cell output from Elman networks developed through standard error back propagation can be simply viewed as a type of extra source of input. Elman network and typical feed-forward networks are contrasted.

Context cells for Jordan networks are bombarded with data from the outcomes layer, whereas context cells for Elman networks are served by the source layer. It displays a partially linked recurrent hidden layer in a partly recurrent neural network. RNNs analogue feed-forward neural networks, which will be tackled in Section 1.4 of this chapter, and must be taught differently. For this to occur, information must diffuse across recurrent connections inside RNN stages. BPTT and RTRL constitute the most popular and widely recognized learning algorithms to facilitate instruction RNNs in time-dependent, supervised training applications. To create a feed forward neural network, the chain of neurons must be unfurled in BPTT. Then, the ensemble set of weights is updated using the extrapolated delta rule.

1.3.2 Training recurrent neural networks

Even though BPTT is by far the most popular method for training recurrent neural networks, RTRL may be the best-known alternative. The method used to determine the weight changes is where BPTT and RTRL diverge most. BPTT and RTRL were combined in the initial formulation of the LSTM-RNNs. We therefore rapidly review both learning algorithms. Training RNNs involves updating the weights in a network's ability to reduce the discrepancy between the intended output and the output predicted by the network. Here is an example of an analytical formula that could be used in the training of RNNs:

Let's say that we have a sequence of input A: $\{a_1, a_2, \ldots, a_N\}$ and corresponding target outputs $B = \{b_1, b_2, \ldots, b_N\}$, where N is the sequence's length. We can represent the output an RNN at 't' using the formula:

$$p_t = f\left(W_p * p_{\{t-1\}} + W_a * a_t + b\right)$$

where p_t is the RNN's concealed state at t, f is activation function, W_p and Wa, b is a bias vector, and weighted matrix.

We can use a loss function, like mean squared error (MSE), to quantify the discrepancy among the desired output and the anticipated output when training the RNN:

$$L = (1/N) * \text{sum}_{\{t=1\}}{}^{N} (y - o_s)^2$$

where o_s is the RNN's outcome at time t.

To reduce the loss function, it can update the RNN's weights using an optimization approach like stochastic gradient descent (SGD). The following formula can be used to express the SGD update rule:

$$W_n = W_o - \alpha * \text{grad}$$

where W_0 is the old weight, α is the learning rate, and grad is the gradient in weight loss function with regard to weight.

During the course of the training phase of RNNs, we must consider the impact of the prior concealed state on the present hidden state. To BPTT an equation is used to accomplish this, which unrolls the RNN over time and propagates the gradients from the output to the input.

The BPTT algorithm involves computing a loss function's pathways with consideration to the weights and biases at each time step, and then summing these gradients over the entire sequence. This can be represented using the formula:

$$\text{grad} = \text{sum}_{\{t=1\}}{}^{N} \left(dL/dh_t * dh_t/dW\right)$$

where dL/dh_t is the loss function's pathways with consideration to the weights state at hidden time t, and dh_t/dW is the gradient of the hidden state at time t with scepticism and weights.

1.3.3 LSTM networks are superior to RNN

LSTM networks, which are a type of RNN, incorporate at least one cycle in their inter-neuronal connections. They were initially introduced by Hochreiter and Schmidhuber and have since been refined over time. Unlike traditional RNNs, LSTM networks focus on learning long-term dependencies and address the issues of vanishing and exploding gradients.

LSTM networks typically consist of an initial layer, a set of memory cells, and an output layer. The input layer has the same number of neurons as the explanatory variables. However, the distinguishing feature of LSTM networks lies in the hidden layer, also known as the buried layer, which is comprised of memory cells. These memory cells enable LSTM networks to effectively capture and retain information over extended sequences.

1.3.4 Cells and their variants

Long short-term memory is a RNN framework that is frequently used with deep learning for speech recognition, study of time series using statistics, and the processing of natural languages. The LSTM cell, which serves as the foundation of this architecture, is intended to assist in resolving the vanishing gradient issue that plagues conventional RNNs.

The basic LSTM cell has three gates:

- INPUT GATE: The extent to which the fresh input should be incorporated into the cell is decided by the input gate. It takes as input the current input and the previous hidden state; output is a value range from 0 to 1 for each element of the memory, where 1 indicates high importance.
- FORGET GATE: The forget gate determines which information from the prior cell state should be omitted. By considering the previous hidden state and current input, it assigns weight ranges 0 and 1 to each element of the cell. A weight approaching 0 indicates an increased likelihood of discarding that particular element.
- OUTPUT GATE: The output gate governs the selection of components within the cell that contribute to the ultimate prediction. It takes into account the present input, previous hidden state, and current state as inputs, assigning weight ranges 0 and 1 to each element of the cell. A weight close to 1 signifies a strong significance towards the final output.

There are several variants of the LSTM cell, including:

1. Peephole LSTM: This version increases the number of connections between the cell state and the gates such that the gates have direct access to the cell state. By enabling these gates to better manage the information flow, this can enhance the model's performance.
   a. Input gate: $i_t = \text{sigmoid}(W_i * [x_t, h_{\{t-1\}}, c_{\{t-1\}}] + b_i)$
   b. Forget gate: $f_t = \text{sigmoid}(W_f * [x_t, h_{\{t-1\}}, c_{\{t-1\}}] + b_f)$
   c. Output gate: $o_t = \text{sigmoid}(W_o * [x_t, h_{\{t-1\}}, c_t] + b_o)$
   d. Memory cell: $c_t = f_t * c_{\{t-1\}} + i_t * \tanh(W_c * [x_t, h_{\{t-1\}}, c_{\{t-1\}}] + b_c)$
   e. Hidden state: $h_t = o_t * \tanh(c_t)$

   Here, c_t represents the memory cell at time t, and the weights and biases have the same meaning as in the basic LSTM.
2. Gated Recurrent Unit: It is a compressed version of the LSTM in cells that combines the inputs and forget gate, cells gate, and outputs into a single update gate. The architecture improves its computing efficiency as a result.
   a. Reset gate: $r_t = \text{sigmoid}(W_r * [x_t, h_{\{t-1\}}] + b_r)$
   b. Update gate: $z_t = \text{sigmoid}(W_z * [x_t, h_{\{t-1\}}] + b_z)$

c. Candidate hidden state: $h_t' = \tanh(W_h * [x_t, r_t * h_{\{t-1\}}] + b_h)$
d. Hidden state: $h_t = (1 - z_t) * h_{\{t-1\}} + z_t * h_t'$

Both the reset gate and update gate are represented here by r_t and z_t, respectively. Using a reset gate and the prior hidden state, the candidate hidden state represented by h_t' is calculated. The update gate z_t is in charge of computing the real hidden state of the system at time t as an ordered combination of the candidate's hidden state and the previous hidden state.

3. Bidirectional LSTM: This variant processes the input sequence both forwards and backwards, combining the outputs from each direction. This can be useful for tasks where the context of a sequence is important.
4. Depth Gated LSTM: In this variation, the LSTM architecture is given more layers, enabling the system to learn more intricate correlations among inputs and outputs.
5. Tree-LSTM: This variant is designed to handle structured input data, such as parse trees, by recursively applying the LSTM operation to the nodes of the tree.

The selection of which variants to utilize will rely on the specific needs of the issue at hand as each of these variations has advantages and disadvantages of their own.

1.3.5 LSTM–dominated neural networks

An LSTM has memory cells that can retain data for a long time and gates that control the information flow in and out of these cells. The gates are controlled by sigmoid activation functions and are trained to decide which information to keep, which to discard, and which to output. LSTMs have been very successful in a wide range of tasks, including natural language processing, speech recognition, and image captioning [5]. They are often used in situations where the input sequence is long or variable in length, and where it is important to remember past information to make accurate predictions.

Overall, an LSTM–dominant neural network would be a neural network architecture that heavily utilizes LSTMs as its primary building blocks, rather than other types of recurrent or feed-forward layers.

1.3.6 Working of LSTMs in RNN

The vanishing gradient problem refers to the issue where the gradients calculated during back propagation become exponentially small as they propagate through time, making the model learn dependencies. The exploding gradient problem is the opposite issue, where the gradients become extremely large and the model can't converge. The LSTM architecture addresses these problems by introducing a set of specialized memory cells and gating mechanisms.

The memory cell is designed to store data over a long duration of time and selectively forget or remember certain information [6, 7]. The information entering and leaving the memory cell is managed by the gating mechanism. At every time interval, the LSTM takes as input the current input vector, the output from the early time interval, and the previous memory cell state. The input and output are passed through a set of linear functions and the non-linear activation functions, while the memory cell state is modified by the gating mechanisms to determine what information should be remembered, forgotten, or updated.

Each gate has its own set of learnable parameters, which are optimized during training to adaptively control the flow of information. By selectively forgetting or remembering information, the LSTM can effectively handle long-term dependencies in sequential data and make accurate predictions.

1.4 OPTIMIZATION OF ADAPTIVE METHOD FOR DATA REDUCTION

Optimization of adaptive methods for data reduction in energy-efficient wireless sensor networks can be achieved through application of deep-learning techniques, specifically utilizing a dispersed data mining strategy based on LSTM neural networks. The target of the data reduction in WSNs is to minimize the amount of data transmitted from the sensors to the base station while maintaining a high level of accuracy in the information collected. This is essential for energy efficiency in WSNs, as the transmission of large amounts of data can quickly deplete the limited power resources of the sensor nodes. One approach to data reduction is to use adaptive methods that adjust the sampling rate of the sensor nodes based on the data patterns observed in the network. However, the effectiveness of these methods relies heavily on the accuracy of the data models used to predict future data patterns [8]. Deep-learning techniques, specifically LSTM networks, have promising results in estimating the time series data patterns. By applying a dispersed data mining strategy, the LSTM model can be trained on data from multiple sensors in the network, allowing for more accurate predictions of future data patterns. The dispersed data mining strategy involves training the LSTM model on data from a subset of sensors in the network, with each subset representing a different region of the network. This allows the model to learn the unique characteristics of each region and make more accurate predictions based on the combined data from all sensors.

The optimized adaptive method for data reduction in energy-efficient WSNs using deep learning and a dispersed data mining strategy based on LSTM would involve the following steps:

1. Collect data from sensors in the network.
2. Divide the network into regions and create subsets of data from sensors in each region.

3. Train an LSTM model on the data from each subset.
4. Combine the LSTM models to make predictions on future data patterns for the entire network.
5. Adjust the sampling rate of the sensor nodes based on the predicted data patterns to minimize the amount of data transmitted while maintaining a high level of accuracy.

By utilizing this optimized approach, WSNs can achieve energy efficiency while still collecting accurate and meaningful data.

1.5 EXPERIMENTAL ANALYSIS

1.5.1 Analysis of energy usage

An essential topic of research is energy-efficient WSNs, which allow for effective and cost-effective data collecting. LSTM networks and other deep learning (DL) algorithms may be used to examine the data gathered by WSNs and derive insightful conclusions. It is possible to increase the effectiveness of data collecting and analysis in WSNs by using a scattered data mining method. By spreading out the data gathering and analysis activities among a number of network nodes, this tactic minimizes the quantity of data that needs to be sent to a central processing unit. As a result, the network will last longer and use less energy. To implement this strategy, each node in the WSN can be equipped with an LSTM network that is responsible for analyzing the data collected by that node. The results of the analysis can be transmitted to a central node, which aggregates the results and generates a comprehensive report on energy usage across the entire network.

Analyzing the energy use of different techniques is an essential phase in the design and optimizing of wireless sensor systems. This involves calculating the quantity of energy necessary by a proposed technique to generate energy-effective networks that can function efficiently in circumstances where sources of electricity may be scarce or nonexistent. The energy consumption can be calculated using the following formula:

$$\text{Energy Consumption} = \text{Power} * \text{Time}$$

For instance, if the power consumed by a wireless sensor is 10 watts and it is operating for 10 hours, then the energy consumption can be calculated as follows:

$$\text{Energy Consumption} = 10\,\text{watts} * 10\,\text{hours} = 100\,\text{Wh}$$

The energy consumption can then be compared with the energy consumed by traditional approaches to determine the effectiveness of the proposed approach in reducing energy usage.

1.5.2 Transmission ratio analysis

WSNs are gaining traction in different applications such as environmental monitoring, industrial automation, and smart homes. However, the energy consumption of WSNs is a major concern since most of the nodes in these networks are battery powered and have limited energy resources. One approach to address this issue is to develop energy-efficient algorithms for data processing and transmission in WSNs. Computational intelligence methodologies, including LSTM networks, have shown promising results in various applications, including WSNs. RNNs further have potential for learning long-term dependencies include LSTM networks in sequential data. A scattered data mining approach can be utilized to lessen the amount of data that must be sent overseas in order to increase the energy efficiency of WSNs. This strategy involves processing the data locally at each sensor node and transmitting only a summary of the data to the base station. The LSTM network can be used to generate the summary by learning the patterns in the data and predicting the future values. The effectiveness of the suggested strategy can be researched using the propagation ratio analysis. This analysis involves comparing the amount of data transmitted using the dispersed data mining strategy with that of a traditional approach where all the data is transmitted to the base station. The analysis' findings can shed light on how well the suggested strategy works to lower WSNs' energy consumption.

Transmission ratio analysis involves evaluating the effectiveness of the proposed approach in reducing the amount of data transmitted from the sensors to the base station. The transmission ratio can be calculated using the following formula:

Transmission Ratio = Data transmitted/Total data collected

For example, if a sensor network collects 1000 data points and transmits only 100 data points to the base station, then the transmission ratio can be calculated as follows:

Transmission Ratio = 100/1000 = 0.1

A low transmission ratio indicates that the proposed approach is effective in reducing the amount of data transmitted, which leads to reduced energy consumption and extended network lifetime.

In conclusion, by lowering the amount of data broadcast, a distributed data mining technique built on LSTM networks can improve the computational efficiency of WSNs. The performance of the suggested strategy can be assessed using the routing ratio analysis, which can also offer suggestions for further improvement.

1.5.3 Network lifetime ratio analysis

The slightly longer lithium-ion battery of the fog nodes makes the energy efficiency of these networks a serious problem. This problem can be solved by optimizing the network lifetime ratio analysis utilizing a DL–based approach and a dispersed data mining positivistic approach on LSTM. RNNs with lengthy interconnections can be captured using LSTMs, which can be applied to time-series data analysis. In the context of WSNs, LSTM can be used to predict the future behaviour of the sensor nodes based on their past data. This prediction can help in making energy-efficient decisions, such as turning off some of the nodes or reducing their sensing frequency, thereby extending the network lifetime. A dispersed data mining strategy involves collecting data from multiple sensor nodes, processing it, and sending the results back to the nodes for decision-making. This approach can help in reducing the amount of data transmitted and the energy consumed by each node, thereby increasing the network lifetime.

Network lifetime ratio analysis involves evaluating the extent to which the proposed approach can extend the lifetime of the wireless sensor network. The network lifetime ratio can be calculated using the following formula:

$$\text{Network Lifetime Ratio} = \text{Network lifetime with proposed approach} / \text{Network lifetime without proposed approach}$$

For example, if a wireless sensor network has a lifetime of 2 years without the proposed approach and a lifetime of 5 years with the proposed approach, then the network lifetime ratio can be calculated as follows:

$$\text{Network Lifetime Ratio} = 5/2 = 2.5$$

A higher network lifetime ratio indicates that the proposed approach is effective in extending the lifetime of the network, which reduces maintenance costs and increases network reliability.

In order to calculate the internet backbone lifespan ratio, the sensor nodes' remaining energy must be compared to their starting energy. This ratio can be used to estimate the network lifetime and to identify the nodes that consume more energy. By using the dispersed data mining strategy based on LSTM, it is possible to optimize this ratio and increase the network lifetime. In conclusion, deep learning-based approaches using a dispersed data mining strategy based on LSTM can be applied to optimize the energy efficiency of WSNs. This approach can help in extending the network lifetime and reducing how much strength the sensor nodes use, thereby enabling the widespread adoption of WSNs in various applications.

1.5.4 Efficiency analysis

LSTM, yet another deep-learning technique, has shown promise in increasing the energy efficiency of WSNs. In this context, a dispersed data mining strategy based on LSTM can be employed to analyse the efficiency of

WSNs. The dispersed data mining strategy involves collecting data from different sensors in the network and processing it locally using lightweight algorithms. The processed data is then transmitted to a central server or gateway for further analysis using deep-learning techniques. This strategy reduces its density or sensitive data that must be transmitted over the network, thereby reducing energy consumption. Recurrent artificial neural networks with LSTM capabilities can detect temporal dependencies in elapsed time, making it well-suited for analyzing sensor data. It can learn patterns in the data and predict future values, which can be used to optimize the energy consumption of WSNs.

To apply LSTM in the efficiency analysis of WSNs, the following steps can be taken:

Data collection: Data from different sensors in the network is collected and processed locally using lightweight algorithms.

Data transmission: The processed data is transmitted to a central server or gateway for further analysis using LSTM.

LSTM training: The LSTM model is trained using the collected data to learn patterns and predict future values.

Energy optimization: The trained LSTM model is used to optimize the energy consumption of WSNs by predicting the optimal time for sensors to transmit data and adjusting their sampling rates.

Efficiency analysis involves evaluating the overall performance of the proposed approach. Efficiency can be calculated using different metrics, such as accuracy, throughput, or energy efficiency. In this case, the efficiency can be evaluated based on the energy consumption per data point collected, which can be calculated using the following formula:

Energy consumption per data point = Energy consumption/Total data collected

For example, if a sensor network consumes 100 Wh of energy to collect 1000 data points, then the energy consumption per data point can be calculated as follows:

$$\text{Energy consumption per data point} = 100\,\text{Wh} / 1000 = 0.1\,\text{Wh per data point}$$

Lower energy consumption per data point indicates higher efficiency and better performance of the proposed approach.

By using a dispersed data mining strategy based on LSTM, energy consumption can be reduced while maintaining the accuracy and reliability of the WSNs. This approach can be applied to various WSN applications, such as environmental monitoring, healthcare, and smart cities, to improve their energy efficiency and prolong their lifespan.

1.6 APPLICATION OF LSTM OVER RNN

1.6.1 Early education activities

Early research has indicated that LSTM networks have the ability to learn tasks that were once believed to be impossible, including tasks that require precise timing and counting, learning context-free languages, and processing high-precision real numbers over noisy sequences. Moreover, researchers have successfully established meta-learning in LSTM networks by using a podcast search task to approximate a learning approach for quadratic functions and have applied reinforcement learning to address non-Markovian learning challenges involving persistent dependencies. These findings underscore the potential of LSTM networks to tackle complex machine learning problems that involve long-term dependencies and noisy input sequences, enabling the development of more effective models and algorithms that can be used to solve a wide range of real-world problems [9].

1.6.2 Cognitive education exercises

LSTM-RNNs were exceptional in completing a variety of cognitive learning tasks. The most prevalent audio and handwriting recognition technologies, as well as, more recently, machine translation, are employed in literature. Other cognitive learning challenges include constituency parsing, conversational modelling, text categorization, handwriting generation, sentiment identification from speech, and text production.

A preliminary assessment of the capabilities of neural networks in situations requiring natural language was made using a conventional neural modelling test. In 2003, effective results were achieved when standard LSTM with a combination of long short-term and polynomial kernel units were implemented for speech-recognition tasks. Better results were obtained after multimodal training with bidirectional long short-term (BLSTM) compared to secret hidden Markov model (HMM)-based systems. Last but not least, a variation known as BLSTM-CTC (connectionist temporal classification) outperformed HMMs; subsequent advances are covered and excellent results were obtained both in 2013 and after using a deep form of both the stacked BLSTM-CTC as well as the reference of a customized CTC objective function.

An LSTM/HMM hybrid design yielded the best outcomes, according to research on the efficiency of different LSTM topologies for big lexicon voice recognition tasks. The performance of LSTM has recently been enhanced using attention-based learning and the sequence-to-sequence framework. A customized speech synthesis framework with two features—listener and attend and spell—was introduced in 2015. The "listener" function uses BLSTM with a pyramidal structure and is based on the previously reported clockwork RNNs (pBLSTM). A newly developed attention-based LSTM transducer is used in the other functionality, "attend and spell." Techniques from the attention-based learning framework and the sequence-to-sequence framework are used to train both functions [10].

1.6.3 Comprehension of handwriting

BLSTM-CTC was first used to detect virtual manuscripts in 2007 and ultimately surpassed hidden Markov methods. Then, combining BLSTM-CTC and a probability language model, a system that could instantly translate online handwritten data was created. This system demonstrated an incredibly high SCADA rates with a rate of errors similar to a human's on this type of activity in a real-world usage scenario. In an offline handwriting tracking task, a different method that combines BLSTM-CTC and multidimensional LSTM outperformed classifiers based on the hidden Markov model. Dropout using the highly effective regularization approach from 2013 was advised.

1.6.4 Automatic interpreting

The authors of a 2014 study leveraged the RNN encoder-decoder neural network architecture to enhance a stochastic translation system's performance in machine translation, resulting in improved accuracy and fluency. The RNN transceiver layout is built on the methodology presented. These results were confirmed, and the pattern-learning deep LSTM architecture was examined. The ability to translate terms that weren't in the lexicon was enhanced by the usage of sequence-to-sequence, which also addressed the problem with rare words. The design was greatly improved by addressing issues with the translation of long phrases by incorporating an attention mechanism into the decoder. The use of sequence-to-sequence improved the capacity to translate terms that weren't in the lexicon and addressed the issue with rare words. By adding an attention mechanism to the decoder, problems with the translation of lengthy words were addressed, substantially enhancing the architecture.

1.6.5 Photo editing

The ability of bidirectional long short-term memory can distinguish between multiple forms of data in textual data, such as texts, mathematics, schematics, and illustrations, allowing it to surpass HMMs and SVMs in 2012 for the identification of keywords and the recognition of mode. The classification of slightly elevated images from the top-ranked collection was also studied by researchers at roughly the same time, with considerably better results than previous approaches. In order to create genuine downright English sentences expressing photos, the more recent sequence-to-sequence–based

LSTM version was successfully trained in 2015. In addition, the authors created a model in 2015 by fusing LSTMs with a hierarchical high-level feature extractor for image classification and interpretation tasks like activity detection and image/video description.

1.7 CONCLUSION

In conclusion, the use of deep-learning techniques such as LSTM can significantly improve the energy efficiency of wireless sensor networks. The dispersed data mining strategy proposed in this study also allows for reduced data transmission and prolongs the network lifetime, particularly in remote locations where maintenance is challenging. Experimental analysis has shown that the suggested approach exceeds traditional techniques in requirements of energy usage, transmission ratio, network lifetime, and efficiency. This research provides a promising solution to the challenges faced in creating sustainable wireless sensor networks, especially in remote areas.

REFERENCES

[1] P. Rawat, K.D. Singh, H. Chaouchi, J.M. Bonnin, Wireless sensor networks: a survey on recent developments and potential synergies. *Journal of Supercomputing* 68(1), 1–48 (2014).

[2] C.P. Chen, S.C. Mukhopadhyay, C.L. Chuang, M.Y. Liu, J.A. Jiang, Efficient coverage and connectivity preservation With Load Balance for Wireless Sensor Networks. *Sensors Journal IEEE* 15(1), 48–62 (2015).

[3] A. Sinha, D.K. Lobiyal, Prediction models for energy efficient data aggregation in wireless sensor network. *Wireless Personal Communications* 84(2), 1325–1343 (2015).

[4] T. Railt, A. Bouabdallah, Y. Challal, Energy efficiency in wireless sensor networks: a top-down survey. *Computer Networks* 67(8), 104–122 (2014).

[5] Y. Song, J. Luo, C. Liu, W. He, *Periodicity-and-Linear-Based Data Suppression Mechanism for WSN* (2015 IEEE Trustcom/BigDataSE/ISPA, Helsinki, 2015), pp. 1267–1271.

[6] Q. Liu, D. Jin, J. Shen, Z. Fu, N. Linge, *A WSN-Based Prediction Model of Microclimate in a Greenhouse Using an Extreme Learning Approach* (2016 18th International Conference on Advanced Communication Technology ICACT), Pyeongchang, (2016).

[7] H. Xiao, S. Lei, Y. Chen, H. Zhou, *WX-MAC: An energy efficient MAC protocol for wireless sensor networks* (2013 IEEE 10th International Conference on Mobile Ad-Hoc and Sensor Systems, Hangzhou, 2013), pp. 423–424.

[8] K.J. Yang, Y.R. Tsai, *WSN18-2: Link Stability Prediction for mobile ad hoc networks in shadowed environments* (IEEE Globecom 2006, San Francisco, 2003).

[9] J. Kolodziej, F. Xhafa, *Utilization of Markov model and non-parametric belief propagation for activity-based indoor mobility prediction in wireless networks* (2011 International Conference on Complex, Intelligent, and Software Intensive Systems, Seoul, 2011), pp. 513–518.

[10] M.A. Razzaque, C. Bleakley, S. Dobson, Compression in wireless sensor networks. *Acm Transactions on Sensor Networks* 10(1), 1–44 (2013).

Chapter 2

Learning-based intelligent energy efficient routing protocols in WSN

Vidya Sagar, G. Bhanu Chander, and Koppala Guravaiah
IIIT Kottayam, Kottayam, India

2.1 INTRODUCTION

Wireless sensor networks (WSNs) are becoming increasingly popular due to their many benefits when compared to wired networks. WSNs are easy to deploy, have low installation costs, are highly mobile, and require no cabling. As a result, WSNs are an attractive technology for smart infrastructures, such as building, factory automation, and process control applications Willig et al. (2005). WSNs are also established as low-cost systems, and they provide Internet of Things (IoT) applications with better sensing and actuation capabilities. A sensor network consists of numerous low-cost sensor nodes that cover a specific Region of Interest (ROI) to measure data using various sensing capabilities and transmit it to the base station (BS) (Mohamed et al. 2018). It has limitations on traditional routing techniques to produce high-performance networks (Yang et al. 2022). The downsides of classical routing algorithms are their poor convergence speed, lengthy training time, and high computing complexity. As a result, dealing with today's Internet's massive traffic increase and diverse consumer demand is difficult. Furthermore, improper routing decisions will cause transmission delay and, in extreme cases, lead to network congestion or packet loss (Dai et al. 2021).

In recent years, artificial intelligence (AI)–based methodologies developed rapidly and became hugely applied in domains such as natural language processing (Lai Zhou and Liu 2020), computer vision (Wu Zhong and Liu 2018), game strategy computation (Silver et al. 2017), and many other areas. AI–based deep learning model research and the development of computer hardware such as the central processing unit (CPU) and graphical processing unit (GPU) have made the techniques that AI models may learn increasingly complicated, and the training and execution efficiency is increasing. With the advancement of processing power and model expression capacity, the AI model has a strong learning ability and good generalization. It is becoming possible to utilize AI models to solve routing optimization challenges and provide intelligence to the network layer.

Machine learning techniques are required for the following reasons according to Alsheikh et al. (2014).

DOI: 10.1201/9781003474524-2

- Wireless sensor networks are often deployed in dynamic environments where the nodes may need to move due to factors like soil erosion or sea turbulence. As a result, these networks must be able to adjust to these changes in order to function effectively. To address this, designers prefer to use reliable machine learning algorithms that can adapt to new data and improve their accuracy over time (Paradis and Han 2007).
- WSNs are often placed in complex environments, making it difficult for researchers to construct precise mathematical models to describe their behavior. Even with simple mathematical models, certain tasks in WSNs, such as the routing problem, require complex algorithms to be solved. In such scenarios, machine learning offers a low-complexity method to estimate the system model (Krishnamachari et al. 2002; Al-Karaki and Kamal 2004). Designers of sensor networks may have access to a vast amount of data, but they may find it challenging to identify significant correlations within the data. Along with ensuring communication connectivity and energy sustainability, sensor network applications often require a minimum data coverage that must be met with limited sensor hardware resources (Romer and Mattern 2004). To ensure maximum data coverage, machine learning methods can be utilized to discover significant correlations in the sensor data and suggest an improved sensor deployment.
- WSNs are being used in new ways and integrated into various technologies such as cyber-physical systems (CPS), machine-to-machine (M2M) communication, and IoT to support better decision-making and autonomous control (Wan et al. 2013). Machine learning plays a crucial role in extracting the necessary levels of abstraction to perform AI tasks with minimal human intervention (Bengio et al. 2009).

2.2 WIRELESS SENSOR NETWORKS

In a WSN, the sensor nodes have several constraints, such as limited energy, processing capacity, and storage capability (Gharajeh and Khanmohammadi 2016). One of the primary objectives for WSNs is the development of an energy-efficient and dependable routing (Gama et al. 2015). The network connectivity must be maintained throughout time, which necessitates balanced node energy consumption. The energy-efficient routing process is shown in Figure 2.1. The nodes at the end collect data and send it to the base station using the correct path. Energy nodes are represented by full circles, while low energy nodes are represented by dotted circles. The solid line indicates a reliable link, while the dotted line indicates an unreliable link (Acar and Adams, 2006). The cluster heads as shown as filled circles. Routing protocols select a route that is energy efficient and reliable to transfer data from the sensor node to the base station.

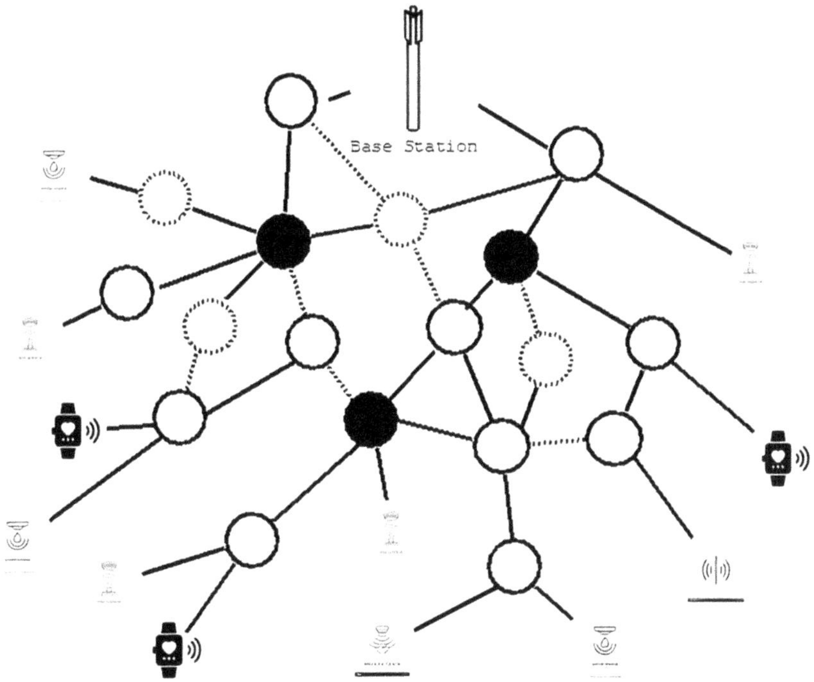

Figure 2.1 Wireless sensor network routing model.

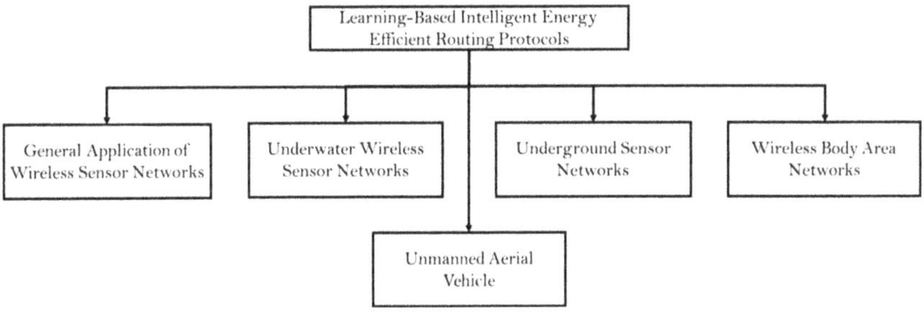

Figure 2.2 Survey considered on WSN and its sister areas.

This chapter mainly concentrates on different protocols implemented and applied to WSN; underwater, underground, unmanned aerial vehicles; and multimedia networks as shown in Figure 2.2.

The energy-efficient routing protocols depend on different properties of the network (Wang and Wang 2006).

- *Routing path length*: To reduce energy consumption, it is important to deliver the collected data to the base station using the shortest path possible. This is done by minimizing the routing path length.
- *Load balance*: Some sensors located on the shortest paths of several data sources may be overused. To prolong the network lifetime, one may want to conserve these sensors that have low residual energy.
- *Link reliability*: Wireless sensor communication can be unstable. Even if bypassing unreliable links results in a shorter path, it may be better to avoid them because the cost of retransmission can outweigh the benefits of a shorter path.
- *Aggregation*: If the collected data from different sources is highly correlated, it can be aggregated before being delivered to the base station. This reduces the packet size and transmission cost.
- *Packet Delivery Ratio*: The packet delivery rate (PDR) is the ratio of delivered data packets to the total number of packets sent by the source. It is important for routing protocols to have a high PDR. Poorly designed routing protocols or paths with routing loops can negatively impact the PDR.

(Mittal and Iwendi 2019)

2.3 PROTOCOLS IN WIRELESS SENSOR NETWORKS

The protocol ELDC (Mehmood et al. 2017) uses artificial neural networks (ANN) to maximize network objectives. The ANN contain typical mathematical operations performed between input and output by utilizing the supervised training data or working on unsupervised input. Using ANN makes it easy to select cluster heads with energy less than the threshold. Further, it also provides efficiency in time consumption, storage, computation and communication, and so on. It makes the network more intelligent regarding energy efficiency, robustness, and mean time; it also increases the network's lifespan and minimizes the latency between nodes. The network life span also depends on the traffic of the network. To handle the traffic of the network, the dynamic three-dimensional fuzzy routing protocol based on traffic control (DFRTP) was proposed (Gharajeh and Khanmohammadi 2016). This works on the transmission path to transmit the packets from source to destination by applying fuzzy decision-making. The fuzzy approach balances the energy consumption by balancing the minimum number of hops with the lowest traffic load and the highest remaining energy (RE) with the minimum number of intermediate nodes. The authors Srivastava and Sudarshan (2015) proposed a genetic fuzzy (GFS) approach to find the optimized clustering that reduces the network's data transmission. The fuzzy method with a genetic approach was applied to the zone-based energy efficient routing protocol for cluster head selection. The fuzzy inference engine is utilized for selecting a node cluster head by considering different node characteristics, that is, energy, shortest path between nodes, node density, and node mobility. A genetic algorithm uses the selected cluster nodes to produce the best nodes of the network that consume less energy in the network. The support vector machine hierarchical routing algorithm proposed for clustering the network reduces the consumption of energy by communicating with the nearest cluster heads (Khan et al. 2016). The Naïve Bayes approach is used to select the cluster head node in WSN, which produces minimum energy consumption, which increases the network lifetime (Jafarizadeh et al. 2017). A new adaptive integrated routing framework based on a Bayesian technique has been presented in Liu Zhang and Cui (2014) for data collection. In the projected technique, an adaptive projection vector is constructed in each routing iteration by introducing a new target node selection. The routing prediction model (Kazemeyni et al. 2014), which can be used for both decentralized and centralized versions, utilizes the Bayesian learning approach. This method ensures balanced energy consumption by scheduling data transmission through the route. It is particularly effective for decentralized methods.

The K-means classification algorithm is used (El Mezouary et al. 2016) for routing, which provides better control of traffic overhead, throughput, and packet delivery ratio with minimum energy consumption. The K-means classification algorithm is used for routing, which provides better control of traffic overhead, throughput, and packet delivery ratio. The energy-efficient clustering protocol using the k-means (EECPK-means) algorithm proposed (Ray and De 2016) to find the optimal cluster head of a cluster from a randomly chosen center node. The cluster head selection depends on the Euclidean distance and residual energy of the node in the network. The EECPK-means provides a multi-hop shortest path to base stations. This algorithm avoids data loss and balances the energy consumption of the sensor nodes. A technique called Energy-Efficient K-Means (EKMT) was introduced by Jain Brar and Malhotra (2018) to identify the best cluster heads (CH) that are in close proximity to both the member nodes and the base station, while also being energy efficient. EKMT calculates the total distance between nodes and selects CHs based on the minimum distance, resulting in improved throughput and reduced delay by dynamically reselecting CHs.

The hybrid IDS approach proposed (Duraisamy et al. 2019) using a rules and extreme machine (RELM) learning algorithm to detect cross-layer attacks and reduce energy consumption. The approach is implemented at each node to detect cross-layer attacks. The rule of the algorithm filters the attacks at each layer and sends normal packets to the base station. The algorithm insertion in each sensor node reduces the data transmission to the base station. In the meantime, the energy consumption is also reduced. Still, if attacks are detected, the alert is given to the administrator; otherwise, data is accepted. The proposed clustering and reinforcement learning (C-SARSA) (Aslam et al. 2019) data routing approach reduces and balances data routing energy consumption, improving network stability. In the same way, the shortest path for wireless chargers is with intelligent routing for minimizing energy consumption among renewable wireless sensor networks (RWSN). The multipath link routing protocol (MLRP) and a hybrid-based TEEN (H-TEEN) protocol have enhanced the security and energy of the signal transmission. The transmitted signals have been classified using recurrent neural networks (RNN), which reduces the data retransmission and provides energy balancing (Ramesh et al. 2023).

A deep Q-network (DQN)–based packet scheduling algorithm (Fu and Kim 2023) proposed for energy-efficient routing. In this approach, one primary node and others are considered secondary nodes where the transmission data is done through the primary node. The primary node is also an agent that controls all the secondary nodes. The controller imposes the quality constraints for packet transmission with a predetermined time. Hence, data transmission is controlled automatically and energy consumption is reduced. The reinforcement learning–based energy-efficient routing protocol (Abadi et al. 2022) minimizes the routing policies and maximizes the reward of each node. Further, it provides the sleep scheduling approach to limit the data transmission rate of nodes with low energy conditions. A deep reinforcement learning algorithm (deep Q-learning), a lightweight distributed reinforcement learning technique (Todorovic 2022), manages the sensor node's data transmission power with minimum energy consumption along with the network's reliability. Meantime DQL also increases the network lifetime. The multi-agent Q-learning routing schema (Okine et al. 2022) with the learning agent, that is, the sensor node, adjusts its routing policies based on Q-values of the route path. The Q-values calculate the route's quickness, reliability, and energy efficiency based on the network features such as number of hops to sink, energy consumption cost for data transmission, and neighbor's packet loss ratio. The approach (Su et al. 2022) utilizes Q-learning and is designed to conserve energy. The wireless sensor collects information about its neighboring sensors including their remaining energy, energy consumption for information transmission, and information transmission direction and distance. This information is then used to update the Q-values of the neighboring sensors. The wireless sensor can then select the best neighboring sensor based on its Q-values for efficient information transmission. The Q-learning reinforcement technique is used to perform the data transmission on cluster nodes, which improves the energy utilization of the network (Kiani 2015). River Formation Dynamics (Guravaiah and Leela velusamy 2015) are used to create multi-hop routing protocols using hop count and remaining energy. This algorithm was extended for hierarchical routing protocol–proposed HCCRFD (Guravaiah & Leela Velusamy 2017), where they used basic LEACH protocol for CH selection and routing will be with RFD. Later the Balanced Energy Adaptive Cluster Head (BEACH) (Guravaiah and Leela Velusamy 2018) selection algorithm is proposed to balance the energy among the cluster heads. All the above algorithms use the RFD–based concept to enhance the lifetime of the network.

The two forms of conserving energy in sensor nodes are proposed (Kang et al. 2020). First, utilize N1-energy saving, a network-level energy-saving strategy that identifies the minimum number of sensor nodes necessary to maintain WSN performance. The first approach is done through a hybrid filter-wrapper feature selection technique, a common machine learning method, to determine the most effective feature subsets. Second, we employ N2-energy saving, which involves adjusting the sampling rate and transmission interval of sensor nodes to conserve energy in WSNs at the node level.

The RL-LEACH protocol is a type of reinforcement learning that selects a cluster head based on environmental observations, such as energy consumption, coverage, and distance between cluster nodes and the base station. By optimizing cluster selection, the neighboring nodes can improve network lifetime by reducing energy consumption. Ultimately, the RL approach selects the cluster head with the highest reward point for data communication. The hybrid routing protocol (Kumar 2023) uses the fuzzy density peak clustering (FDPC) approach to perform clustering on the network. Further, an Adaptive Donkey Theorem Optimization (ADTO) method was applied to determine the CH necessary for effective data transfer. The data transferring using the cluster head to the sink node is done through an efficient hybrid deep marine reinforcement learning (H-DMRL) approach. The proposed hybrid routing protocol earns reduction in energy consumption and improvement in network lifetime.

A Centralized Routing Protocol for Lifetime and Energy Optimization using genetic algorithms and LSPI (CRPLEOGALSPI) (Obi et al. 2023) for WSN, which comprises a collection of sensor nodes and a sink, is initially represented as a weighted graph in order to carry out the suggested protocol's design. The sink builds the potential minimum spanning trees that are utilized as the routing tables (RTs), using a GA after network startup. After a sensor node or nodes die during a round of data transmission, the sink keeps building MSTs until the network graph is severed. In order to maximize network lifespan while reducing network energy consumption, the sink employs LSPI to learn the best or nearly best MST(s) during the cycle of data transmission. To get more insights see Table 2.1.

2.4 PROTOCOLS IN UNDERWATER WIRELESS SENSOR NETWORKS

There is a wide range of aquatic applications using the underwater wireless sensor networks (Akyildiz et al. 2006), such as navigation assistance, disaster management, surveillance of coastal area (Rice 2005) and detecting mines (Freitag et al. 2005).

The Q-learning–based topology-aware routing (QTAR) protocol (Nandyala et al. 2023) considers network topology for determining the route path's next forwarder (NF) node candidate. It utilizes the Q-learning

Table 2.1 Wireless sensor networks literature

Reference	*Name of the protocol*	*Learning method*	*Metrics simulated*
Mehmood et al. (2017)	An artificial neural network–based energy-efficient and robust routing	ANN	Energy efficiency with network reliability and network lifetime
Gharajeh and Khanmohammadi (2016)	Dynamic three-dimensional fuzzy routing based on traffic probability	Fuzzy	Network lifetime and packet delivery ratio
Srivastava and Sudarshan (2015)	A genetic fuzzy system–based optimized zone–based energy efficient routing protocol	Fuzzy	Energy efficiency along with load balancing and network lifetime
Khan et al. (2016)	Support vector machine–based energy aware routing	SVM	Energy dissipation
Jafarizadeh et al. (2017)	Naïve Bayes method	Naïve Bayes	Network lifetime
Liu et al. (2014)	An adaptive data collection algorithm	Bayesian	Energy efficiency
Kazemeyni et al. (2014)	Formal modeling and analysis of learning-based routing	Bayesian	Energy efficiency
El Mezouary et al. (2016)	An energy-aware clustering approach based on the K-means method	K-means	Energy efficiency along with network lifetime
Ray and De (2016)	An energy efficient clustering protocol based on K-means algorithm	K-means	Energy efficiency along with load balancing and increasing the network lifetime
Jain et al. (2018)	An energy efficient K-means clustering algorithm	K-means	Energy efficiency, throughput and delay
Duraisamy et al. (2019)	The rules and extreme learning machine algorithm	Extreme learning machine (ELM) algorithm	Energy efficiency
Aslam et al. (2019)	Optimal wireless charging inclusive of intellectual routing based on SARSA learning	SARSA learning	Energy efficiency and network lifetime

Ramesh et al. (2023)	Classification-based signal processing	RNN	Energy efficiency and lifespan
Fu and Kim (2023)	A practical deep Q-network (DQN)–based packet scheduling algorithm	Q-learning	Energy efficiency, lifespan, and QoS
Abadi et al. (2022)	Reinforcement-learning–based energy efficient control and routing protocol	Q-learning	Energy efficiency and network lifetime
Todorovic (2022)	A lightweight distributed reinforcement learning	Deep Q-learning	Energy efficiency and network life time
Okine et al. (2022)	The multi-agent Q-learning routing schema	Q-learning	Energy efficiency, data transmission, and packet delivery ratio
Su et al. (2022)	A Q-learning–based routing approach	Q-learning	Energy efficiency and network life time
Kiani et al. (2015)	Efficient intelligent energy routing protocol	Q-learning	Energy efficiency and data transmission
Kang et al. (2020)	Machine learning—based energy-saving framework	Fuzzy c-means and simulated annealing	Energy efficiency, data transmission and packet delivery ratio
Kumar (2023)	Hybrid deep marine reinforcement learning–based routing protocol	Deep marine reinforcement learning	Energy efficiency and data transmission
Obi et al. (2023)	A Centralized Routing Protocol for Lifetime and Energy Optimization using GA and LSPI	Q-learning	Energy efficiency, network lifetime, and data transmission

approach to decide NF from the NF node candidates. Further, implicit cut-vertex is used to select an optimal number of NF nodes, reducing the energy waste of data transmission. The DQN–based intelligent routing (DQIR) protocol approach (Geng and Zhang 2023) uses a Markov decision process (MDP) model for routing decision problems. The MDP is utilized for finding the next hop with the highest reward for forwarding data by applying DQN. The agent is chosen based on residual energy balancing with minimum routing distance, which improves the network lifetime with minimum average delay. A new approach called Q-learning–based hierarchical routing protocol with unequal clustering (QHUC) has been proposed (Yuan 2023) to improve data forwarding while extending network lifetime. The protocol builds a hierarchical network structure initially, then uses unequal clustering and Q-learning to distribute remaining energy evenly throughout the network. With the Q-learning algorithm, the optimal CH and next hop can be determined without incurring any additional costs. This approach results in better routing decisions for the network. QELAR is a routing protocol based on reinforcement learning that is adaptive, energy efficient, and lifetime aware (Hu and Fei 2010). It increases the lifespan of the network by distributing residual energy of sensor nodes evenly. The protocol calculates the reward function to choose the best path for packet forwarding, considering the group of nodes available in the routing path. Khan et al. (2021) introduce the Q-learning–based energy-efficient and balanced data gathering (QL-EEBDG) routing protocol. QL-EEBDG selects FNs based on their remaining energy and groups them according to their neighboring nodes' energy levels. This approach optimizes energy consumption and prolongs the network's lifespan. However, the void node recovery process may fail when the network's topology changes. To prevent this issue, the QL-EEBDG-ADN scheme identifies alternative neighbor routes for packet transmission, ensuring uninterrupted communication in the network. A Q-learning–based application-adaptive routing protocol (QAAR) approach (Han et al. 2022) is used to improve the end-to-end delay, energy consumption, and packet delivery ratio. In this approach, Q-learning reward function considers factors such as residual energy, energy distribution, distance, and density that influence network performance when nodes select the next-hop forwarder. Further, the analytic hierarchy process (AHP) model is used to assign weights to these parameters for different underwater network applications. By using distance information and choosing forwarding nodes with the highest Q-value, the QMCR protocol applies Q-learning to multi-hop cooperative routing. This results in lower energy consumption and higher data transmission efficiency (Chen et al. 2021).

The multi-agent reinforcement algorithm is used to design an adaptive cluster in routing protocol (Sun 2022). The multi-agent system creates a network as a multi-agent and selects the optimal global route collaboratively by all the nodes. It also uses the adaptive approach for cluster head selection without communication overhead and minimizes the probability of hotspot generation. It does not consider the consensus from neighbor nodes. Hence, nodes autonomously determine they can act as cluster heads depending on the network information. Further, a biased reward function approach provides feedback on the routing performance for the adaptive cluster head selection. Also, it motivates nodes to select cluster heads as relays. Hence, adaptive clustering routing protocol accomplishes more heightened routing efficiency, lower energy consumption, and longer network lifetime. A novel underwater clustering–based hybrid routing protocol (UC-HRP) (Sathish Kumar et al. 2023) works in three phases; phase one is initializing clustering parameters such as Doppler spread, noise, path loss, and multipath using the fuzzy-ELM method. The second phase is called Cluster Centre Cluster Head Selection (C3HS) for selecting the cluster head based on link quality, node degree, distance, and residual energy. The last phase of the protocol is a Hybrid Artificial Bee Colony (HABC) to provide an optimal route based on reliability, average path loss, average transmission latency, and bandwidth effectiveness. By using a learning-based approach (Wang et al. 2023), the DROR protocol achieves energy efficiency and real-time data transmission. It avoids the void nodes by using the void recovery mechanism and continuous forwarding. It also uses the Q–based dynamic scheduling procedure to send data packets through the best routing path. The mix of opportunistic routing with reinforcement learning provides a new protocol (RLOR) (Zhang et al. 2021), a distributed routing method that depends on the peripheral status of nodes to select them as relay nodes. The RLOR also has a recovery mechanism to avoid the void area efficiently for forwarding the packets within the sparse network packet delivery ratio. The Q-learning assisted ant colony routing protocol (QLACO) (Fang et al. 2020) uses the reward mechanism and the optimal global route selection provided by artificial ant colony to address the challenges of link instability and energy efficiency. The reward function improves the packet delivery ratio, and an anti-void mechanism handles the void region.

In Su (2019), the advanced routing protocol, called deep Q-network–based energy and latency-aware routing protocol (DQELR), is designed to extend the lifespan of universal autonomous system numbers by considering both energy and latency concerns. DQELR employs a deep Q-network algorithm that utilizes both off-policy and on-policy methods to make optimal routing decisions. By considering the energy and depth levels of nodes at different communication stages, the protocol intelligently selects the nodes with the highest Q-value as forwarders. DQELR also employs a hybrid approach of broadcast and unicast communication to reduce network overhead. Additionally, the protocol is able to adapt to changes in network topology through an on-policy method that reroutes traffic when the current route is compromised. With its energy efficiency and strict latency requirements, DQELR can significantly prolong the lifespan of UASNs. An innovative algorithm called energy-efficient depth-based opportunistic routing algorithm with Q-learning (EDORQ) (Lu et al. 2020) uses depth-based opportunistic routing and Q-learning to improve energy efficiency and data transmission reliability in UWSNs. Combining the strengths of these two techniques results in enhanced network performance in terms of energy consumption, packet delivery ratio, and average network overhead. EDORQ factors in the void detection rate, residual energy, and depth information of candidate nodes to proactively identify void nodes and reduce energy consumption. Authors also proposed a simple and scalable void node recovery mode to recover packets stuck in void nodes. Additionally, we have designed a new method to set the holding time for packet forwarding schedules based on Q-value, which reduces packet collision and redundant transmission.

Authors in Li et al. (2020) developed a new routing protocol for underwater wireless sensor networks (UOWSNs) called DMARL. This protocol uses multi-agent reinforcement learning and considers factors like residual energy and link quality to improve the network's ability to adapt to changes and prolong its lifespan. They also used two optimization strategies to speed up the learning process and included a reward mechanism for the distributed system. Overall, this approach is designed to be efficient, effective, and sustainable. The literature (Wang and Shin 2019) explores resource management in hierarchical networks using a Q-learning routing protocol which is efficient. In this, the authors define the relation between distance and energy as single hopping bonus metrics and use them to derive an updating formula for the routing algorithm. They also establish the relationship between energy priority and distance priority. Furthermore, a regulatory factor is introduced to adjust the balance between energy saving and low delay, catering to diverse requirements. See Table 2.2 for more detailed information.

2.5 PROTOCOLS IN WIRELESS UNDERGROUND SENSOR NETWORKS

The underground is a setting for a variety of sectors, including seismic mapping, agriculture, oil and gas, and border surveillance (Table 2.3). These applications need to collect pertinent data from the installed underground objects. To transport information between the surface and subterranean objects, a single communication method cannot be used due to the hard subsurface propagation environment, which includes sand, rock, and watersheds (Saeed et al. 2019). Wireless communication in wireless autonomous system numbers (WUSNs) is far more difficult than it is in terrestrial over-the-air wireless sensor networks because of the precipitation and harsh weather that regularly cause the subterranean surroundings to undergo changes (Zhao et al. 2020). In order to make their deployment cost effective, WUSN devices should have a lifespan of at least several years, depending on the intended use. The lossy subterranean channel makes this problem more difficult and necessitates that WUSN devices have radios

Table 2.2 Protocols in underwater wireless sensor networks

Reference	*Name of the protocol*	*Learning method*	*Metrics simulated*
Nandyala et al. (2023)	The Q-learning–based topology-aware routing (QTAR) protocol	Q-learning	Energy efficiency and data transmission
Geng and Zhang (2023)	The deep Q-network (DQN)–based intelligent routing (DQIR) protocol	Q-learning	Energy efficiency, network life time, and average delay
Yuan et al. (2023)	Q-learning–based hierarchical routing protocol with unequal clustering	Q-learning	Energy efficiency, network lifetime, and cost reduction
Hu and Fei (2010)	An adaptive, energy-efficient, and lifetime-awarerouting protocol based on reinforcement learning	Q-learning	Energy efficiency and network lifetime
Khan et al. (2021)	The Q-learning–based energy-efficient and balanced data gathering	Q-learning	Energy efficiency, network lifetime, and alternative routing paths
Han et al. (2022)	A Q-learning-–based application-adaptive routing protocol	Q-learning	Energy efficiency, packet delivery ratio, end-to-end delay
Chen et al. (2021)	A Q-learning to multi-hop cooperative routing	Q-learning	Energy efficiency and data transmission
Sun et al. (2022)	Adaptive clustering routing protocol	Q-learning	Energy efficiency, best routing paths and network lifetime
Sathish Kumar et al. (2023)	A novel underwater clustering–based hybrid routing protocol	Fuzzy-ELM and Hybrid Artificial Bee Colony	Energy efficiency, data transmission, reliability
Wang et al. (2023)	Reinforcement learning–based opportunistic routing protocol	Q-learning	Energy efficiency, end-to-end delay and reliability
Zhang et al. (2021)	The mix of opportunistic routing with reinforcement learning protocol	Q-learning	Energy efficiency and packet delivery ratio
Fang et al. (2020)	The Q-learning assisted ant colony routing protocol	Q-learning	Energy efficiency, link stability and packet delivery ratio

(*Continued*)

Table 2.2 (Continued)

Reference	*Name of the protocol*	*Learning method*	*Metrics simulated*
Su et al. (2019)	Deep Q-network–based energy and latency-aware routing protocol	Q-Network algorithm	Energy efficiency, delay, and network lifetime
Lu et al. (2020)	An energy-efficient depth-based opportunistic routing algorithm with Q-learning	Q-learning	Energy efficiency, packet delivery ratio, and average network overhead
Li et al. (2020)	An efficient routing protocol based on multi-agent reinforcement learning	Multi-agent reinforcement learning	Energy efficiency, link quality, and network lifetime
Wang and Shin (2019)	A resource management in hierarchical networks using a Q-learning routing protocol	Q-learning	Energy efficiency, delay, and priority on route path

Table 2.3 Protocols in wireless underground sensor networks

Reference	*Name of the protocol*	*Learning method*	*Metrics simulated*
Radhakrishnan et al. (2022)	The deep learning (DL)–based cooperative communication protocol	Deep learning	Energy efficiency, cost effectiveness, and reliability
Zhao et al. (2023)	The multi-agent RL (MARL) algorithm	multi-agent RL	Energy efficiency
Soundararajan et al. (2023)	Deep learning–based multi-channel learning and protection model	Multi-channel learning	Energy efficiency, throughput, delay, and packet delivery ratio and data transmission

with more transmission power than terrestrial WSN devices. As a result, energy conservation is a top priority while designing WUSNs (Akyildiz and Stuntebeck 2006).

The deep learning (DL)–based cooperative communication protocol (Radhakrishnan et al. 2022) aims for the productive utilization of a cluster-based cooperative model using relay nodes to provide reliability and cost-effectiveness by communicating through inter-cluster cooperative transmission to increase energy efficiency. By considering packet collisions, energy usage, and connection quality, the multi-agent RL (MARL) algorithm (Zhao et al. 2023) increases network energy efficiency. Second, a reward system is suggested in order to better the proposed algorithm's capacity to adapt to a dynamic subterranean environment by defining an independent state and action for each node. An algorithmic framework for channel attribute categorization is the deep learning–based multi-channel learning and protection model (DMCAP) (Soundararajan et al. 2023). For assessing the channel conditions, it makes use of the multi-channel ensemble model, ensemble multilayer perceptron (EMLP) classifiers, nonlinear channel regression models, nonlinear entropy analysis models, and ensemble nonlinear support vector machine (ENLSVM). Additionally, the distributed environment's intrusion detection techniques are created by the variable generative adversarial network (VGAN) engine. First, it identifies the channels and uses EMLP and ENLSVM to perform SNIR (signal to noise interference ratio) and channel entropy distortions of multiple channels. Nonlinear regression is used to forecast the channel behavior based on training data.

2.6 PROTOCOLS IN WIRELESS BODY AREA NETWORKS (WBANS)

The wireless body area networks (WBANs) applications have provided smart sensor node–based health-care monitoring systems (Table 2.4). These programmes offer sophisticated monitoring systems that can manage and keep track of patient health indicators through early detection (Qureshi et al. 2020). The sensor nodes are applied to the patient's body (inside or outside) to detect the early symptoms and communicate data for additional processing to the washbasin node or other devices (Ullah et al. 2010, Hasan et al. 2017). Battery life, processing speed, and data storage on sensor nodes are all finite. Operating networks with insufficient resources can result in node exhaustion, overhead, data loss, and latency, among other major consequences. Additionally, the network's routing is complicated by patient migration. Another important problem is sensor node battery life, which has a substantial influence on data transmission since it prevents replacement of batteries and charging, particularly for implanted sensor nodes within the patient's body (Anwar et al. 2018a, 2018b). A system called energy aware routing (EAR) (Qureshi et al. 2020) assesses the connection quality of sensor nodes in order to save energy consumption and choose the best next hop. To balance the load, reduce energy consumption, and improve data transmission, the suggested protocol examines the energy level, link quality, and remaining energy level. The QL-CLUSTER (Kiani 2017) is a reinforcement learning method that offers an optimized routing channel for data transmission to cut down on costs and power use. It is a routing system based on Q-learning that employs a Q-value at each node

Table 2.4 Protocols in wireless body area networks (WBANs)

Reference	*Name of the protocol*	*Learning method*	*Metrics simulated*
Qureshi et al. (2020)	Energy aware routing (EAR)	Next forwarder selection	Energy efficiency, load balancing, and data transmission
Kiani (2017)	QL-CLUSTER	Q-learning	Energy efficiency, data transmission, and cost reduction
Anand et al. (2017)	A distributed Q-learning agent	Q-learning	Energy efficiency and data transmission
Mohammadi and Shirmohammadi (2023)	The duty cycle can be determined using deep reinforcement learning (DRDC) protocol	Deep Q-network	Energy efficiency and packet delivery ratio
Kim and Kim (2022)	An energy delay-efficient mechanism	Reinforcement learning	Energy efficiency and network lifetime

and a unique reward for each packet. To enable the distributed Q-learning agent to automatically make an optimized choice, each node chooses a path based on the cumulative reward acquired from the transition of the packets from one node to another node (Anand et al. 2017). Alqahtani et al. (2023) proposed a machine learning method that dynamically indicates the path for efficient data transmission between network intermediates and servers. Their approach takes into account the estimation of energy harvesting, the standard deviation of energy usage, and the incorporation of multiple sensors in a WBAN. The duty cycle can be determined using deep reinforcement learning (DRDC) protocol (Mohammadi and Shirmohammadi 2023). DRDC has some unique features, such as considering the change rate of data sensed by BN along with its energy to prevent emergency packet loss and unnecessary sleep/wake cycles, utilizing -DQN with a lightweight neural network to accurately determine BN's duty cycle, implementing a three-layer communication architectural model when there are extreme limitations in BN resources to maintain EH-BN's memory constraints and computational power, executing the algorithm on a local server, and transmitting only the trained policy to EH-BN. Additionally, a reward function is designed to ensure the appropriate performance of the DQN algorithm.

The proposed techniques in Kim and Kim (2022) include a reinforcement learning–based process for selecting CH and super-frame extension to improve energy efficiency and network lifespan. Moreover, the authors suggest a quick collection and delivery of vital information to medical personnel for accurate decision-making in case of critical signals.

2.7 PROTOCOLS IN UNMANNED AERIAL VEHICLES

UAVs, or unmanned aerial vehicles, are aircraft that are lightweight and can be controlled either remotely or through preprogrammed instructions (Table 2.5). They are typically outfitted with a range of sensors, computational units, cameras, GPS, and transceivers (Nazib and Moh 2020). Efficiently routing data in flying ad hoc networks (FANETs) with UAVs presents significant challenges due to the high dynamic topology, rapid motion of nodes, and frequent link failures. As such, designing a reliable energy efficient routing protocol is a crucial area of research for these networks (Lansky et al. 2022).

The study addresses the problem of allocating resources for UAV communications using downlink rate-splitting multiple access (RSMA) technology (Xiao et al. 2023). By measuring traffic, the UAV is able to determine user requirements for achievable rates and optimize its deployment, beamforming, rate allocation, and subcarrier allocation to maximize energy efficiency while meeting those requirements. To solve the non-convex problem,

Table 2.5 Protocols in unmanned aerial vehicle

Reference	*Name of the protocol*	*Learning method*	*Metrics simulated*
Liu et al. (2020)	QMR:Q-learning–based multi-objective optimization routing protocol	Q-learning and multi-objective optimization	Energy efficiency with optimal routing path
Yang et al. (2020)	Q-learning–based fuzzy logic system	Q-learning and fuzzy logic	Energy efficiency, data transmission, and link metrics
Khan et al. (2022)	A deep Q-network (DQN)–based vertical routing	Deep Q-network	Energy efficiency, network lifetime, and routing path
Rovira-Sugranes et al. (2021)	A self-adaptive learning rate and simulated annealing optimization	Q-learning	Energy efficiency
Xiao et al. (2023)	The downlink rate-splitting multiple access (RSMA) technology		Energy efficiency
Souto et al. (2023)	Path planning optimization	Q-learning and K-means	Energy efficiency
Guo et al. (2023)	Intelligent clustering routing approach	Reinforcement learning	Energy efficiency, packet delivery ratio, and end-to-end delay

a joint optimization approach is proposed. First, a heuristic approach is used to determine the UAV's location. Then, a successive convex approximation method is used to optimize RSMA parameters. Finally, a swap matching algorithm is employed to solve the subcarrier allocation problem as a many-to-one two-sided matching game, introducing a new routing protocol for FANETs with UAVs that utilizes Q-learning and multiobjective optimization to guarantee low-delay and low-energy services (Liu et al. 2020). Unlike existing Q-learning–based protocols that use fixed Q-learning parameters, this protocol adapts to the dynamic nature of FANETs by allowing the parameters to be adjusted accordingly. Additionally, the protocol introduces a new exploration and exploitation mechanism to discover optimal routing paths while also utilizing acquired knowledge. Rather than relying on past neighbor relationships, the protocol re-estimates these relationships during the routing decision process to ensure a more reliable next hop is selected. The authors in Yang (2020) developed a Q-learning–based fuzzy logic system for the FANET routing protocol. This system makes it easier to select the best routing paths by considering both link and overall path performance. Each UAV determines the optimal routing path to the destination using a fuzzy system that takes into account both link-level and path-level parameters. The link-level parameters include transmission rate, energy state, and flight status between neighbor UAVs, while the path-level parameters include hop count and successful packet delivery time. The reinforcement learning method is used to dynamically update the path-level parameters. A new routing method called DQN–based vertical routing (Khan et al. 2022) selects routes with higher residual energy and lower mobility rates across network planes (macro-plane, pico-plane, and femto-plane). The aim is to improve network performance by reducing frequent link disconnections and network partitions. In the 5G access network, there are distributed controllers (DCs) and a central controller (CC) in different network planes. The proposed routing method is a hybrid approach that allows a CC and DCs to exchange information and handle global and local information respectively. This routing method is suitable for highly dynamic ad hoc FANETs, and it enables data communication between UAVs in different applications such as monitoring borders and object tracking. The routing is performed over a clustered network, and clusters are formed across different network planes to provide inter-plane and inter-cluster communications. This helps to offload data traffic across different network planes to enhance network performance and lifetime. In Liu et al. (2020), the authors proposed a routing protocol for FANETs based on Q-learning and aim to provide reliable and efficient service. Unlike most existing Q-learning protocols, it adjusts its parameters to account for the ever-changing environment of FANETs. Additionally, it uses a novel exploration and exploitation mechanism to discover optimal routing paths and use the acquired knowledge. To ensure the reliability of the route, the method re-evaluates neighbor relationships during the routing decision process. It provides efficient energy consumption and packet delivery ratio. The Q-routing algorithm has been enhanced with a self-adaptive learning rate and simulated annealing optimization (Rovira-Sugranes et al. 2021). This allows the algorithm to adjust its exploration rate based on the temperature decline rate, which is controlled by the Q-values' experienced variation rate. The protocol can adapt to network changes without requiring manual re-initialization, even during abrupt topology changes. Furthermore, this results in improved energy efficiency. Yang (2020) proposed a Q-learning fuzzy logic system for the FANET routing protocol that makes it easier to select routing paths based on their overall performance and link performance. Each UAV uses a fuzzy system to determine the best routing path to the destination, which considers link-level parameters such as transmission rate, energy state, and flight status between neighboring UAVs, as well as path-level parameters such as hop count and successful packet delivery time. The reinforcement learning method dynamically updates the path-level parameters. To determine the node positions and potential actions for UAVs navigating through randomly placed urban obstacles and varying wind speeds, a combination of K-means and Q-learning methods were utilized (Souto et al. 2023).

The intelligent clustering routing approach (ICRA) (Guo et al. 2023) is made up of three parts: the clustering module, the clustering strategy adjustment module, and the routing module. During the clustering process, each node must calculate its utility to maintain stability and longevity in different network states. The clustering strategy adjustment module uses reinforcement learning to continuously learn the benefits of different strategies for calculating node utility in specific network states. With this knowledge, the module can determine the best clustering strategy for the current network state. During the routing phase, the proposed scheme introduces inter-cluster forwarding nodes to forward messages among different clusters, reducing end-to-end delay and improving packet delivery rate. Additionally, better energy efficiency is provided.

2.8 PROTOCOLS IN WIRELESS MULTIMEDIA SENSOR NETWORKS

Wireless multimedia sensor networks (WMSN) are made up of a variety of devices, including cameras, microphones, and other tools for producing multimedia output. These gadgets have the ability to record sounds and take pictures and motion pictures (Nagalingayya and Mathpati 2021). The WMSN is also utilized for traffic avoidance, environmental monitoring, law enforcement, and industrial process control. The nature of resource restrictions and the QoS needs of WMSN should be considered by the routing protocols modelled for WSMN (Akila and Venkatesan 2016).

An energy-efficient distributed adaptive cooperative routing (EDACR) for WMSN was designed by considering QoS constraints and energy consumption. This protocol uses a reinforcement learning–based mechanism with QoS and making routing energy efficient by considering the reliability and also delay (Wang et al. 2020). The long short-term memory model–based packet priority based the protocol that works depending on the classification approach, which considers the word embedding of a packet. The semantics of the packet header is used for the packet's priority. Additionally, a hybrid meta-heuristic algorithm called butterfly-based rider optimization combines the rider optimization algorithm and butterfly optimization algorithm to provide an optimized route for data transmission along with packet priority and other parameters such as energy consumption, packet delivery ratio, and others (Chiwariro and Thangadurai 2023).

2.9 RESEARCH CHALLENGES

The literature on intelligent wireless sensor networks energy efficient routing protocols mostly concentrated on the nearest single hop nodes to provide better results in the network. There is a necessity to concentrate on the entire path of the route. The environment of underwater wireless sensor networks is unpredictable, and the mobility of sensor nodes is high. There is a need for a more intelligent understanding of the behavior of the sensor network. Therefore, the design of routing protocols without packet loss, energy efficiency, and maximum network lifetime is achieved.

Data transfer is one of the challenges in wireless body sensor networks. It is critical if it is a health monitoring system. So, finding a reliable path requires more intelligence in routing protocols.

The data in multimedia sensor networks are sometimes secure. There is a need for secure data transfer routing protocols.

When it comes to routing methods in FANETs that rely on RL, exchanging control messages at set intervals is necessary to gather information about nearby nodes.

Unfortunately, this leads to higher bandwidth consumption, greater routing overhead, and more network congestion. To combat these issues, it's important to adjust broadcast intervals of these messages based on network dynamics in order to create more efficient RL–based routing methods.

The data in multimedia sensor networks are sometimes secure. There is a need for secure data transfer routing protocols.

REFERENCES

Abadi, A. F. E., Asghari, S. A., Marvasti, M. B., Abaei, G., Nabavi, M., and Savaria, Y. (2022). Rlbeep: Reinforcement-learning-based energy efficient control and routing protocol for wireless sensor networks. *IEEE Access*, 10, 44123–44135.

Acar, G. and Adams, A. (2006). Acmenet: An underwater acoustic sensor network protocol for real-time environmental monitoring in coastal areas. *IEE Proceedings-Radar, Sonar and Navigation*, 153(4), 365–380.

Akila, I. and Venkatesan, R. (2016). A fuzzy based energy-aware clustering architecture for cooperative communication in wsn. *The Computer Journal*, 59(10), 1551–1562.

Akyildiz, I. F., Pompili, D., and Melodia, T. (2006). State-of-the-art in protocol research for underwater acoustic sensor networks. In *Proceedings of the 1st International Workshop on Underwater Networks*, pages 7–16.

Akyildiz, I. F. and Stuntebeck, E. P. (2006). Wireless underground sensor networks: Research challenges. *Ad Hoc Networks*, 4(6), 669–686.

Al-Karaki, J. N. and Kamal, A. E. (2004). Routing techniques in wireless sensor networks: A survey. *IEEE Wireless Communications*, 11(6), 6–28.

Alqahtani, A. S., Changalasetty, S. B., Parthasarathy, P., Thota, L. S., and Mubarakali, A. (2023). Effective spectrum sensing using cognitive radios in 5G and wireless body area networks. *Computers and Electrical Engineering*, 105, 108493.

Alsheikh, M. A., Lin, S., Niyato, D., and Tan, H.-P. (2014). Machine learning in wireless sensor networks: Algorithms, strategies, and applications. *IEEE Communication Surveys and Tutorials*, 16(4), 1996–2018.

Anand, J., Perinbam, J. R. P., and Meganathan, D. (2017). Q-learning-based optimized routing in biomedical wireless sensor networks. *IETE Journal of Research*, 63(1), 89–97.

Anwar, M., Abdullah, A. H., Altameem, A., Qureshi, K. N., Masud, F., Faheem, M., Cao, Y., and Kharel, R. (2018a). Green communication for wireless body area networks: Energy aware link efficient routing approach. *Sensors*, 18(10), 3237.

Anwar, M., Abdullah, A. H., Butt, R. A., Ashraf, M. W., Qureshi, K. N., and Ullah, F. (2018b). Securing data communication in wireless body area networks using digital signatures. *Technical Journal*, 23(02), 50–55.

Aslam, N., Xia, K., and Hadi, M. U. (2019). Optimal wireless charging inclusive of intellectual routing based on sarsa learning in renewable wireless sensor networks. *IEEE Sensors Journal*, 19(18), 8340–8351.

Bengio, Y. et al. (2009). Learning deep architectures for ai. *Foundations and Trends® in Machine Learning*, 2(1), 1–127.

Chen, Y., Zheng, K., Fang, X., Wan, L., and Xu, X. (2021). Qmcr: A q-learning-based multihop cooperative routing protocol for underwater acoustic sensor networks. *China Communications*, 18(8), 224–236.

Chiwariro, R. and Thangadurai, N. (2023). Deep learning-based prioritized packet classification and optimal route selection in wireless multimedia sensor networks. *IETE Journal of Research*, 14, 1–19.

Dai, B., Cao, Y., Wu, Z., Dai, Z., Yao, R., and Xu, Y. (2021). Routing optimization meets machine intelligence: A perspective for the future network. *Neurocomputing*, 459, 44–58.

Duraisamy, S., Pugalendhi, G. K., and Balaji, P. (2019). Reducing energy consumption of wireless sensor networks using rules and extreme learning machine algorithm. *The Journal of Engineering*, 2019(9), 5443–5448.

El Mezouary, R., Choukri, A., Kobbane, A., and El Koutbi, M. (2016). An energy-aware clustering approach based on the k-means method for wireless sensor networks. In *Advances in Ubiquitous Networking: Proceedings of the UNet'15* 1, pages 325–337. Springer.

Fang, Z., Wang, J., Jiang, C., Zhang, B., Qin, C., and Ren, Y. (2020). Qlaco: Q-learning aided ant colony routing protocol for underwater acoustic sensor networks. In *2020 IEEE Wireless Communications and Networking Conference (WCNC)*, pages 1–6. IEEE.

Freitag, L., Grund, M., von Alt, C., Stokey, R., and Austin, T. (2005). A shallow water acoustic network for mine countermeasures operations with autonomous underwater vehicles. *Underwater Defense Technology (UDT)*, pages 1–6.

Fu, X. and Kim, J. G. (2023). Deep-q-network-based packet scheduling in an IoT environment. *Sensors*, 23(3), 1339.

Gama, S., Walingo, T., and Takawira, F. (2015). Energy analysis for the distributed receiver based cooperative medium access control for wireless sensor networks. *IET Wireless Sensor Systems*, 5(4), 193–203.

Geng, X. and Zhang, B. (2023). Deep q-network-based intelligent routing protocol for underwater acoustic sensor network. *IEEE Sensors Journal*, 16, 457–472.

Gharajeh, M. S. and Khanmohammadi, S. (2016). DFRTP: Dynamic 3d fuzzy routing based on traffic probability in wireless sensor networks. *IET Wireless Sensor Systems*, 6(6), 211–219.

Guo, J., Gao, H., Liu, Z., Huang, F., Zhang, J., Li, X., and Ma, J. (2023). ICRA: An intelligent clustering routing approach for UAV Ad Hoc networks. *IEEE Transactions on Intelligent Transportation Systems*, 24(2), 2447–2460.

Guravaiah, K. and Leela Velusamy, R. (2017). Energy efficient clustering algorithm using RFD based multi-hop communication in wireless sensor networks. *Wireless Personal Communications*, 95, 3557–3584.

Guravaiah, K. and Leela Velusamy, R. 2015. RFDMRP: River formation dynamics based multi-hop routing protocol for data collection in wireless sensor networks. *Procedia Computer Science*, 54, 31–36.

Guravaiah, K. and Leela Velusamy, R. (2018). BEACH: Balanced energy and adaptive cluster head selection algorithm for wireless sensor networks. *Adhoc & Sensor Wireless Networks*, 42, 286–302.

Han, C., Xu, C., Song, S., Liu, J., Yang, T., and Cui, J.-H. (2022). Qaar: An application adaptive routing protocol based on q-learning in underwater sensor networks. In *2022 IEEE/CIC International Conference on Communications in China (ICCC)*, pages 162–167. IEEE.

Hu, T. and Fei, Y. (2010). Qelar: A machine-learning-based adaptive routing protocol for energy-efficient and lifetime-extended underwater sensor networks. *IEEE Transactions on Mobile Computing*, 9(6), 796–809.

Jafarizadeh, V., Keshavarzi, A., and Derikvand, T. (2017). Efficient cluster head selection using na¨ıve bayes classifier for wireless sensor networks. *Wireless Networks*, 23, 779–785.

Jain, B., Brar, G., and Malhotra, J. (2018). Ekmt-k-means clustering algorithmic solution for low energy consumption for wireless sensor networks based on minimum mean distance from base station. In *Networking Communication and Data Knowledge Engineering*: Volume 1, pages 113–123. Springer.

Kang, J., Kim, J., Kim, M., and Sohn, M. (2020). Machine learning-based energy-saving framework for environmental states-adaptive wireless sensor network. *IEEE Access*, 8, 69359–69367.

Kazemeyni, F., Owe, O., Johnsen, E. B., and Balasingham, I. (2014). Formal modeling and analysis of learning-based routing in mobile wireless sensor networks. *Integration of Reusable Systems*, 8, 127–150.

Khan, F., Memon, S., and Jokhio, S. H. (2016). Support vector machine based energy aware routing in wireless sensor networks. In *2016 2nd International Conference on Robotics and Artificial Intelligence (ICRAI)*, pages 1–4. IEEE.

Khan, M. F., Yau, K.-L. A., Ling, M. H., Imran, M. A., and Chong, Y.-W. (2022). An intelligent cluster-based routing scheme in 5g flying ad hoc networks. *Applied Sciences*, 12(7), 3665.

Khan, Z. A., Karim, O. A., Abbas, S., Javaid, N., Zikria, Y. B., and Tariq, U. (2021). Qlearning based energy-efficient and void avoidance routing protocol for underwater acoustic sensor networks. *Computer Networks*, 197, 108309.

Kiani, F. (2017). Reinforcement learning based routing protocol for wireless body sensor networks. In *2017 IEEE 7th International Symposium on Cloud and Service Computing (SC2)*, pages 71–78. IEEE.

Kiani, F., Amiri, E., Zamani, M., Khodadadi, T., and Abdul Manaf, A. (2015). Efficient intelligent energy routing protocol in wireless sensor networks. *International Journal of Distributed Sensor Networks*, 11(3), 618072.

Kim, S. C. and Kim, H. Y. (2022). An energy- and delay-efficient transmission mechanism using reinforcement learning in wireless body sensor networks. *Journal of Southwest Jiaotong University*, 57(5), 555–561.

Krishnamachari, L., Estrin, D., and Wicker, S. (2002). The impact of data aggregation in wireless sensor networks. In *Proceedings 22nd International Conference on Distributed Computing Systems Workshops*, pages 575–578. IEEE.

Kumar, A. (2023). Hybrid deep marine reinforcement learning-based routing protocol in wireless sensor networks. In *International Conference on Mathematical and Statistical Physics, Computational Science, Education, and Communication (ICMSCE 2022)*, volume 12616, pages 133–150. SPIE.

Lai, Q., Zhou, Z., and Liu, S. (2020). Joint entity-relation extraction via improved graph attention networks. *Symmetry*, 12(10), 1746.

Lansky, J., Ali, S., Rahmani, A. M., Yousefpoor, M. S., Yousefpoor, E., Khan, F., and Hosseinzadeh, M. (2022). Reinforcement learning-based routing protocols in flying ad hoc networks (fanet): A review. *Mathematics*, 10(16), 3017.

Liu, J., Wang, Q., He, C., Jaffrès-Runser, K., Xu, Y., Li, Z., and Xu, Y. (2020). Qmr: Q-learning based multi-objective optimization routing protocol for flying ad hoc networks. *Computer Communications*, 150, 304–316.

Liu, Z., Zhang, M., and Cui, J. (2014). An adaptive data collection algorithm based on a bayesian compressed sensing framework. *Sensors*, 14(5), 8330–8349.

Lu, Y., He, R., Chen, X., Lin, B., and Yu, C. (2020). Energy-efficient depth-based opportunistic routing with Q-Learning for underwater wireless sensor networks. *Sensors*, 20(4), 1025.

Mehmood, A., Lv, Z., Lloret, J., and Umar, M. M. (2017). Eldc: An artificial neural network based energy-efficient and robust routing scheme for pollution monitoring in wsns. *IEEE Transactions on Emerging Topics in Computing*, 8(1), 106–114.

Mittal, M. and Iwendi, C. (2019). A survey on energy-aware wireless sensor routing protocols. *EAI Endorsed Transactions on Energy Web*, 6(24), 1022–1046.

Mohamed, R. E., Saleh, A. I., Abdelrazzak, M., and Samra, A. S. (2018). Survey on wireless sensor network applications and energy efficient routing protocols. *Wireless Personal Communications*, 101(2), 1019–1055.

Mohammadi, R. and Shirmohammadi, Z. (2023). DRDC: Deep reinforcement learning based duty cycle for energy harvesting body sensor node. *Energy Reports*, 9, 1707–1719.

Nagalingayya, M. and Mathpati, B. S. (2021). A comprehensive review on energy efficient routing in wireless multimedia sensor networks. In *2021 6th International Conference on Inventive Computation Technologies (ICICT)*, pages 144–151. IEEE.

Nandyala, C. S., Kim, H.-W., and Cho, H.-S. (2023). Qtar: A q-learning-based topology-aware routing protocol for underwater wireless sensor networks. *Computer Networks*, 17, 109562.

Nazib, R. A. and Moh, S. (2020). Routing protocols for unmanned aerial vehicle-aided vehicular ad hoc networks: A survey. *IEEE Access*, 8, 77535–77560.

Obi, E., Mammeri, Z., and Ochia, O. E. (2023). A centralized routing for lifetime and energy optimization in wsns using genetic algorithm and least-square policy iteration. *Compute*, 12(2), 22.

Okine, A. A., Adam, N., and Kaddoum, G. (2022). Reinforcement learning aided routing in tactical wireless sensor networks. In *International Symposium on Ubi quitous Networking*, pages 211–224. Springer.

Paradis, L. and Han, Q. (2007). A survey of fault management in wireless sensor networks. *Journal of Network and Systems Management*, 15, 171–190.

Qureshi, K. N., Din, S., Jeon, G., and Piccialli, F. (2020). Link quality and energy utilization based preferable next hop selection routing for wireless body area networks. *Computer Communications*, 149, 382–392.

Radhakrishnan, K., Ramakrishnan, D., Khalaf, O. I., Uddin, M., Chen, C.-L., and Wu, C.-M. (2022). A novel deep learning-based cooperative communication channel model for wireless underground sensor networks. *Sensors*, 22(12), 4475.

Ramesh, S., Nirmalraj, S., Murugan, S., Manikandan, R., and Al-Turjman, F. (2023). Optimization of energy and security in mobile sensor network using classification based signal processing in heterogeneous network. *Journal of Signal Processing Systems*, 95(2–3), 153–160.

Ray, A. and De, D. (2016). Energy efficient clustering protocol based on k-means (eecpkmeans)- midpoint algorithm for enhanced network lifetime in wireless sensor network. *IET Wireless Sensor Systems*, 6(6), 181–191.

Rice, J. (2005). Seaweb acoustic communication and navigation networks. In *Proceedings of the International Conference on Underwater Acoustic Measurements: Technologies and Results*. Citeseer.

Romer, K. and Mattern, F. (2004). The design space of wireless sensor networks. *IEEE Wireless Communications*, 11(6), 54–61.

Rovira-Sugranes, A., Afghah, F., Qu, J., and Razi, A. (2021). Fully-echoed Q-routing with simulated annealing inference for flying Adhoc networks. *IEEE Transactions on Network Science and Engineering*, 8(3), 2223–2234.

Saeed, N., Alouini, M.-S., and Al-Naffouri, T. Y. (2019). Toward the internet of underground things: A systematic survey. *IEEE Communication Surveys and Tutorials*, 21(4), 3443–3466.

Sathish Kumar, P., Ponnusamy, M., Radhika, R., and Dhurgadevi, M. (2023). Underwater clustering based hybrid routing protocol using fuzzy elm and hybrid abc techniques. *Journal of Intelligent Fuzzy Systems*, (Preprint), 1–13, doi:10.3233/JIFS-230172

Silver, D., Schrittwieser, J., Simonyan, K., Antonoglou, I., Huang, A., Guez, A., Hubert, T., Baker, L., Lai, M., Bolton, A., et al. (2017). Mastering the game of go without human knowledge. *Nature*, 550(7676), 354–359.

Soundararajan, R., Stanislaus, P. M., Ramasamy, S. G., Dhabliya, D., Deshpande, V., Sehar, S., and Bavirisetti, D. P. (2023). Multi-channel assessment policies for energy-efficient data transmission in wireless underground sensor networks. *Energies*, 16(5), 2285.

Souto, A., Alfaia, R., Cardoso, E., Araújo, J., and Francês, C. (2023). UAV path planning optimization strategy: Considerations of urban morphology, microclimate, and energy efficiency using Q-learning algorithm. *Drones*, 7(2), 123.

Srivastava, J. R. and Sudarshan, T. (2015). A genetic fuzzy system based optimized zone based energy efficient routing protocol for mobile sensor networks (ozeep). *Applied Soft Computing*, 37, 863–886.

Su, X., Ren, Y., Cai, Z., Liang, Y., and Guo, L. (2022). A q-learning based routing approach for energy efficient information transmission in wireless sensor network. *IEEE Transactions on Network and Service Management*, 15, 489–511.

Su, Y., Fan, R., Fu, X., and Jin, Z. (2019). DQELR: An adaptive deep Q-network-based energy- and latency-aware routing protocol design for underwater acoustic sensor networks. *IEEE Access*, 7, 9091–9104.

Sun, Y., Zheng, M., Han, X., Li, S., and Yin, J. (2022). Adaptive clustering routing protocol for underwater sensor networks. *Ad Hoc Networks*, 136, 102953.

Todorovic, D. (2022). Bringing Intelligence to Wireless Sensor Nodes: Improving Energy Efficiency and Communication Reliability in Sensor Nodes. B.S. thesis, University of Twente.

ul Hasan, N., Ejaz, W., Atiq, M. K., and Kim, H. S. (2017). Energy-efficient error coding and transmission for cognitive wireless body area network. *International Journal of Communication Systems*, 30(7), e2985.

Ullah, S., Higgins, H., Shen, B., and Kwak, K. S. (2010). On the implant communication and mac protocols for wban. *International Journal of Communication Systems*, 23(8), 982–999.

Wan, J., Chen, M., Xia, F., Di, L., and Zhou, K. (2013). From machine-to-machine communications towards cyber-physical systems. *Computer Science and Informa_ tion Systems*, 10(3), 1105–1128.

Wang, C., Shen, X., Wang, H., Zhang, H., and Mei, H. (2023). Reinforcement learning based opportunistic routing protocol using depth information for energy-efficient underwater wireless sensor networks. *IEEE Sensors Journal*, 14, 1246–1261.

Wang, D., Liu, J., Yao, D., and Member, I. (2020). An energy-efficient distributed adaptive cooperative routing based on reinforcement learning in wireless multimedia sensor networks. *Computer Networks*, 178, 107313.

Wang, P. and Wang, T. (2006). Adaptive routing for sensor networks using reinforcement learning. In *The Sixth IEEE International Conference on Computer and Information Technology (CIT'06)*, pages 219–219. IEEE.

Wang, S. and Shin, Y. (2019). Efficient routing protocol based on reinforcement learning for magnetic induction underwater sensor networks. *IEEE Access*, 7, 82027–82037.

Willig, A., Matheus, K., and Wolisz, A. (2005). Wireless technology in industrial networks. *Proceedings of the IEEE*, 93(6), 1130–1151.

Wu, S., Zhong, S., and Liu, Y. (2018). Deep residual learning for image steganalysis. *Multimedia Tools and Applications*, 77, 10437–10453.

Li, X., Hu, X., Zhang, R., and Yang, L. (2020). Routing protocol design for underwater optical wireless sensor networks: A multiagent reinforcement learning approach. *IEEE Internet of Things Journal*, 7(10), 9805–9818. doi:10.1109/JIOT.2020.2989924

Xiao, M., Cui, H., Huang, D., Zhao, Z., Cao, X., and Wu, D. O. (2023). Traffic-aware energy efficient resource allocation for rsma based uav communications. *IEEE Transactions on Network Science and Engineering*, 19(3), 382–398.

Yang, Q., Jang, S.-J., and Yoo, S.-J. (2020). Q-learning-based fuzzy logic for multi-objective routing algorithm in flying Ad Hoc networks. *Wireless Personal Communications*, 113(1), 115–138.

Yang, S., Tan, C., Madsen, D. Ø., Xiang, H., Li, Y., Khan, I., and Choi, B. J. (2022). Comparative analysis of routing schemes based on machine learning. *Mobile Information Systems*, 2022, 1189–1208.

Yuan, Y., Liu, M., Zhuo, X., Wei, Y., Tu, X., and Qu, F. (2023). A q-learning-based hierarchical routing protocol with unequal clustering for underwater acoustic sensor networks. *IEEE Sensors Journal*, 22, 624–639.

Zhang, Y., Zhang, Z., Chen, L., and Wang, X. (2021). Reinforcement learning-based opportunistic routing protocol for underwater acoustic sensor networks. *IEEE Transactions on Vehicular Technology*, 70(3), 2756–2770.

Zhao, D., Zhou, Z.,Wang, S., Liu, B., and Gaaloul, W. (2020). Reinforcement learning–enabled efficient data gathering in underground wireless sensor networks. *Personal and Ubiquitous Computing*, 20, 1–18.

Zhao, G., Lin, K., Chapman, D., Metje, N., and Hao, T. (2023). Optimizing energy efficiency of lora-wan-based wireless underground sensor networks: A multi-agent reinforcement learning approach. *Internet of Things*, 22, 100776.

Chapter 3

Optimizing secure routing for mobile ad-hoc and WSN in IoT through dynamic adaption and energy efficiency

J. Viji Gripsy and A. Jayanthiladevi
Institute of Computer Science and Information Science,
Srinivas University, Mangalore, India

3.1 INTRODUCTION

3.1.1 Wireless sensor network

WSNs are systems of spatially dispersed, specialized sensors that track and record environmental variables and transmit the acquired information to a centralized point [1]. Temperature, levels of pollution, sound, humidity, and wind are among the variables that WSNs can monitor. Moreover, WSNs have gotten a lot of attention lately, due to the advent of micro-electromechanical systems (MEMS) advanced technologies, which make it possible to create sensor nodes that can perform multiple functions while also being small, cheap, and only requiring a small amount of processing power. They can detect physical amounts as well as ambient variables, which allows them to analyze information locally, wirelessly communicate, and collaborate [2]. This is an overview of WSNs' overall structure, as seen in Figure 3.1. A sensor node is invented consisting of a radio transceiver, a sensing device, a microcontroller, and a power supply, which is generally a battery, as well as other components. Sensor nodes may cost anything from a few dollars to several hundred dollars, depending on their capabilities. As a result of the sensor nodes' price and size limitations, the system's available resources, including energy, transmission, and computing, were also limited [3].

In addition to being an energy source, a sink node is a resourceful node that has unlimited communication and computing capabilities. It may be either fixed or dynamic, and it serves as an interface between a sensor network and a central administration hub. Based on the WSN application, the activity under observation may be either fixed or mobile. One use for movable sensor nodes is to mount them on wild animals to observe their behavior when they move in an unconventional way. When compared to sensor nodes, fixed and well-known sites are preferable [4].

A WSN is a kind of wireless network in which a large number of sensor nodes (also known as motes) circulate freely, self-direct, and use minimal power. While these networks undoubtedly include many small,

DOI: 10.1201/9781003474524-3

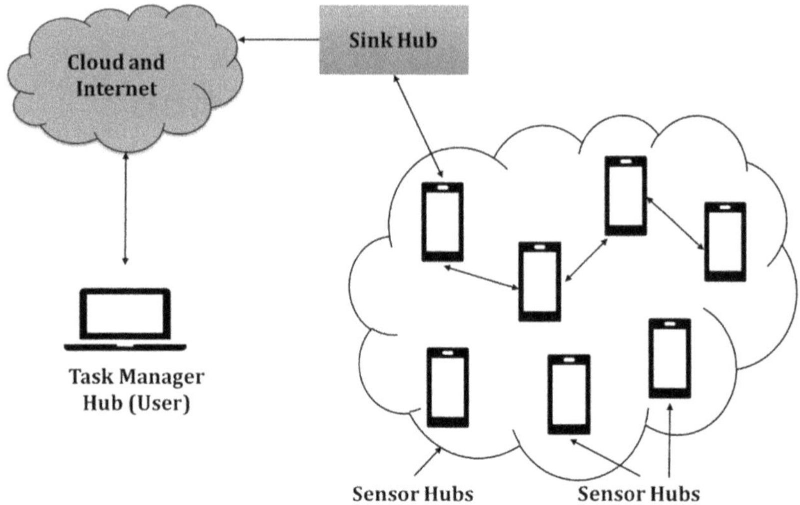

Figure 3.1 Structure of WSN.

battery-operated, geographically dispersed embedded devices that are networked for the purpose of meticulously capturing, processing, and transferring data to the users, they have also controlled the computing and processing capacities. Small computers known as nodes are used in networks to connect other nodes [5].

The sensor node is an energy-efficient wireless sensor node with a wide range of functions. Motes have a broad range of industrial uses. To meet particular application goals, a network of sensor nodes gathers data from the environment. Transceivers are used by motes to communicate with one another [6]. The number of motes in a WSN may be in the hundreds or even thousands. Ad hoc networks, in contrast to sensor networks, will have fewer nodes but will lack any organization. As a result of their widespread use, wireless sensor networks offer a slew of new conceptual and optimization challenges, including deployment, location, and tracking [7]. For the most part, coverage provides answers to concerns regarding a sensor network's ability to offer quality service (surveillance).

3.1.1.1 Motivation

The zone-based location-aided routing technique reduces energy consumption. Mobile nodes in this network have limited battery power. The large mobility nodes and zone size utilize more energy. Reduce energy usage with zone-based location-aided routing protocol. It uses zone routing and location-assisted node selection. The CZLAR also targets big zones and mobile nodes. CZLAR improves packet delivery ratio in low- and high-density networks. Health care, commerce, and sports stadiums can also consider real-life random node deployment and distribution scenarios.

3.1.2 Ad hoc networks (AN)

Ad hoc networks (AN) refer to a class of peer-to-peer wireless communication network in which direct communication takes place between wireless devices without the assistance of any access points (APs). As and when the need for communication arises, the WN is spontaneously established, subject to the condition that the participating nodes should be in close proximity to each other. In cases where the two devices are not in close range, multi-hop communication comes into play [8]. However, as the size and number of devices increase, the connection quality and communication speed deteriorate. On the contrary, the self-supporting nature of AN is quite handy in dealing with situations such as natural calamities, emergency situations, military operations, and other scenarios where a prompt response for data communication is required. Nevertheless, in spite of being easy to deploy and having quick response times, there are few practical limitations to AN, especially in terms of security and reliability.

A. Categories of AN

   AN can be categorized as follows:

   - MANET: This involves of a pool of nodes, which are mobile and self-sustaining. MANET has introduced a new category of network establishment that is best suited for an environment where an instant and cost-effective communication solution is required.
   - Wireless mesh networks (WMN): This is an intranet of various clients, that is, laptops, wireless nodes, and others, structured and connected in the form of a mesh topology. The data generated in WMN is shared within the mesh network and is not transmitted to the outside world.
   - WSN: This encompasses many sensor-based nodes and is deployed to sense and observe environmental data. The application of WSNs is widespread and covers areas such as smart cities, vehicle detection, weather monitoring, and so on [9].

3.1.3 Mobile ad hoc networks (MANET)

MANET is moving between different locations. In a general wireless network, the whole geographic region has some base stations, and each of them has fixed coverage in the geographic area. The mobile transmits whatever signals the nodes can hear at the base station. Any mobile node will always be coming to some base station, and the base stations deliver whatever the signal or data [10]. This environment is a generic view of wireless networks. The MANET nodes, in contrast, participate in cooperative packet transmission, with every mobile node participating for both transmission and reception, with the goal of transmitting the packet to a different node. A customized routing operation does not need to be carried out through the wireless router. To transmit the packet to the intended node, each mobile node can execute routing and route selection [11].

Mobile ad hoc networks require cooperative transmission of packets. The reason behind the cooperative communication is that, unlike other types of wireless networks, there is no base station to coordinate the routing process and to transfer the packet [12]. Because the MANET lacks a topology, collaborative transmission is necessary for packet forwarding. Figure 3.2 depicts a general mobile ad hoc network.

Similar to topology, there are other constraints on mobile nodes in this kind of network. The most important one is that the nodes come up with a single radio that has a fixed transmission range. The radio's transmission range determines how far the signal transmitted by the mobile node can be heard [13]. In addition, each node has some constant power in watts, which is required to activate the radio for the transmission and reception of signals. Whatever the process is, whether transmission or reception of the signal, the node loses some energy, and the lifetime of the node is only as long as the energy it has.

The mobile node has other properties like mobility speed; the node can move at any speed and in any direction. This property changes the topology of the network in a more frequent manner, which affects the transmission of packets highly. The factors that influence the functioning of the mobile ad hoc network and routing are discussed in detail in this section.

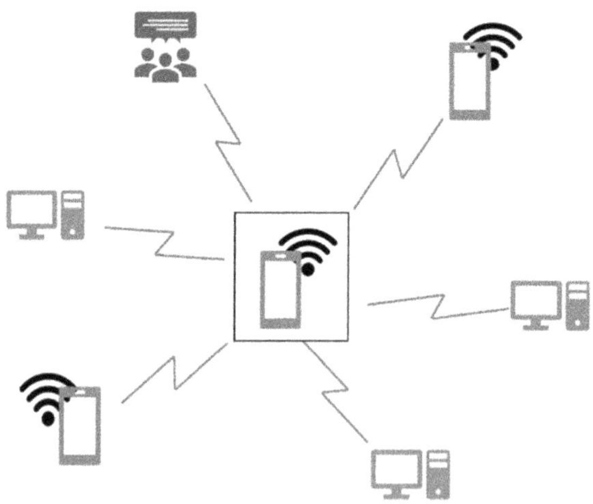

Figure 3.2 Mobile ad hoc network.

3.1.4 Overview of IoT

The IoT is largely reliant on sensors, when it comes to real-time data. Using these electronic gadgets in every sector would result in a huge flood of big data as they expand across the world's population. When designing electrical systems, the inverse aspects of low battery usage and high performance are important considerations to take into account [14]. The overview of the IoT is illustrated in Figure 3.3. There are two approaches to developing IoT:

1. Create a distinct network that is comprised only of physical things.
2. Expand the Internet's reach even farther, but this will require the use of hard-core innovations, including stringent cloud computing, large data storage, and many others (all expensive).

There is an incredible expanse of data produced by IoT devices that must be kept on a trustworthy storage server. Cloud technology comes into play in this situation. We can learn more about the system as the data is processed and analyzed, allowing us to better understand where things like electrical malfunctions are located inside it [15].

An internet connection is required in order to communicate, since each physical item is referenced by an IP address on the internet network [16]. The IP naming scheme, on the other hand, only allows for a finite number

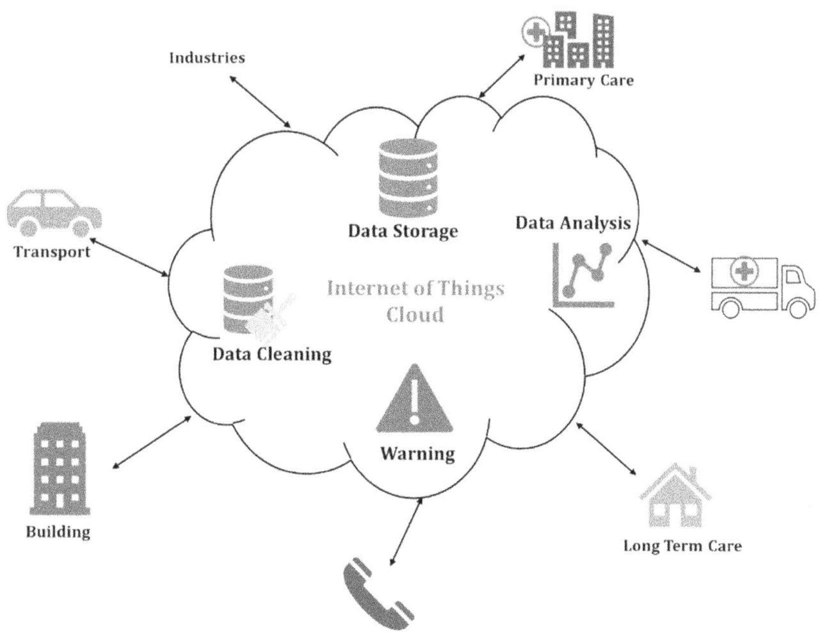

Figure 3.3 Overview of IoT.

of addresses to be used. This naming scheme will become unworkable if the number of devices continues to increase. Therefore, researchers are searching for an alternate naming system to describe each physical item in order to save time and resources [17].

3.2 DESIGN OF WSN IN IOT

This chapter presents a thorough technique of the IoT and its architecture used in this inquiry. Furthermore, the application formulations used in the calculations of diverse systems are examined. Detailed application-based architecture design and development is covered in this section, shown in Figure 3.4. Sensor network architecture is utilized in WSN. In addition to hospitals, highways, schools, and buildings, this kind of architecture may be used in a variety of settings, including disaster management, security management, and crisis management [4]. Because WSN is an application-based system, describing the designs in terms of application requirements and constraints is straightforward. Despite rapidly developing technology, society is moving toward a paradigm that is constantly linked [18]. Fixed and mobile networks are commonplace; open standards have been developed and made available for certain addressing procedures [5].

Future Internet concepts are being studied, developed, and constantly adapted to everyday life. IoT is a novel idea linked to the "future internet." By incorporating IoT, we envision a future in which everyday objects are integrated into the internet, their location and status are recognized, and intelligence is added in order to effectively expand the internet. This seamless integration of the virtual environment with the physical realm has an impact on people's individual and interpersonal environments [19].

IoT refers to the networking of real-world objects that have electronics as part of their construction [20]. This gives these objects the ability to communicate with one another and to understand how they are interacting

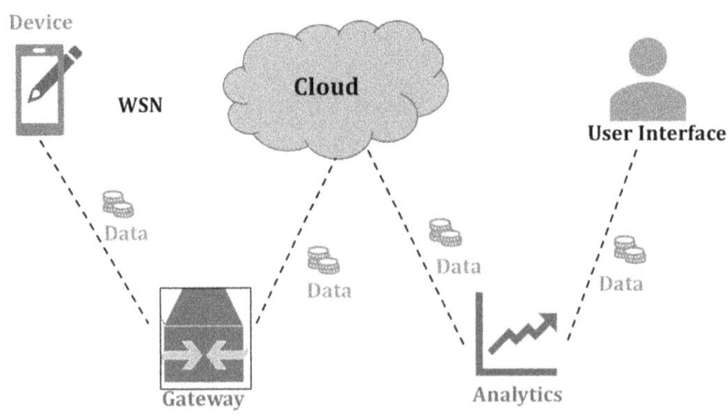

Figure 3.4 Structure of IoT in WSN.

with their environment. In the coming years, "technology based on the Internet of Things will provide advanced levels of service and drastically change the way people go about the business of living their everyday lives" [21]. IoT has become well established in a variety of domains, including but not limited to the following: the fields of medicine, agribusiness, gene therapies, smart cities, electric power, and smart homes, to name just a few. When it comes to industrial IoT automation, installing big local wireless sensor networks with extended battery life and the best operating system for their needs is a good first step. Developing and extending IoT networks includes using different kinds of devices and developing and improving wireless communication protocols and layering systems to better suit the WSN market [6].

3.2.1 Integration of WSN into IoT

In recent years, the Internet has transitioned from being a network of people to becoming an IoT. The integration of WSNs into the Internet of Things has resulted in high degrees of complexity in the management of interoperability. The IoT Gateway is essential for IoT applications because it connects wireless sensor networks to conventional communication networks or the Internet [22]. It enables the management and control of wireless sensor networks in addition to the seamless integration of WSNs with mobile communication networks or the Internet Figure 3.5.

WSNs have been opened up to the Internet; attackers may now conduct their nefarious operations from any location in the world. As a result, the WSNs should make every effort to resolve the problems that have arisen as a result of this internet connection, such as malware and other threats. In order to guarantee effective protection by the existing WSNs, they have been supplied with a central, strong gateway that is unique and powerful. Because of the scarcity of computing resources, the need for energy, and the need to save memory, it is difficult to repurpose the current security mechanism. Using WSN, users can monitor and manage system environments.

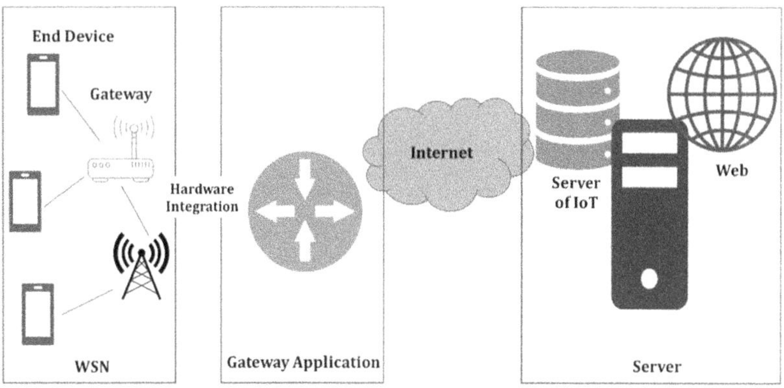

Figure 3.5 Structure of WSN with IoT system.

3.3 APPLICATIONS OF WSN AND IOT

The IoT project conducted a study in 2010 and found that IoT scenario applications could be categorized into 14 different categories, such as transportation and smart cities, smart homes, lifestyle, agriculture and retail, and smart factories. The most familiar applications are articulated, such as:

- Smart cities
- Smart grids and energy saving
- Health care
- Smartphone detection
- Water flow monitoring
- Wearables
- Smart homes
- Detection of radiation/hazardous
- Earthquake detection
- Traffic monitoring

The majority of IoT applications are found in enterprises that use sensors and other IoT devices, such as industry, commerce, and utilities. It's also found uses in agriculture, architecture, and building automation, prompting a few of these businesses to go on a digitalization path.

Figure 3.6 shows that WSNs and IoT have many uses. Mobility, heterogeneity, coverage, topology, and QoS classify these applications. The Internet of Things may help farmers streamline their work. Sensors can measure temperature, rainfall, humidity, soil content, and other elements to automate agricultural methods. Sensors measure soil content. The IoT can also monitor

Figure 3.6 Applications of WSN.

infrastructure-related activities. Sensors can monitor building, road, and infrastructure structural stability [12]. This can save money, boost productivity, improve quality of life, and streamline workflows. The Internet of Things affects every area, including health care, banking, retail, and industry.

3.4 SECURE ROUTING IN IOT–BASED MANET

Due to the reliability and security requirements of the acquired data to be transmitted to the base station, secure routing is an essential and difficult issue in MANET. The majority of current routing efforts concentrate on route discovery, route response, and route maintenance tasks for the dependable conveyance of the data gathered by the sensor nodes to the base station. Security considerations are not the main focus of these current algorithms. The attackers use their malicious actions to target the IoT–enabled sensor networks. Consequently, the most difficult design factors to take into account for the optimal design of an IoT–enabled MANET are energy efficiency and security concerns [23].

Ad hoc network routing deviates significantly from typical infrastructure network routing. It must take a number of things into account, including topology, routing path choice, and routing overhead. It needs to locate a path swiftly and effectively. The number of resources available is typically lower than in infrastructure networks, necessitating the use of optimal routing. A route to the destinations must be found among the network's many nodes, which necessitates a substantial and frequent exchange of routing control information. The volume of update traffic is significant, and it increases when the network has high mobility nodes. This can cause routing systems' overhead to crash, leaving no capacity for data packet delivery [24].

A node's transmission range is frequently constrained, and not all nodes can interact with one another directly. In order to communicate with other nodes across the network, a node must transmit packets. As there is no predetermined topology or setup of fixed routes, an ad hoc routing protocol is used to find and keep current routes between connecting nodes [25].

3.4.1 Energy efficiency in secure routing

The ratio of a node's data transmission rate to its overall energy consumption is known as energy efficiency. More energy efficiency indicates that the node can transmit more packets with the same amount of energy reserve. Whatever the packet is, it always travels through an intermediate node in an ad hoc network, consuming energy along the way. Also, it controls the asymmetric power consumption of the nodes. Restricted energy resources, difficulty renewing batteries, a lack of central coordination, restrictions on battery supply, choice of best transmission power, and channel utilization are the main causes of energy consumption in MANET [26].

In the MANETs, when the transmission takes place using the routing protocols, the energy consumption mainly occurs in three phases. They are:

- When a packet is being sent
- When a packet is received,
- When the computer is idle.

The greatest source of energy utilization in each node is the energy consumed while sending a packet. The energy is consumed when it is received by a packet. However, the energy is also used when the node is changed from its idle state during communication.

3.5 CRESENT ZONE–BASED LOCATION AIDED ROUTING PROTOCOL

MANETs help many network connections and applications. MANETs offer flexibility, efficiency, and inexpensive installation and use costs due to their dynamic nature. MANETs are self-configuring, remotely connected mobile device communication networks. Devices are mobile since the networks are wireless. Online learning, emergency response, and other industries benefit from MANET, which requires multipoint-to-multipoint connectivity. Wireless transmitter and receiver patterns in MANETs depend on node positions in an ad hoc topology. Ad hoc topology changes when nodes move. In practice, mobile and wireless networks' reliability characteristics mean that each node in a wireless network has always been susceptible to failure and that the system's reliability depends on its constituents' dependability and the wireless network design's redundancy. MANETs rarely provide QoS and trustworthy services. Unreliability is caused by low memory, processing power, battery capacity, high mobility, and unexpected node behavior. Bandwidth, delay, packet loss, and jitter are crucial for MANET reliability and stability. Multimedia app demand rises. QoS becomes essential. Multicasting supports efficient one-to-many or many-to-many communications. Multicasting communication involves sending a single packet with a list of addresses to the destination, which is subsequently repeated to all accessible destinations. MANET can help researchers improve existing processes see the below table for more information.

MANET routing protocols find a viable path and conserve resources. MANET's properties limit data packet routing. Thus, MANET cannot employ wired network routing techniques, so a protocol that addresses these unique needs is needed. MANET protocols are proposed. The protocol may store stale routes due to dynamic node connection and disconnection, creating routing resource overhead. Routing protocols must balance all issues. Routing relies on a protocol's ability to discover new nodes.

The vast majority of the ongoing research is concerned with the development of fault-tolerant, low-cost, and energy-efficient routing protocols. The goal of MANET research was to lower the node's energy usage. Given its significance, some studies have concentrated on a different element that can help MANET protocols reduce network effects while increasing the number of packets sent to the sender by adding overhead. As a result, transmitting the message indefinitely reduces battery life until shutdown.

3.6 LITERATURE REVIEW (TABLE 3.1)

Table 3.1 Comparison for zone-based location aided routing protocols

Authors	*Methodology*	*Pros*	*Cons*
Michaelraj Kingston Roberts et al. (2022)	An enhanced high-performance cluster-based secure routing protocol, a framework that is dependable and efficient with respect to energy use [27]	Improves the data management quality	High energy consumption
Mohammad Ali Alharb et al. (2021)	Clustering and routing have been improved so that they now provide area-based clustering that is determined by the transmission range of network nodes [28]	Efficient node density management	Low network capacity
Payal Thakrar et al. (2020)	Route selection Algorithm TORA used to find a path can be easy [29]	Low energy consumption	Proper assortment of path cannot be possible
Srivastava (2019)	Routing techniques in MANET utilized for information correspondence layer [30]	Low delay	Low in robustness
Sushma Ghode et al. (2018)	Improved EE-HRP that uses a routing approach to find the optimal balance between minimizing power consumption [31]	Increased network longevity	Low network capacity
Mohamed et al. (2018)	Cluster-based routing monitoring the base station periodically [32]	Minimize loss of nodes	Packet size is large and difficult to handle
Khalid et al. (2017)	LBR protocols such as Focused Beam Routing (FBR) Hop-Hop Vector-based Forwarding, and Vector-based Forwarding (VBF) [33]	Higher number of alive nodes	High E2E and energy consumption
Gripsy et al. (2017)	ArcRect Route Selection	Reduced route size	Consumes lot of energy

(*Continued*)

Table 3.1 (Continued)

Authors	*Methodology*	*Pros*	*Cons*
Lin et al. (2016)	Moving zone-based routing protocol achieves a delivery rate, delivery time, and communication overhead [34]	It has fewer location updates	Overall performance is Low
Selvi & Vijayaraj (2016)	Zone-based routing protocol improved the quality of service [35]	Reduces the control overhead	Security issues
Gripsy et al. (2014)	TWRect routing technique	Lowering the overhead of network routing while simultaneously improving security	Consumes more energy

3.7 METHODOLOGY

The route-finding procedure eliminates intra-zone mobile nodes, as shown in Figure 3.7 This only considered nodes that were located within the crescent-shaped area. For this process, the radius value is computed using the hop count. The radius size in this instance, for example, will be 8. The controller is in charge of communicating data to the target. The wave length can be cut even further if there are fewer mobile nodes.

The algorithm provided below, which details the phases of the protocol's implementation, serves as a representation of the CZLAR.

ALGORITHM 3.1: ALGORITHM FOR CZLER

Step 1: Every node's neighbor and zone are initially determined and kept in node N, destination D.
Step 2: To create a zone (Z), obtain the number of hops as a radius (R).
Step 3: Analyze the transaction log to choose a controller.
Step 4: If (certain node N can receive traffic T,
Step 5: The mode of Node A is active.
Step 6: If (node N = destination D) then
Step 7: Accept and send an acknowledgment (ACK).
Step 8: Else
Step 9: Check that Destination D is located in zone (Z), and then send the packet using the inter-zone routing method if that is the case.
Step 10: Else, employ the intra-zone routing mechanism.
Step 11: End

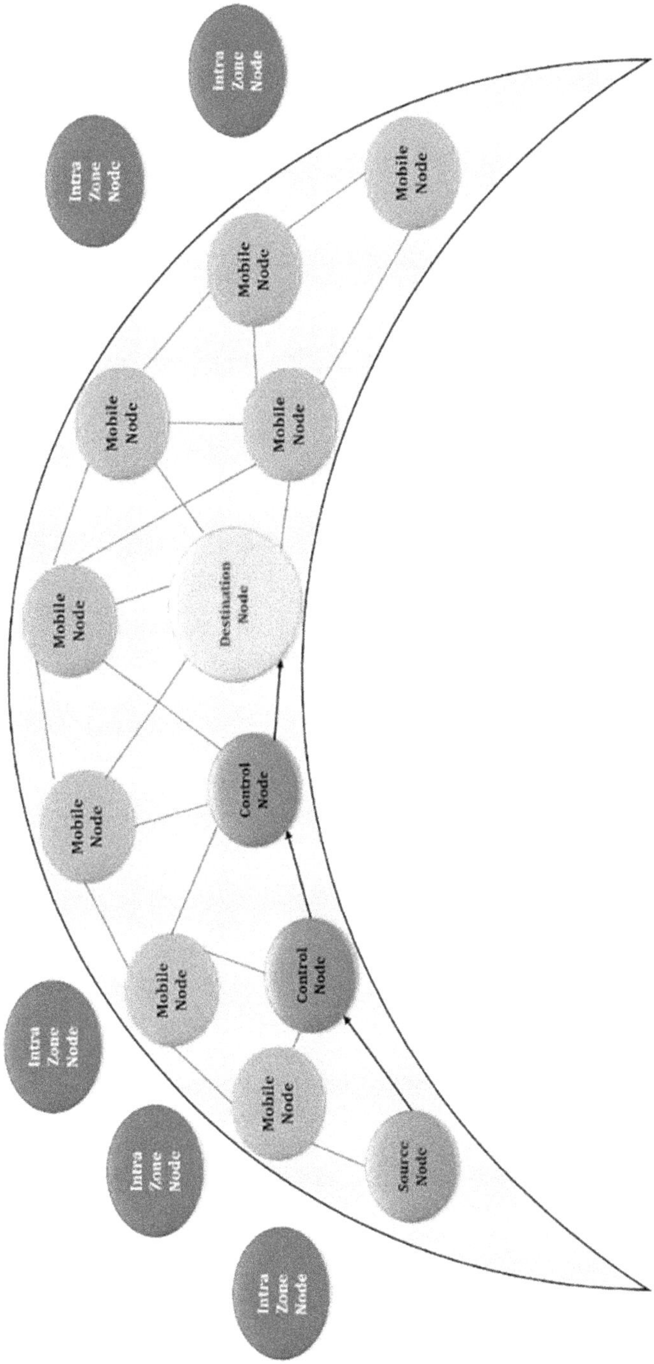

Figure 3.7 Route discovery of CZLAR.

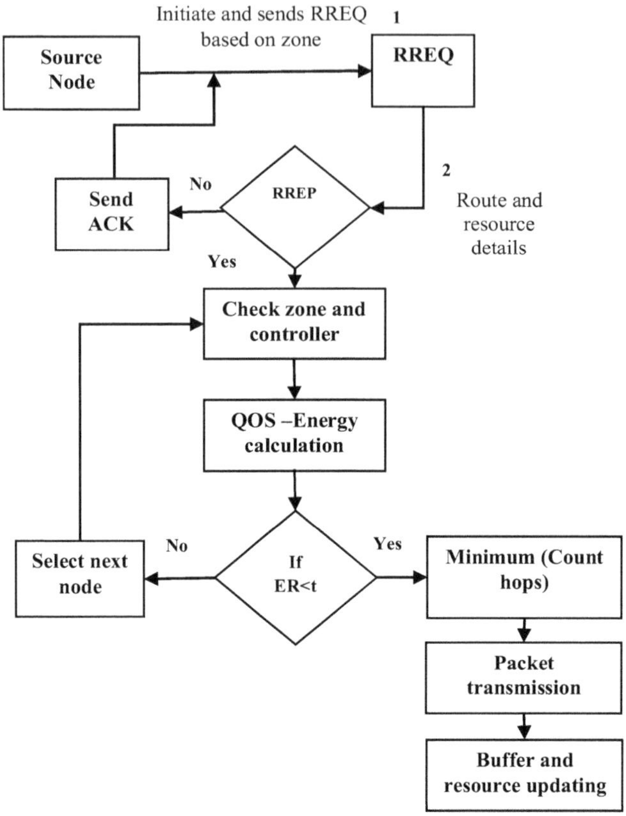

Figure 3.8 Flowchart of CZLAR.

All mobile nodes regularly initiate the controller selection and rejection procedure. The node needs all neighbor data and node utility for this procedure, regardless of whether it has coordinators or not. Every mobile node has its neighbors discovered in step 1 of the process. The remaining energy of that node is also estimated in combination with this data for the selection procedure Figure 3.8. The residual energy of the controller or intermediary node is computed to achieve the QOS in data transmission. The "HELLO" messages, which carry all the network's information, are linked to this data. With this knowledge, a wait period before declaring a coordinator is computed for each of the nodes. The quantity of neighbors and the amount of residual energy are indirectly related to the delay duration.

3.8 RESULTS AND DISCUSSION

The simulation of the proposed work has been done and is shown in following table.

Table 3.2 Simulation parameters

Parameter	*Specifications*
Simulator	NS2.35
Simulation Area	20 × 20
Data Dissemination	2 to 40 seconds
Energy Consumption	2 to 40 seconds
No. of nodes	50 to 100
No. of rounds	180
BS location	150*50
Transmitter/receiver electronic circuit	Eelc = Etx = Erx = 50 nJ/bit
Energy of nodes	50 J

The proposed CZLER is compared with ArcRect and TWRDASR. Performance is evaluated mainly according to the following metrics. End to End Delay, Packet Delivery Ratio, Throughput, and Energy Consumption.

To assess the scalability of the protocols, the number of nodes varies from 20 to 100. It can be observed that the E2D of the proposed CZLER ranges from 12 to 104 milliseconds. Since it was proposed, CZLER performs load balancing, which reduces the queue waiting time during congestion. Hence the delay of the proposed CZLER is less when compared to ArcRect and TWRDASR, which are shown in Tables 3.2, 3.3, and Figure 3.9.

The result graph of throughput for different numbers of nodes is shown in Figure 3.10. It can be observed that the throughput of the proposed CZLER ranges from 432 to 532. Because the proposed CZLER performs load balancing, packet drops caused by network node overload are reduced. Hence, the throughput of CZLER is higher than that of ArcRect and TWRDASR Table 3.4.

Table 3.3 End-to-end delay

Nodes	*ArcRect*	*TWRDASR*	*Proposed CZLER*
20	33	23	12
40	53	43	32
60	83	73	61
80	93	83	70
100	123	113	104

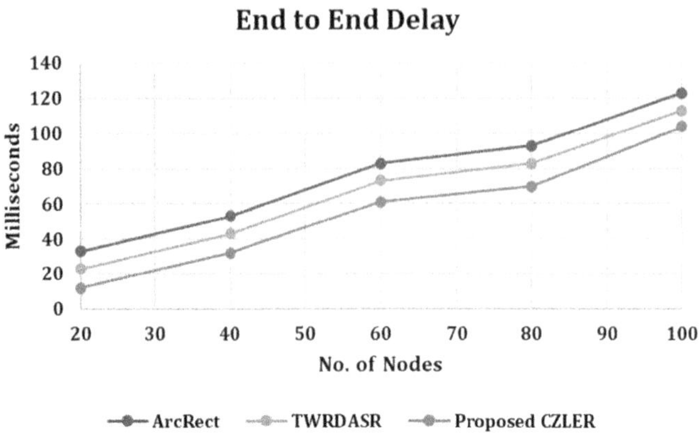

Figure 3.9 End-to-end delay.

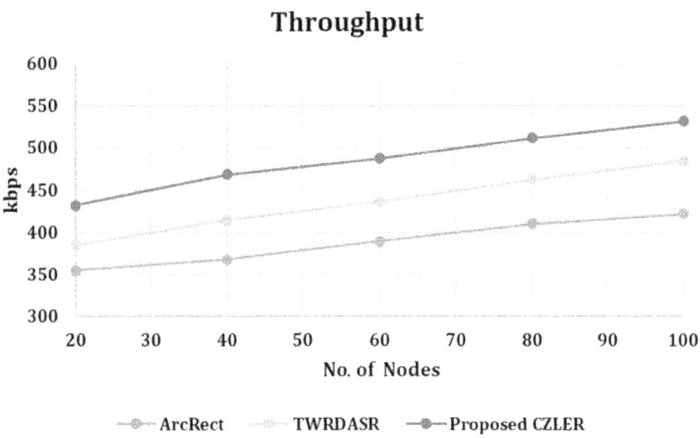

Figure 3.10 Throughput.

Table 3.4 Throughput

Nodes	*ArcRect*	*TWRDASR*	*Proposed CZLER*
20	355	385	432
40	368	414	468
60	389	436	487
80	410	462	512
100	422	484	532

Table 3.5 Packet delivery ratio

Nodes	*ArcRect*	*TWRDASR*	*Proposed CZLER*
20	0.37	0.44	0.62
40	0.39	0.45	0.63
60	0.26	0.29	0.47
80	0.06	0.1	0.28
100	0.04	0.08	0.26

The result graph of PDR for different numbers of nodes is shown in Figure 3.11 see Tables 3.5 and 3.6. It can be observed that the PDR of the proposed CZLER ranges from 0.62 to 0.26. Since the proposed CZLER performs load balancing, the packet drops due to overload are reduced. Hence, the PDR of the proposed CZLER is higher than ArcRect and TWRDASR.

Table 3.6 Energy consumption

Nodes	*ArcRect*	*TWRDASR*	*Proposed CZLER*
20	7.45	8.76	9.55
40	9.16	9.97	10.76
60	10.1	11.32	12.11
80	10.13	11.58	12.37
100	11.05	12.4	13.19

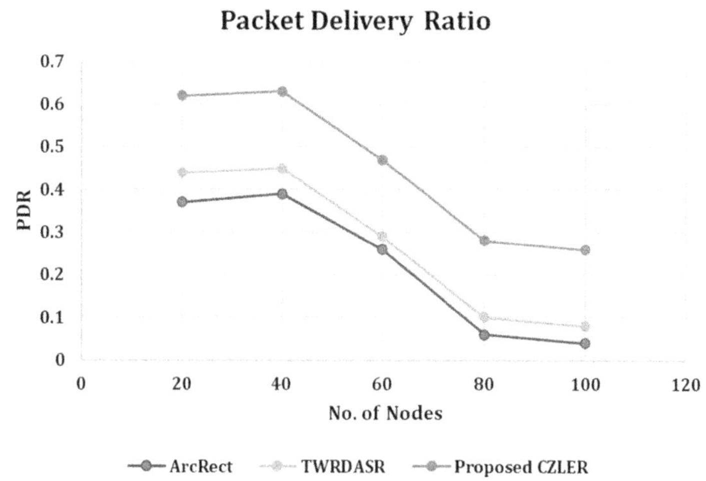

Figure 3.11 Packet delivery ratio.

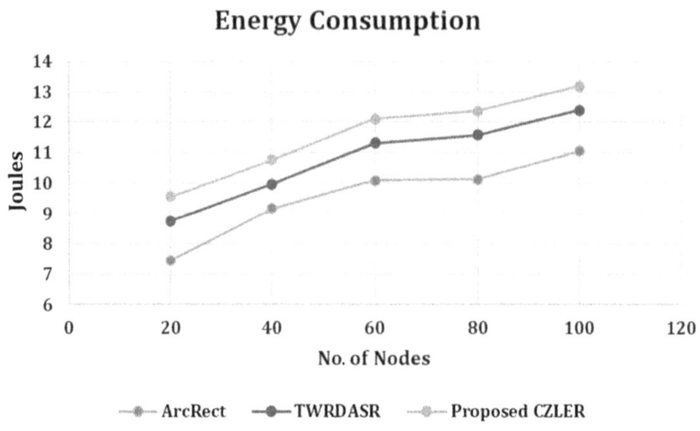

Figure 3.12 Energy consumption.

The result graph of energy consumption for different numbers of nodes is shown in Figure 3.12. It can be observed that the energy consumption of proposed CZLER ranges from 9.55 to 13.19 joules. Since CZLER implements energy-efficient route selection, the energy consumption of CZLER is less when compared to ArcRect and TWRDASR.

3.9 CONCLUSION

Mobile ad-hoc networks have a number of qualities that set them apart from other networks. MANET presents a substantial problem in the implementation of a routing protocol architecture despite having many distinctive features and being portable. In regard to energy use, resource utilization, and packet delivery, this becomes more challenging. With the successful construction of a MANET routing protocol, this is possible. As a result, the objective of this research is to create a routing protocol that is energy efficient while retaining QoS. The CZLER protocol extends the life of the node, uses less energy, performs with a high packet delivery ratio and low packet drop rate, is cost-effective, and allows for quick and secure intra-node communication. The simulation's findings demonstrate that the suggested CZLER uses less energy while achieving an improved throughput and packet transmission ratio. Considerations for real-life random node deployment and random node distribution scenarios can also be made in the context of health-care organizations, shopping zones, and stadiums for sports and this work will be extended to IoT–based medical sensors.

REFERENCES

[1] Nguyen, Thien D Jamil Y Khan, & Duy T Ngo, 'An effective energy-harvesting-aware routing algorithm for WSN-based IoT applications', In *IEEE International Conference on Communications (ICC)*, pp. 1–6.

[2] Ahmad, A, A Ullah, C Feng, M Khan, S Ashraf, M Adnan, S Nazir, & HU Khan 2020, 'Towards an improved energy efficient and end-to- end secure protocol for IoT healthcare applications', *Security and Communication Networks*, https://doi.org/10.1155/2020/8867792

[3] Al Qundus, Jamal, Kosai Dabbour, Shivam Gupta, Régis Meissonier, & Adrian Paschke 2020, 'Wireless sensor network for AI-based flood disaster detection', *Annals of Operations Research*, pp. 1–23, https://doi.org/10.1007/s10479-020-03754-x

[4] Ali, Ahmad, Yu Ming, Sagnik Chakraborty, & Saima Iram 2019, 'A comprehensive survey on real-time applications of WSN', *Future Internet*, vol. 9, no. 4, pp. 77.

[5] Amale, Onkar, & Rupali Patil 2019, 'IoT based rainfall monitoring system using WSN enabled architecture', In *3rd International Conference on Computing Methodologies and Communication (ICCMC).*

[6] Al-Turjman, Fadi 2019, 'Cognitive-node architecture and a deployment strategy for the future WSNs', *Mobile Networks and Applications*, vol. 24, no. 5, pp. 1663–1681

[7] Behera, TM, SK Mohapatra, UC Samal, MS Khan, M Daneshmand, & AH Gandomi 2019, 'I-SEP. An improved routing protocol for heterogeneous WSN for IoT-based environmental monitoring', *IEEE Internet of Things Journal*, vol. 7, no. 1, pp. 710–717.

[8] Elappila, Manu, Suchismita Chinara, & Dayal Ramakrushna Parhi 2020, 'Survivability Aware Channel Allocation in WSN for IoT applications', *Pervasive and Mobile Computing*, vol. 61, pp. 101107.

[9] Haseeb, Khalid, Ikram Ud Din, Ahmad Almogre, & Naveed Islam 2020, 'An energy efficient and secure IoT-based WSN framework: An application to smart agriculture', *Sensors*, vol. 20, no. 7, pp. 2081.

[10] Ahmed, M, M Salleh, & MI Channa 2018, 'Routing protocols based on protocol operations for underwater wireless sensor network: A survey', *Egyptian Informatics Journal*, vol. 19, no. 1, pp. 57–62.

[11] Abella, CS, S Bonina, A Cucuccio, S D'Angelo, G Giustolisi, AD Grasso, A Imbruglia, GS Mauro, GA Nastasi, G Palumbo, & S Pennisi 2019, 'Autonomous energy-efficient wireless sensor network platform for home/office automation', *IEEE Sensors Journal*, vol. 19, no. 9, pp. 3501–3512.

[12] Alghamdi, Turki Ali 2020, 'Energy efficient protocol in wireless sensor network: Optimized cluster head selection model', *Telecommunication Systems: Modelling, Analysis, Design and Management*, vol. 74, no.3, pp. 331–345.

[13] Li, Wen, & Sami Kara 2017, 'Methodology for monitoring manufacturing environment by using wireless sensor networks (WSN) and the internet of things (IoT)', *Procedia CIRP*, vol. 61, pp. 323–328.

[14] Shukla Anurag, & Sarsij Tripathi 2020, 'multi-tier-based clustering framework for scalable and energy efficient WSN-assisted IoT network', *Wireless Networks*, vol. 14, pp. 1–23.

[15] Elijah, Olakunle, Tharek Abdul Rahman, Igbafe Orikumhi, Chee Yen Leow, & MHD Nour Hindia, 'An overview of Internet of Things (IoT) and data analytics in agriculture: Benefits and challenges', *IEEE Internet of Things Journal*, vol. 5, no. 5, pp. 3758–3773.

[16] Ahmed, G, X Zhao, MMS Fareed, MR Asif, & SA Raza 2019, 'Data redundancy-control energy-efficient multi-hop framework for wireless sensor networks', *Wireless Personal Communications*, vol. 108, no. 4, pp. 2559–2583.

[17] Farhan, Laith, Sinan T Shukur, Ali E Alissa, Mohmad Alrweg, Umar Raza, & Rupak Kharel 2017, 'A survey on the challenges and opportunities of the Internet of Things (IoT)', In *Eleventh International Conference on Sensing Technology (ICST)*, IEEE, pp. 1–5.

[18] Gardaševiü, Gordana, Konstantinos Katzis, Dragana Bajiü, & Lazar Berbakov 2020, 'Emerging wireless sensor networks and Internet of Things technologies—Foundations of smart healthcare', *Sensors*, vol. 20, no. 13, pp. 3619.

[19] Preeth, SSL, R Dhanalakshmi, R Kumar, & PM Shakeel 2018, 'An adaptive fuzzy rule-based energy efficient clustering and immune-inspired routing protocol for WSN-assisted IoT system', *Journal of Ambient Intelligence and Humanized Computing*, vol. 22, no. 1, pp. 1–13.

[20] Shahraki, Amin, Amir Taherkordi, Øystein Haugen, & Frank Eliassen 2020, 'A survey and future directions on clustering: From WSNs to IoT and modern networking paradigms', *IEEE Transactions on Network and Service Management*, vol. 8, no. 2, pp. 2242–2274.

[21] Jaiswal, Kavita, & Veena Anand 2019, 'EOMR pp. an energy-efficient optimal multi-path routing protocol to improve QoS in wireless sensor network for IoT applications', *Wireless Personal Communications*, vol. 16, pp. 1–23.

[22] Elappila, Manu, Suchismita Chinara, & Dayal Ramakrushna Parhi 2018, 'Survivable path routing in WSN for IoT applications', *Pervasive and Mobile Computing*, vol. 43, pp. 49–63.

[23] Bajaj, K, B Sharma, & R Singh 2020, Integration of WSN with IoT applications: A vision, architecture & future challenges. *Integration of WSN and IoT for Smart Cities*, pp. 79–102.

[24] Roberts, MK, P Ramasamy 2023, An improved high performance clustering based routing protocol for wireless sensor networks in IoT. *Telecommun Syst*, vol. 82, pp. 45–59. https://doi.org/10.1007/s11235-022-00968-1

[25] Benayache, Ayoub, Azeddine Bilami, Sami Barkat, Pascal Lore, & Hafnaoui Taleb 2019, 'MsM A microservice middleware for smart WSN-based IoT application', *Journal of Network and Computer Applications*, vol. 144, pp. 138–154.

[26] Chaitra, M, & B Sivakumar 2017, 'Disaster debris detection and management system using WSN & IoT', *International Journal of Advanced Networking and Applications*, vol. 9, no. 1, p. 3306.

[27] Alharbi, MA, M Kolberg, & M Zeeshan 2021, Towards improved clustering and routing protocol for wireless sensor networks. *Journal on Wireless Communications and Networking*, vol. 2021, p. 46. https://doi.org/10.1186/s13638-021-01911-9

[28] Payal M Thakrar, Bhushan Bharat, & G Sahoo 2019, 'Routing protocols in wireless sensor networks', In *Computational Intelligence in Sensor Networks*, pp. 215–248.

[29] Srivastava, A, A Prakash, & R Tripathi 2020, 'Location based routing protocols in VANET: Issues and existing solutions', *Vehicular Communications*, vol. 23, pp. 100231.
[30] Srivastava, D, V Sharma, & D Soni 2019, 'Optimization of CSMA (Carrier Sense Multiple Access) over AODV, DSR & WRP routing protocol', In *4th International Conference on Internet of Things: Smart Innovation and Usages (IoT-SIU)*, pp. 1–4.
[31] V Srivastava, S Tripathi, & K Singh 2020, 'Energy efficient optimized rate based congestion control routing in wireless sensor network', *Journal of Ambient Intelligence and Humanized Computing*, vol.11, no.3, pp. 1325–1338.
[32] Mohamed, RE, WR Ghanem, & MA Mohamed 2018, 'TECEAP: Two-tier era-based clustering energy-efficient adaptive and proactive routing protocol for wireless sensor networks', In *35thNational Radio Science Conference (NRSC)*, pp. 187–196.
[33] Khalid, M, Z Ullah, N Ahmad, H Khan, HS Cruickshank, & OU Khan 2017, 'A comparative simulation-based analysis of location-based routing protocols in underwater wireless sensor networks', In *2nd Workshop on Recent Trends in Telecommunications Research (RTTR)*, pp. 1–5.
[34] Lin, D, J Kang, A Squicciarini, Y Wu, S Gurung, & O Tonguz 2016, 'MoZo: A moving zone-based routing protocol using pure V2V communication in VANETs', *IEEE Transactions on Mobile Computing*, vol. 16, no. 5, pp. 1357–1370.
[35] Selvi, SA, & A Vijayaraj 2016, 'Increasing quality of service in video traffic using zone routing protocol in wireless networks', In *World Conference on Futuristic Trends in Research and Innovation for Social Welfare (Startup Conclave)*, pp. 1–5.

Chapter 4

Energy efficient in-network data aggregation in Internet-of-Things

Radhakrishnan Maivizhi
SRM Institute of Science and Technology, Kattankulathur, India

Palanichamy Yogesh
CEG, Anna University, Chennai, India

4.1 INTRODUCTION

Over the last few years, the Internet of Things (IoT) paradigm emerged as an eminent technology and has become an indispensable part of humans' everyday lives. The widespread deployment of IoT technologies promotes the human lifestyle more intelligently by offering flexibility and facility in a range of day-to-day applications, namely smart city (Wang et al. 2018), smart meter (Li et al. 2018), smart home (Jia et al. 2019), intelligent transport (Ramamoorthi and Sangaiah 2019), smart healthcare (Fischer et al. 2020), and production automation (Huang 2020). The IoT is designed as an interconnected network of a large number of things or objects or devices or nodes. The smart things connected with IoT are heterogeneous and hence the architecture of IoT is flexible, which is illustrated in Figure 4.1. It comprises four layers, namely sensing, networking, cloud computing, and application layers. Due to its flexible architecture and communication between the virtual world and the real world, the number of things connected to IoT is increasing every year. The rise in number of things increases the amount of sensory data generated, which results in a high amount of data traffic. In general, the devices of IoT are battery powered with limited energy. As more things (devices) are connected to IoT–based systems, the associated computational and communication overhead poses significant challenges to the life of IoT–based systems. According to Ding et al. 2003, the energy consumed for transmitting information is very high when compared to the energy consumed for computation. The increase in energy utilization therefore decreases the lifespan of IoT devices which in turn degrades the network efficiency. The process of reducing the amount of data traffic thus is of paramount importance, which automatically conserves a significant amount of bandwidth and energy which in turn improves the lifetime of IoT–based systems. Since enhancing the lifetime of IoT is crucial, the researcher community thus turned their attention toward developing techniques that increases the lifetime of IoT networks.

 DOI: 10.1201/9781003474524-4

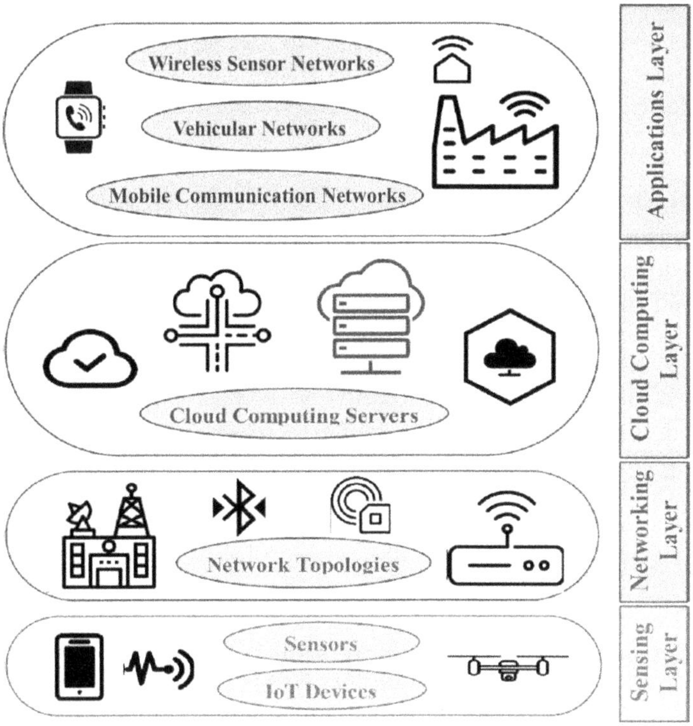

Figure 4.1 Architecture of IoT (Yousefi et al. 2021).

4.1.1 Data reduction techniques

A central and challenging issue in IoT is the energy restriction of smart devices. To tackle this issue, researchers have investigated different data reduction techniques in order to keep down the number of data transmissions taking place in IoT. For instance, the techniques like data aggregation, data compression, data fusion, data elimination, and data filtering help in saving energy of IoT devices.

The main objective of data aggregation is to eliminate data redundancy and reduce the number of data transmissions in IoT whereas data compression techniques use encoding techniques in order to bring down the size of the data. However, more energy will be consumed if the amount of data compressed is large. In data fusion, which is a superset of data aggregation, the focus is on other information rather than data with the use of techniques like machine learning, statistical analysis, and probability. One way of processing data is data elimination, which eliminates some of the data packets based on sampling techniques or temporal aggregation. But the selection of sample size affects the accuracy of data. Although data filtering is a subset of data fusion, which removes the invalid or incorrect data packets, the amount of data traffic will not be reduced to the intended limit.

Among the preceding techniques, an important renowned paradigm that is widely applied to minimize the amount of data transmission in IoT–based systems is data aggregation.

4.1.1.1 Data aggregation

In the IoT–based systems, data aggregation is a kind of data fusion, in which redundant data is eliminated and data of different devices are combined using aggregation functions namely sum, average, count, minimize, maximize, and so on, and forward the single aggregated data packet either to the upstream device or to the sink or base station (BS) / cloud center (CC). It is a distributed approach and follows the many-to-one data processing paradigm. As shown in Figure 4.2, the no-aggregation scenario requires a total of 11 data transmissions for a network of five devices and the aggregation scenario requires only five data transmissions. Data aggregation thus minimizes the amount of data transmissions that take place in the IoT network in order to extend the lifetime of networks. In general, data aggregation prolongs the lifetime of devices and thereby increases the network lifetime. However, the increase in speed of data generation and transmission increases the cost of data aggregation. Therefore, the essential goal in the design of data aggregation protocols is to gather and combine the sensory data efficiently in terms of data traffic, energy utilization, network lifespan, and data accuracy. However, the procedure employed in the data aggregation process depends on the type of network, topology, amount of data generated, and the type of communication. As the data are processed (aggregated) at intermediate nodes, data aggregation is otherwise known as in-network data aggregation. The benefits of performing data aggregation (in-network mode) in IoT–based systems are:

i. Data traffic in the system is minimized
ii. Data latency is reduced
iii. Energy depletion of devices is minimized
iv. Network scalability is increased

Therefore, by employing data aggregation techniques, the lifetime of devices and networks in an IoT ecosystem can be enhanced.

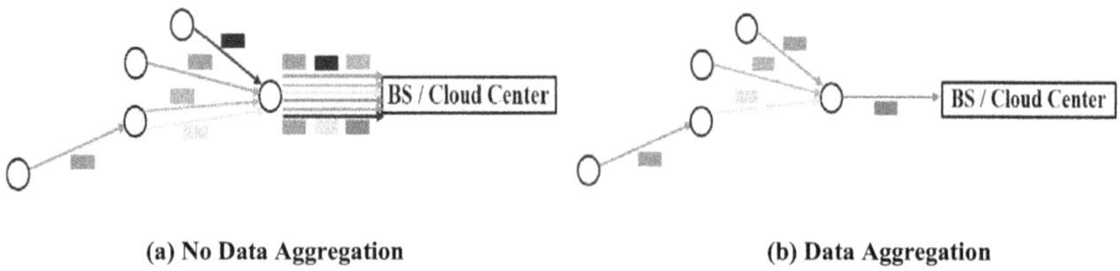

Figure 4.2 Effect of data transmission in IoT.

To minimize data traffic, the routing protocols play an important role in data aggregation techniques. In addition, scheduling of devices minimizes the data aggregation latency to ensure data freshness. Apart from routing and scheduling, ensuring privacy of data is very important during aggregation of data. This chapter focusses on research works carried out by chief experts in the arena of IoT data aggregation in terms of routing, scheduling, and security.

4.2 DATA AGGREGATION ROUTING IN IOT

Though in-network data aggregation algorithms reduce the energy utilization of IoT–based systems, the efficacy of these algorithms depends on the routing structure employed. During data aggregation, the huge volume of data generated by different devices entails the transformation of aggregated outcome into intelligence. Such a transformation is possible only when the aggregated outcome is forwarded to the BS. To accomplish the forwarding process, a routing structure in required. Routing structure thus is the principal component of IoT data aggregation. As most of the devices change their position often, communication among the devices also changes and this results in varying network topology. The frequent changes in network topology and restricted resources make the routing process a challenging task in in-network data aggregation. Several incipient researches have investigated the routing structure problem in networks of IoT–based systems and proposed many energy efficient data aggregation routing algorithms (Yousefi et al. 2021). Depending on the routing structures, the data aggregation approaches are classified into three kinds, namely cluster-based, tree-based, and centralized approaches, which are shown in Figure 4.3. This

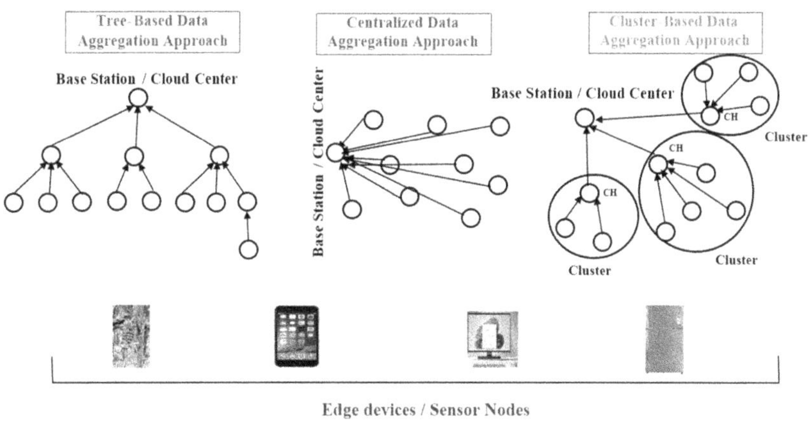

Figure 4.3 Different kinds of in-network data aggregation routing techniques (Khan and Chishti 2022).

section explores the different routing structures to enhance the performance of data aggregation.

4.2.1 Tree-based approaches

In tree-based data aggregation, the network is modeled as a tree, in which the sink initiates the tree construction, starting from itself and proceeding hierarchically toward the leaf nodes. Nodes between the sink and the leaf nodes act as aggregators, aggregating the data from their children.

For complex queries such as skyline and equality joins, the data aggregation tree construction entails connecting a series of aggregation operations (Yin and Wei 2018). These aggregation operations are established by utilizing the aggregation gain concept. For an aggregation operation, the aggregation gain is defined as follows:

$$\text{Aggregation gain} = \frac{\text{Size of data points that is pruned during aggregation}}{\text{Size of data points that is transmitted for aggregation}}$$

Based on the aggregation gain value, the parent nodes are selected and the aggregation operations are constructed. Because of temporal correlations among data, the constructed aggregation tree is updated periodically when queries are issued continuously. The aggregation tree construction process is shown in Figure 4.4. Though this work constructs a communication-efficient data aggregation tree, it does not scale well with a large amount of data.

Learning-based data aggregation routing protocol consists of two objective functions (Homaei et al. 2019). According to the first objective function, an aggregation tree is built, in which the maximum number of children

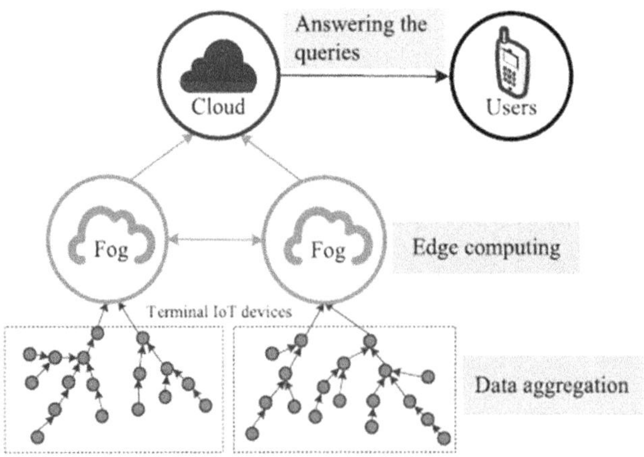

Figure 4.4 Aggregation tree construction for user queries in terminal IoT devices (Yin and Wei 2018).

allocated to a parent node is limited. As a result, there is an increase in the tree height, which decreases the probability of collisions, which in turn reduces network congestion. The second objective function is that the learning automata equipped with each node decides whether to perform data aggregation or directly send the data to the parent node. Although limiting degree decreases the number of control messages to be exchanged, it significantly increases the data aggregation delay by increasing the height of the aggregation tree.

4.2.2 Cluster-based approaches

In a cluster-based system, the whole network is partitioned into several clusters. Generally, each cluster consists of several devices, and by executing an election algorithm, the leader device from each cluster is nominated as cluster head (CH). The CH is accountable for aggregating the data and forwarding the aggregated outcome either directly or in a multi-hop fashion to the sink.

A cluster-based data aggregation technique based on beta dominating set principle is proposed (Khan and Chishti 2022). In an IoT network modelled as a graph, a set of nodes, S is a beta-dominating set, if all nodes in the network are in S or their distance is less than β (threshold) to at least one of the nodes of S. Initially, the number of clusters are fixed. For each cluster, centroids are calculated. Then using the sum of squares metric, the distance between the centroid and the sensory data (represented as data points) is calculated. If the calculated distance is smaller than β (threshold), then the data point becomes a member of the current cluster. Otherwise, the centroids are recomputed using beta-dominating principle. The combination of these two metrics assigns more data points to each centroid resulting in a smaller number of clusters, which in turn improves the overall data aggregation process. Although this method of data aggregation improves the data freshness and abridges the end-to-end delay, its efficiency depends on the initial random selection of clusters. In a hybrid method, fuzzy logic and the Capuchin search optimization algorithm are combined in order to route aggregated data to BS (Mohseni et al. 2022). This hybrid routing technique includes two phases. In the first phase, the fuzzy logic system employs the nodes' residual (remaining) energy and distance from sink for dividing the network nodes into number of clusters. To prevent all the nodes from becoming CHs, a threshold value is predefined. In the second phase, the Capuchin search algorithm finds the routes between CHs and cluster members. In addition, routes from CHs to BS are discovered. On the basis of current position and best position of nodes, the proposed algorithm determines the routes. After finding all the routes, the best routes are selected based on fitness function, which takes energy consumption, number of data packets delivered, delay, and network lifespan as parameters. The proposed aggregation routing protocol alleviates the energy hole problem and is appropriate for scalable IoT–based systems. The efficiency of the routing algorithm can be improved by including temporal information during the cluster formation and route selection process.

4.2.2.1 *Grid-based approach*

A variant of cluster based approach is grid-based data aggregation, in which the network area is partitioned into grids of equal size (Sankar et al. 2020). As shown in Figure 4.5, for each grid, grid head is selected by employing probabilistic method. The grid head is accountable for aggregating sensory data from nodes in the grid. The parent of each grid head is chosen based on the expected transmission count for routing the aggregated data to the sink. Though this work is simple, it incurs high message exchange overhead.

4.2.3 Centralized approaches

In centralized approach, data from all the nodes in the network are sent to a designated node called central node. When compared to other nodes, the central node is equipped with rich resources and features. The nodes in the network send data to the central node, which then aggregates all the data and forwards a single aggregated data packet. If direct transmission is not possible, then each node sends its data to the central node through the shortest path (Uddin et al. 2018). A distributed approach which aggregates data from several nodes in various IoT applications based on request-response model eliminates the data traffic problem (Zhu et al. 2017). This approach solves the heterogeneity problem by utilizing semantic technologies. However, the central node in this work is not easily accessible.

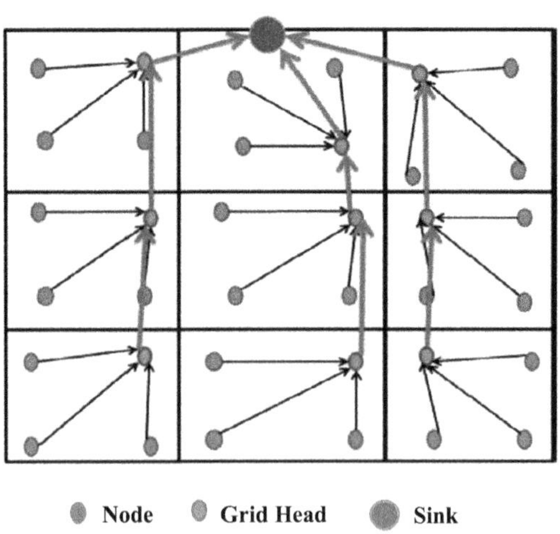

Figure 4.5 Grid-based data aggregation (Sankar et al. 2020).

4.3 IOT DATA AGGREGATION SCHEDULING

Although in-network data aggregation as an inherent paradigm of IoT–based systems saves a significant amount of energy, it still faces an intractable problem: how to perform data aggregation with minimum delay? The data aggregation process increases the overall delay, which is not applicable to practical applications. The total amount of time the nodes should wait indicates the aggregation delay—the higher the waiting time of intermediate nodes, the higher the end-to-end delay. The waiting time of a node may increase due to unbalanced tree structure, congestion, node failure, and loss of data packets. In addition, the rise in delay increases the energy utilization of IoT networks. One solution for this problem is to schedule the nodes effectively with minimum number of time slots. Minimizing delay thus is a cardinal significance in data aggregation with the objective of conserving energy. This section provides techniques for performing data aggregation at the desired level with reduced delay and energy consumption.

4.3.1 Effect of heterogeneity charging rate of devices

In wireless powered IoT (WPIoT), the charging rate of the devices is highly heterogeneous. This heterogeneity of charging rates and the diversity of interference among wireless signals increase the aggregation delay in WPIoT. As data aggregation with minimum delay is critical under the heterogeneity charging rate of IoT devices, many algorithms were developed for WPIoT networks (Jiao et al. 2021a, 2021b). The heterogeneous devices, variety of interference signals, and energy features make the aggregation scheduling problem a complicated one. To fix this problem, the authors proposed a data aggregation scheduling approach based on hypergraph. Since hypergraphs effectively capture the correlations among multiple relations, they are used to process the high-order data in a great manner. In order to minimize the delay of data aggregation procedure, the proposed approach divides the problem into two subproblems. First, an aggregation tree is constructed considering the energy features and forwarding time slots of IoT devices. In the tree construction process, devices with high residual energy and more charging rates are chosen as forwarding nodes (parents). Second, links with sufficient residual energy and that are signal interference free are identified, and by leveraging hypergraph theory, transmission time slots are assigned to these links.

4.3.2 Effect of duty cycling on aggregation delay

Since the energy-consuming rate of sensors is much higher than their recharging rate (Chen et al. 2019), the energy efficacy of sensors can be upgraded by employing a prevalent method called duty cycling (Buettner et al. 2006). In this technique, the sensors wake up only for receiving and/or transmitting

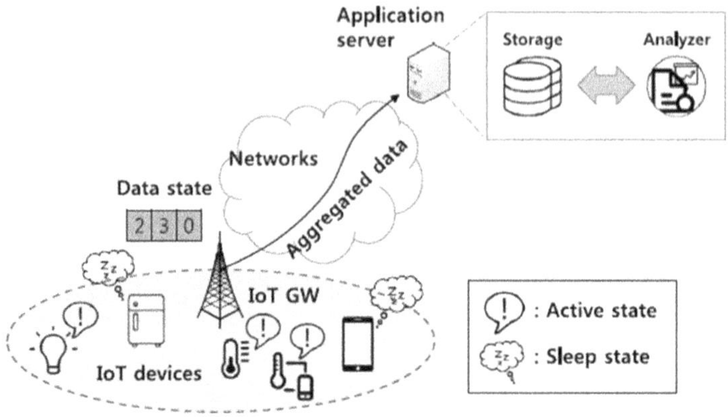

Figure 4.6 Data aggregation scheduling in duty cycling scenario (Ko et al. 2019).

data in a working time period. Otherwise, the sensors are put into sleep state in order to save energy. As the data aggregation is an intermittent process, duty cycling techniques, shown in Figure 4.6, are suitable for aggregating data in IoT networks. Nevertheless, the sleeping time compromises the delay caused by aggregation, especially when the sleeping time of sensors is very high when compared to the active time. Many researchers have investigated the minimum delay data aggregation problem in a duty cycling scenario.

The use of predefined routing structure in existing data aggregation scheduling limits the use of active slots of sensors. In addition, the traditional bottom-up and two-phase methods lead to higher aggregation delay. These problems are studied in Nguyen et al. (2021), in which the active time slots of sensors are efficiently utilized by simultaneously building the aggregation tree and allocating a transmission schedule for the nodes being added to the tree. The proposed period-driven pipeline scheduling approach builds the tree by determining the senders from the set of unselected nodes to the selected receivers already in the tree. Simultaneously, for each pair of nodes, the time slots for data transmission are assigned. The sender-receiver pairs are selected using the multilevel ranking strategy (based on two metrics, namely set-based neighbor count and link delay) on a bipartite graph of candidate nodes. Though the proposed approach performance is good in terms of aggregation delay and energy utilization, limiting the number of children assigned for each node is a major downside of this work.

4.3.3 Impact of waiting time on data accuracy

During active states, IoT devices consume significant amounts of energy even if they are not sensing and/or transmitting data. In such cases, the energy efficiency of these devices can be improved by turning off the device's radio so that they go into a sleep state. Furthermore, data aggregation minimizes the volume of data to be transmitted to save energy of IoT devices.

Nonetheless, if the IoT device is idle for a long period of time and/or the sink node does not receive any information for a long period of time, the data may be inaccurate and is not useful for further analysis. Data accuracy thus is important in the data aggregation scheduling process.

To resolve this issue, the sink node decides the sleep duration of IoT devices. Each device has different sleep duration, as the IoT devices connected in the network are sensing and transmitting different kinds of phenomena. As the required degree of data accuracy varies with the type of IoT devices (type of data sensed), the sleep duration assigned to the IoT devices varies accordingly. For instance, for usual data like temperature, pressure, humidity, and so on, data accuracy can be kept at high level even if the aggregated data is not instantly sent to the server, whereas for emergency data like heartbeat rate, fire detection, and so on, data should be sent to the server for immediate action, as data accuracy is very important. Hence, the duration of sleep mode of devices and data aggregation duration varies with respect to the type of IoT devices. To achieve desired level of accuracy of aggregated data, the sink node jointly determines the IoT device's sleep time and aggregation time. This problem is expressed as a Markov decision process (MDP), where the ideal policy jointly determines the sleep and aggregation duration of IoT devices (Ko et al. 2019). The proposed work defines when a device goes into sleep state and when it performs data aggregation/transmission. The joint policy is optimal, as it considers the energy conserved during the sleep state and the amount of data and data accuracy for scheduling IoT devices. Although this joint optimal policy outperforms other existing algorithms in terms of energy, accuracy and volume, it will work only in a homogeneous network environment. This is a major drawback, as almost all IoT–based systems consist of heterogeneous devices.

4.3.4 Effect of content type

In many applications, the data generated are highly correlated and data aggregation combines the data before forwarding to the sink. Consequently, the amount of data sent decreases, which results in network efficiency. On the other hand, different applications generate different data and they cannot be simply aggregated. As a result, hot spot and congestion problems take place frequently at nodes nearer to the sink node in the network. A new cross-layer paradigm employing 6TiSCH for data aggregation scheduling scheme is proposed (Jin et al. 2017). Based on the contents of data packets, this approach creates routing structures and scheduling schemes. As different devices generate different contents, overlapped routing trees and many schedulers are created and are shown in Figure 4.7. That is, for each content type, a separate routing tree and a MAC schedule is allocated based on two modules. The first module, which is responsible for finding the routing structure for each application, consists of two functions: by employing a greedy algorithm, the first function determines the routing structure that

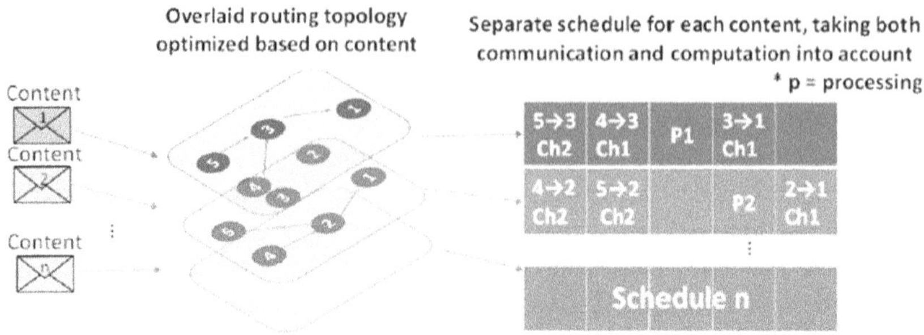

Figure 4.7 Data aggregation routing and scheduling (Jin et al. 2017). For each content type, a separate routing structure and scheduler are created.

best fits each application requirement. That is, for the same source node, different relay nodes with different data packets are selected. Distinctive routing structures are also formed. The second function balances the overall data traffic among all applications. For instance, if too many applications use a single node, then this module suggests an alternative path for routing data, thereby congestion is reduced in the network. The second module takes the routing structures created by the first module and assigns time slots to each node of each routing structure. In addition, special time slots are reserved for critical messages. These special timeslots are retained at fixed intervals in a slot frame. Though the proposed work reduces network traffic and end-to-end delay in a significant manner, it has its own practical implementation difficulties due to hardware memory limitations.

4.3.5 Link-delay aware data aggregation

Performing data aggregation at link level is one way of decreasing the end-to-delay. In a massive IoT, the use of multiple channels minimizes the data aggregation delay (Vo et al. 2022). By leveraging the active time slots of sensor nodes, an aggregation tree containing pipeline links is constructed in a top-down fashion. Initially, the routing tree consists of only the sink node. Nodes are then added to the tree on the basis of delay value. The candidate nodes having minimum link-delay value with the selected node is included in the tree. This procedure is iterated until all the nodes are added in the aggregation tree. Once the aggregation tree is constructed, time slots are assigned to all nodes except the sink node in a bottom-up fashion starting from the leaf nodes. The proposed reinforcement scheduling approach enhances the node scheduling by employing the unused time slots and channels that are not used by the main scheduling algorithm. The complete reinforcement aggregation schedule for a model network of 12 nodes is shown in Figure 4.8. Though the proposed scheme produces a minimum-length aggregation schedule, energy, an important and limited resource directly affecting the lifetime of IoT networks, is not within the scope of this work.

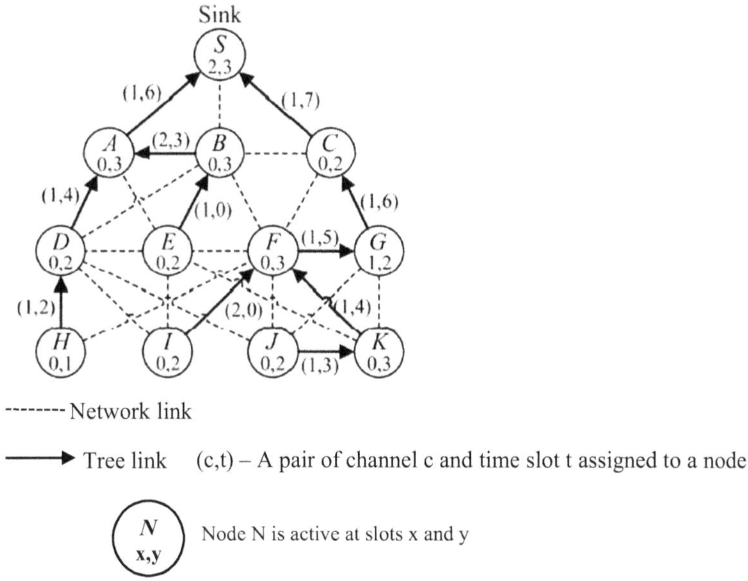

Figure 4.8 Complete aggregation schedule on a sample tree (Vo et al. 2022).

4.4 SECURE DATA AGGREGATION IN IOT

During data aggregation, devices connected with IoT are assumed to be honest and trustworthy. However, the open nature of IoT and wireless medium allow adversaries to launch a series of security attacks while aggregating data. These attacks necessitate the implementation of a security mechanism with data aggregation. This section describes the research works related to secure data aggregation in IoT.

4.4.1 Encrypted data aggregation

One way of securely aggregating the data is to encrypt the data. As homomorphic encryption schemes perform aggregation on encrypted data, it is possible to do data aggregation securely. With encryption, authentication, and signature techniques, the privacy and integrity of the aggregated data can be achieved.

A secure data aggregation scheme for aggregating the sensory data of IoT devices is modelled by Zhang et al. (2020). In order to protect data confidentiality, the sensory data comprising the user's sensitive information are encrypted using the public key of Paillier homomorphic encryption scheme. The resulting ciphertext is attached with signature and is sent to the edge server. The edge server acts as aggregator. It checks the signature of all ciphertext in order to ensure data integrity, and if all signatures are valid, it aggregates the ciphertext as shown in Figure 4.9. After aggregation, the edge

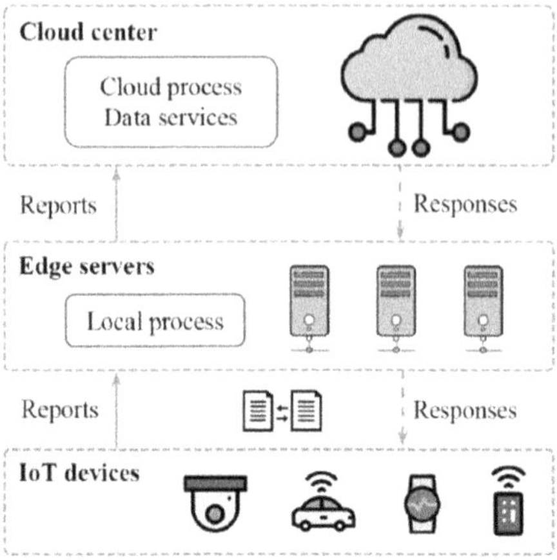

Figure 4.9 Privacy preserving data aggregation (Zhang et al. 2020). The data generated by IoT devices are encrypted and forwarded to edge server for aggregation and the aggregated data is forwarded to cloud server for further analysis and decision making.

server attaches a signature with the aggregated ciphertext and forwards it to the CC. The CC decrypts the aggregated ciphertext using the private key of Paillier cryptosystem for making intelligent decisions. Although the suggested scheme is lightweight, verifiable, and secure against chosen message attack, it is still vulnerable to collusion attack.

The healthcare devices connected with IoT generate a huge volume of data. As these devices have limited resources, data aggregation plays a central role in conserving the resources of health-care devices. However, the data aggregation process creates security issues such as compromising the integrity and privacy of the health data (Sándor et al. 2015, Mineraud et al. 2016). A novel mechanism in which the health data produced by the IoT devices are securely aggregated by combining a public key cryptosystem with secret sharing techniques is proposed. Initially, the health data is signed by the patients using private keys and who send the signed data to health centers. This signature enables the health centers to determine the duplicate data and award the patients accordingly. The health centers encrypt the data using Boneh-Goh-Nissim cryptosystem and aggregate the ciphertext. Before transmitting the data to the cloud center, the health centers add noise with aggregated data. The addition of noise prevents the users and/or cloud servers from inferring the health data by launching differential attacks. The cloud center, in turn, aggregates all the aggregated data transmitted from health centers and sends the final aggregated result to the users for community health control or research investigation (Tang et al. 2019).

Several researches on secure data aggregation consider only one-dimensional data. Nonetheless, the IoT devices generate multidimensional data. As shown in Figure 4.10, to protect the privacy of multidimensional

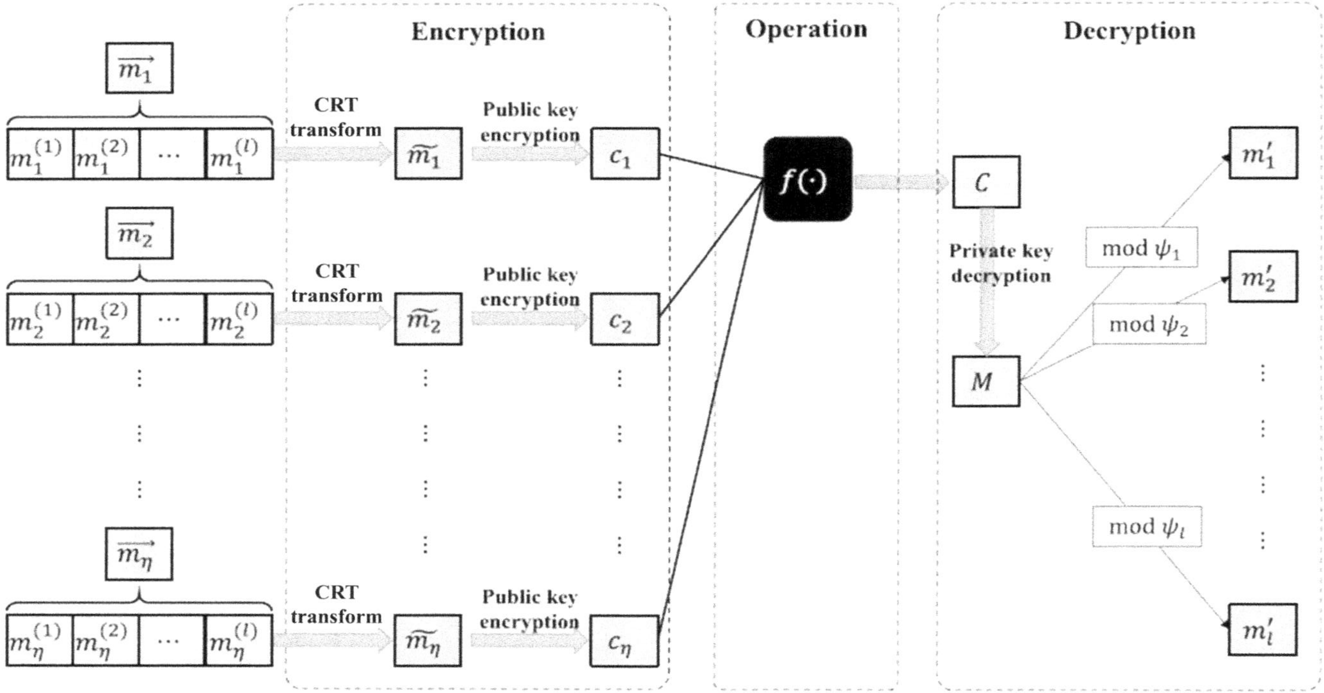

Figure 4.10 Multidimensional data aggregation (Peng et al. 2021). The CRT technique transforms the multidimensional data $m_i^1, m_i^2, \ldots, m_i^l$, $1 \leq i \leq n$ into a single entity, which is then encrypted using public key, then aggregated, and the aggregated data is decrypted using private key.

data during aggregation, a homomorphic encryption scheme is developed based on Chinese Remainder Theorem (CRT) (Peng et al. 2021). The multi-dimensional data is encrypted with a CRT–based homomorphic encryption scheme and the ciphertext is then attached with a signature before sending it to edge centers. The edge center verifies all the signatures and if they are valid, it aggregates the ciphertext and sends the aggregated data to the cloud for further analysis. When compared to existing solutions in terms of communication and computing costs, the suggested method meets the acceptable level of security.

Another work which aggregates data based on the query-response paradigm is developed in González-Manzano et al. (2016). The basic working idea behind the proposed model is that the network nodes are organized into clusters. Initially, the sink node issued the multi-attribute queries into the system. The cluster heads receiving the queries forward them to cluster members. In this model, actual cluster members generate data. By employing Paillier homomorphic encryption techniques, the cluster members encrypt the data and transmit the ciphertext to the cluster head. After receiving the ciphertext from all of its members, the cluster head aggregates the ciphertext by enabling correlation between multiple attributes. In addition, aggregation is performed on the assumption of preorder combination of attributes. Each cluster head in turn sends the aggregated ciphertext to the sink, where the aggregated data is decrypted and the responses are sent to the user for making intelligent decisions. To make the data aggregation task secure, this model depends on trusted authority, which makes the proposed model less efficient. Furthermore, the number of attributes aggregated by the model is limited.

The security mechanism (Liu et al. 2022) tries to thwart the attackers from eavesdropping on private data during data aggregation. Data are aggregated based on cluster topology, where the CHs are elected depending on residual energy and distance between the smart device and the BS. The cluster members employ the elliptic curve encryption scheme to encrypt the sensory data. It then generates a homomorphic message authentication code (MAC) and attaches it with the ciphertext and sends the ciphertext along with authentication code to the cluster header. The CH verifies the MAC and if it is valid, it aggregates the ciphertext of its children and forwards the single aggregated ciphertext to the sink node. After that, the sink node decrypts the aggregated data to recover the original data. Though this mechanism increases the accuracy, the energy consumption of smart devices significantly increases.

4.4.2 Data aggregation under node/device compromise attack

One way of securing in-network data aggregation is to defend against the most destructive attack, the node/device compromise attack. If the IoT devices are deployed in unattended environments (especially in environmental surveillance), these attacks are easily launched by the adversaries in order to take control of the devices. Once the devices are compromised, the attacker launches malicious attacks to make the aggregated result useless and in some cases dangerous. Of these attacks, the most devastating attack is the false data injection attack (FDI). By acquiring the secrets, the devices are compromised and are ordered to inject false data, so that the base station could mislead to erroneous determination and wastes network resources. Therefore, it is significant to detect the compromised devices and false data at an initial stage to protect the network from FDI attacks.

A robust technique for detecting FDI attacks even with low frequency and intensity is developed in Yang et al. (2017). To combat FDI attacks, a Bayesian model is adopted which describes the temporal and spatial attributes of sensory data. By exploiting the statistical attributes of the sensory data, an anomaly detection method detects the compromised nodes at an early stage. Further, the suspected devices are identified by hypothesis testing. Threshold-based mechanisms are designed for protecting the integrity of the aggregated data, which analyzes the uncertainties in the aggregated data. By modeling the FDI attacks using game theoretical approach, robust strategies are devised to prove that the gain acquired by adversaries are limited by the defenders. Although the proposed method prevents the injection of false data, the assumption of predefined cluster structure and an aggregation protocol are major pitfalls of this work.

4.4.3 Blockchain-based data aggregation

To make the IoT devices interact in an auditable manner with trust, researchers have combined blockchain technology with IoT. As blockchain technology protects the privacy of users' data in an end-to-end manner, it is used in the IoT data aggregation process to secure the data being aggregated.

Based on block chain technology, the data aggregation process can be secured (Wang et al. 2020). To preserve the privacy, the blockchain is modified by adding a security label in the block header. This security label comprises of two pieces of information: the level of security and the requirements to be completed. Due to the addition of the security label, new rules are designed to generate the blocks. These rules improve the computation capability of blocks which in turn decreases the latency and increases the throughput of the data aggregation process. In addition, to protect privacy, the transactions and transaction receivers are decomposed into groups. To make the scheme energy efficient, routes are constructed by employing deep reinforcement learning techniques. Though this work is strong against collusion attack, it is not scalable in terms of latency.

The blockchain technology also evades the problem of single point of trust during the data aggregation process (Loukil et al. 2021) and is shown in Figure 4.11. To protect the privacy of aggregated data, the blockchain is combined with homomorphic encryption. The raw sensory data is encrypted

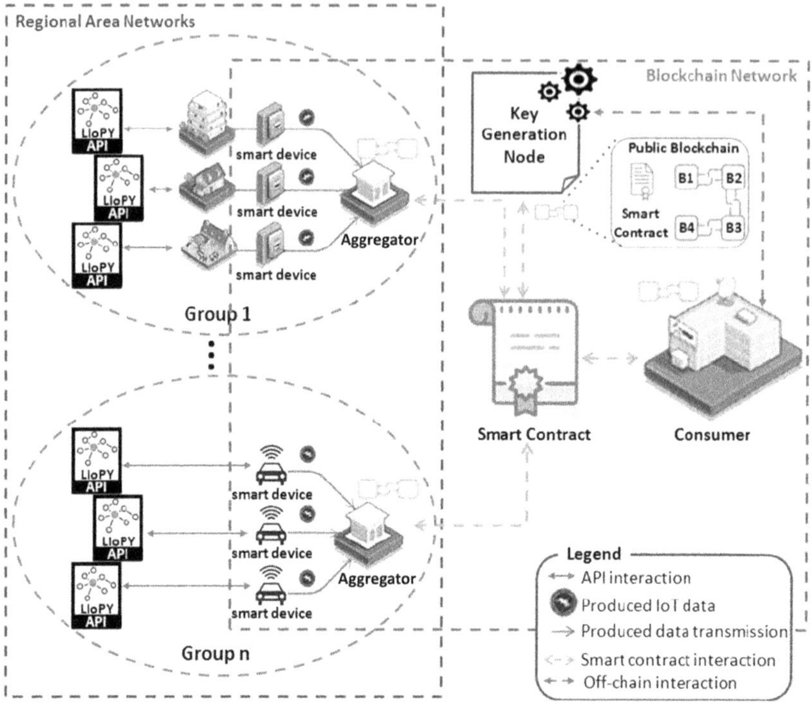

Figure 4.11 Blockchain-based data aggregation (Loukil et al. 2021). The data of smart devices in a regional area network are aggregated by aggregators and the aggregated data is sent to the block chain as a response to the request made by blockchain. All the nodes in the blockchain network have at least one blockchain address in order to interact with the blockchain, which serve as a distributed tamper protection.

with homomorphic encryption scheme which is then stored in blockchain which acts as a distributed data storage and eventually avoids the trust problem. The data aggregation process is controlled by smart contracts, which accept the request, group the smart devices based on this request for the response, and publish the aggregation result. Though blockchain overcomes trust issues, it incurs storage overhead. Since the IoT devices are resource constrained, the use of blockchain technology restricts the number of applications that perform data aggregation using block chain technology.

4.5 CONCLUSION

In-network data aggregation in IoT has received a lot of interest from both industry and academia because of its potential for conserving significant amounts of energy of resource-constrained smart devices. This chapter described how in-network data aggregation approaches minimize the amount of data transmissions among the smart devices in order to enhance the lifetime of IoT–based systems. This chapter explained several techniques describing how to perform data aggregation and how it impacts data traffic, latency, energy, scalability, and security. Though in-network aggregation solved many issues related to IoT, still there are many issues to be solved. Some of the open issues are:

- In most of the IoT data aggregation approaches, the monitoring area is static. Performing data aggregation in a dynamic environment is an interesting research problem.
- Though heterogeneous devices are connected with IoT, most of the research dealt with homogeneous devices. An open issue is performing data aggregation in heterogeneous devices.
- Reducing the computational overhead of mobile devices is another open problem.
- Another research problem is performing data aggregation for complex operations like median, quantile, rank, mode, and so on.
- Most of the data aggregation approaches discussed in this chapter use single sink. Performing data aggregation in the presence of multiple sink is another research issue.

REFERENCES

Buettner, Michael, Gary V. Yee, Eric Anderson, and Richard Han. "X-MAC: a short preamble MAC protocol for duty-cycled wireless sensor networks." In *Proceedings of the 4th International Conference on Embedded Networked Sensor Systems*, pp. 307–320. 2006.

Chen, Kunyi, Hong Gao, Zhipeng Cai, Quan Chen, and Jianzhong Li. "Distributed energy-adaptive aggregation scheduling with coverage guarantee for battery-free wireless sensor networks." In *IEEE INFOCOM 2019-IEEE Conference on Computer Communications*, pp. 1018–1026. IEEE, 2019.

Ding, Min, Xiuzhen Cheng, and Guoliang Xue. "Aggregation tree construction in sensor networks." In *2003 IEEE 58th Vehicular Technology Conference. VTC 2003-Fall (IEEE Cat. No. 03CH37484)*, vol. 4, pp. 2168–2172. IEEE, 2003.

Fischer, Gabriel Souto, Rodrigo da Rosa Righi, Vinicius Facco Rodrigues, and Cristiano André da Costa. "Use of Internet of Things with data prediction on healthcare environments: a survey." *International Journal of E-Health and Medical Communications (IJEHMC)* 11, no. 2 (2020): 1–19.

González-Manzano, Lorena, José M. de Fuentes, Sergio Pastrana, Pedro Peris-Lopez, and Luis Hernández-Encinas. "PAgIoT–Privacy-preserving aggregation protocol for Internet of Things." *Journal of Network and Computer Applications* 71 (2016): 59–71.

Homaei, Mohammad Hossein, Ely Salwana, and Shahaboddin Shamshirband. "An enhanced distributed data aggregation method in the Internet of Things." *Sensors* 19, no. 14 (2019): 3173.

Huang, Xianming. "Intelligent remote monitoring and manufacturing system of production line based on industrial Internet of Things." *Computer Communications* 150 (2020): 421–428.

Jia, Mengda, Ali Komeily, Yueren Wang, and Ravi S. Srinivasan. "Adopting Internet of Things for the development of smart buildings: A review of enabling technologies and applications." *Automation in Construction* 101 (2019): 111–126.

Jiao, Xianlong, Wei Lou, Songtao Guo, Yong Li, Junmei Yao, Fuqiang Gu, and Junchao Ma. "Energy-aware concurrent data aggregation scheduling for wireless powered IoT leveraging hypergraph theory." *IEEE Wireless Communications Letters* 10, no. 11 (2021a): 2464–2468.

Jiao, Xianlong, Wei Lou, Songtao Guo, Ning Wang, Chao Chen, and Kai Liu. "Hypergraph-based active minimum delay data aggregation scheduling in wireless-powered IoT." *IEEE Internet of Things Journal* 9, no. 11 (2021b): 8786–8799.

Jin, Yichao, Usman Raza, Adnan Aijaz, Mahesh Sooriyabandara, and Sedat Gormus. "Content centric cross-layer scheduling for industrial IoT applications using 6TiSCH." *IEEE Access* 6 (2017): 234–244.

Khan, Ab Rouf, and Mohammad Ahsan Chishti. "βDSC 2 DAM: beta-dominating set centered Cluster-Based Data Aggregation mechanism for the Internet of Things." *Journal of Ambient Intelligence and Humanized Computing* 13 (2022): 4279–4296.

Ko, Haneul, Jaewook Lee, and Sangheon Pack. "CG-E2S2: Consistency-guaranteed and energy-efficient sleep scheduling algorithm with data aggregation for IoT." *Future Generation Computer Systems* 92 (2019): 1093–1102.

Li, Ruinian, Carl Sturtivant, Jiguo Yu, and Xiuzhen Cheng. "A novel secure and efficient data aggregation scheme for IoT." *IEEE Internet of Things Journal* 6, no. 2 (2018): 1551–1560.

Liu, X., X. Wang, K. Yu, X. Yang, W. Ma, G. Li, and X. Zhao. "Secure data aggregation aided by privacy preserving in Internet of Things." *Wireless Communications and Mobile Computing* 2022 (2022): 1–14.

Loukil, Faiza, Chirine Ghedira-Guegan, Khouloud Boukadi, and Aïcha-Nabila Benharkat. "Privacy-preserving IoT data aggregation based on blockchain and homomorphic encryption." *Sensors* 21, no. 7 (2021): 2452.

Mineraud, Julien, Oleksiy Mazhelis, Xiang Su, and Sasu Tarkoma. "A gap analysis of Internet-of-Things platforms." *Computer Communications* 89 (2016): 5–16.

Mohseni, Milad, Fatemeh Amirghafouri, and Behrouz Pourghebleh. "CEDAR: A cluster-based energy-aware data aggregation routing protocol in the internet of things using capuchin search algorithm and fuzzy logic." *Peer-to-Peer Networking and Applications* (2022): 1–21.

Nguyen, Tien-Dung, Duc-Tai Le, and Hyunseung Choo. "Sensory data aggregation in Internet of Things: Period-driven pipeline scheduling approach." *IEEE Transactions on Mobile Computing* 21, no. 9 (2021): 3326–3341.

Peng, Cong, Min Luo, Huaqun Wang, Muhammad Khurram Khan, and Debiao He. "An efficient privacy-preserving aggregation scheme for multidimensional data in IoT." *IEEE Internet of Things Journal* 9, no. 1 (2021): 589–600.

Ramamoorthi, Jaya Subalakshmi, and Arun Kumar Sangaiah. "SCGR: Self-configuring greedy routing for minimizing routing interrupts in vehicular communication networks." *Internet of Things* 8 (2019): 100108.

Sándor, Hunor, Béla Genge, and G. Á. L. Zoltán. "Security assessment of modern data aggregation platforms in the internet of things." *International Journal of Information Security Science* 4, no. 3 (2015): 92–103.

Sankar, S., P. Srinivasan, Ashish Kr Luhach, Ramasubbareddy Somula, and Naveen Chilamkurti. "Energy-aware grid-based data aggregation scheme in routing protocol for agricultural internet of things." *Sustainable Computing: Informatics and Systems* 28 (2020): 100422.

Tang, Wenjuan, Ju Ren, Kun Deng, and Yaoxue Zhang. "Secure data aggregation of lightweight E-healthcare IoT devices with fair incentives." *IEEE Internet of Things Journal* 6, no. 5 (2019): 8714–8726.

Uddin, Mostafa, Sarit Mukherjee, Hyunseok Chang, and T. V. Lakshman. "SDN-based multi-protocol edge switching for IoT service automation." *IEEE Journal on Selected Areas in Communications* 36, no. 12 (2018): 2775–2786.

Vo, Van-Vi, Tien-Dung Nguyen, Duc-Tai Le, Moonseong Kim, and Hyunseung Choo. "Link-delay-aware reinforcement scheduling for data aggregation in massive IoT." *IEEE Transactions on Communications* 70, no. 8 (2022): 5353–5367.

Wang, Tian, Guangxue Zhang, Anfeng Liu, Md Zakirul Alam Bhuiyan, and Qun Jin. "A secure IoT service architecture with an efficient balance dynamics based on cloud and edge computing." *IEEE Internet of Things Journal* 6, no. 3 (2018): 4831–4843.

Wang, Xiaoding, Sahil Garg, Hui Lin, Georges Kaddoum, Jia Hu, and M. Shamim Hossain. "A secure data aggregation strategy in edge computing and blockchain-empowered Internet of Things." *IEEE Internet of Things Journal* 9, no. 16 (2020): 14237–14246.

Yang, Lijun, Chao Ding, Meng Wu, and Kun Wang. "Robust detection of false data injection attacks for data aggregation in an Internet of Things-based environmental surveillance." *Computer Networks* 129 (2017): 410–428.

Yin, Bo, and Xuetao Wei. "Communication-efficient data aggregation tree construction for complex queries in IoT applications." *IEEE Internet of Things Journal* 6, no. 2 (2018): 3352–3363.

Yousefi, Shamim, Hadis Karimipour, and Farnaz Derakhshan. "Data aggregation mechanisms on the internet of things: a systematic literature review." *Internet of Things* 15 (2021): 100427.

Zhang, Jiale, Yanchao Zhao, Jie Wu, and Bing Chen. "LVPDA: A lightweight and verifiable privacy-preserving data aggregation scheme for edge-enabled IoT." *IEEE Internet of Things Journal* 7, no. 5 (2020): 4016–4027.

Zhu, Tao, Sahraoui Dhelim, Zhihao Zhou, Shunkun Yang, and Huansheng Ning. "An architecture for aggregating information from distributed data nodes for industrial internet of things." *Computers & Electrical Engineering* 58 (2017): 337–349.

Chapter 5

Future 6G approaches

Integrating intelligent security, sensing, and communication into a green network's architecture

P. Jayadharshini, G. V. Kamalam, and T. Abirami
Kongu Engineering College, Erode, India

5.1 INTRODUCTION

Over the past two decades, the need for wireless communication has grown significantly. The 5G model will soon be widely used and have significantly better features than 4G. A new distance communicating architecture termed 6G featuring AI support is predicted to be in use between 2027 and 2030. Industry and academia were interested in 6G networks during the global implementation of fifth generation communications. In order to meet a variety of network requirements, 6G offers a larger frequency range, a higher transmission rate, enhanced robust, increased capacity for connection setup, and improved bandwidth efficiency. It also has reduced latency and wider coverage (Luo et al., 2020). For the purpose of fully comprehending and altering the physical world, information acquisition is a crucial precursor to E2E information processing. In general, it describes how to behave when sensing objects and gathering information about them through data interfaces, analysis, and sensing devices. With the global adoption of 5G network technologies increasing, 6G connections are beginning to draw the attention of academia and business (Luo et al., 2020). The following are the impacts and history of 6G:

When compared to 5G, 6G has more exacting requirements regarding energy use, delay, dependability, and privacy, among other things. Additionally, it offers greater coverage and higher quality of service (QoS) than previous wireless communications. Technology advancement gives intelligence; consequently 6G is the newest and most creative generation of distance translation. 6G is anticipated to be based on an intelligent system in order to provide statistics learning approaches to massively parallel networks with various topologies. Machine learning systems, on the other hand, need centralized data collection and server processing and create a constraint for application areas that meet daily demands.

The most challenging problems in wireless communication networks will be solved simultaneously by ML and AI, which will make use of enormous amounts of data and computer power. In the future, 6G systems, localization,

 DOI: 10.1201/9781003474524-5

sensing, and communication coexist while utilizing the same time-frequency-spatial resources. In addition to omnipresent communication, with the use of 6G, a cognitive network of communications, sensing services can now identify objects with greater accuracy and clarity. There has been a significant increase in the various IoT gadgets, an advanced communication system that strives for high energy and spectral efficiency, low latency, and support for numerous connections. IoT gadgets offer premium services that enhance transport, track behavioural intentions, and manage it in connected autonomous vehicles and drones. Furthermore, they make it possible for full high definition (HD) video transmission, virtual reality (VR), and telemedicine. From 2025, there will be an estimated 25 billion IoT devices in use, which will make it difficult to handle large IoT devices with the existing various access strategies. Large IoT devices cannot even be supported by the current fifth generation (5G) connectivity systems worldwide.

5.2 EVOLUTION OF MULTIPLE COMMUNICATIONS GENERATIONS

Everyone now has access to wireless technology thanks to the establishment of the first generation of mobile telecommunications in 1979 (Kulkarni & Kulkarni, 2021). International roaming was introduced in the fall, enabling wireless communication among individuals in multiple nations. Digital or analogue modulation can be used to create 1G signals. Analog signals were used in 1G to transport information (Figure 5.1). This was limited to voice conversations with a 2.4 Kbps to 150 MHz frequency range, producing extensive coverage, high delay, and effective energy consumption. However, the musicianship was subpar. It may be moved in the typical manner.

2G: The second generation of mobile telecommunications, which incorporates GSM, was introduced in 1991 (Kulkarni & Kulkarni, 2021). In 2G, packet switching takes the place of analogue modulation. To offer multitasking, 2G leverages TDMA and CDMA. Its frequency was 900 MHz. VoIP services could now support text messages, which enhanced the call quality. To send data to other mobile phones, mobile phones require a BTS. MSC, who oversaw the entire network, had MSC in charge of all BSC. It made use of its own databases and communication channels. Due to its fast transmission speed, 2G gave birth to multimedia messaging services that let users transfer large files. In 2003, G Edge was published.

3G: In order to speed up the network, the third generation was introduced in 1998 (Kulkarni & Kulkarni, 2021). The first 3G standard was WCDMA-UMTS. At first, three frequency bands—850, 1900, and 2100 MHz—were employed for CDMA and packet switching. UMTS is a brand new standard, while WCDMA is based on GSM. Between 384 Kbps and 2 Mbps, data is sent at various rates. Instead of basing charges

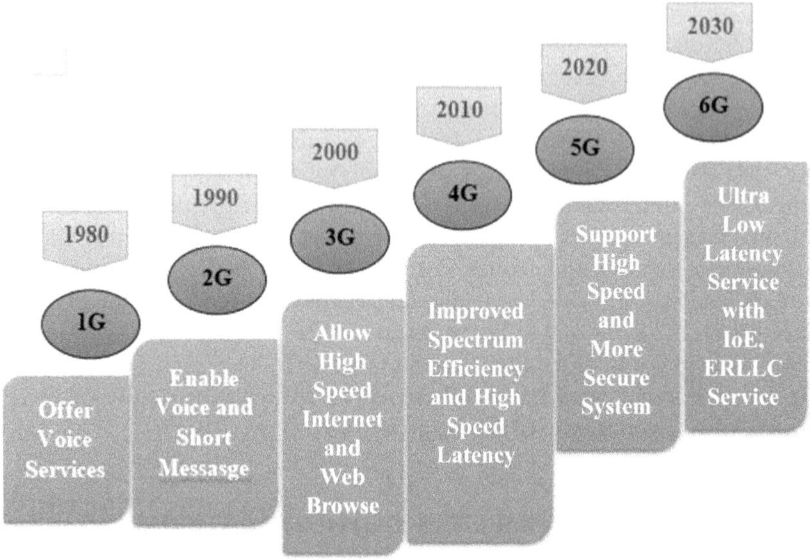

Figure 5.1 Evolution of 6G.

on transmission time, 3G customers were paid based on the volume of data sent. The following years saw the release of HSPA and HSPA+, two further standards.

4G: The International Telecommunication Union (ITU), which created the fourth generation of mobile technology in 2004 and commercialized it in 2009, is known as LTE (Kulkarni & Kulkarni, 2021). LTE and WiMax are two major standards. LTE employs OFDM-MIMO that is entirely packet switched over IP. It offers a 100 Mbps to 1 Gbps speed range and a 2 to 8 GHz frequency range. The combination of low latency and high speed has sped up data transmission.

5G: The interconnect of the fifth generation, which was slated to launch in 2020, uses a new radio standard (Kulkarni & Kulkarni, 2021). It uses physical partition duplexing in place of FTD and allots one uplink slot for every three downlinks. In comparison to 4G, 5G is faster, has a larger capacity, and uses less energy. It also supports interactive streaming of video and audio.

6G: Cellular data networks will be made possible by the sixth generation of wireless communications technology standard, which is currently being created. A strategy has been put up to deploy 100 trillion internet-connected sensors by 2030 in order to revolutionize 6G. In order to eventually develop a society that is wiser, a 1000-fold price reduction is necessary. Less delay, long wavelength and data transfer rate, speedier mobility, and increased high scalability are all requirements of 6G in comparison with 5G.

5.3 KEY COMPONENTS OF 6G

The major elements of 6G that make it the most advanced technology (Figure 5.2) available today are,

- Increased data rate.
- Reduced bandwidth.
- Reliable and accurate networking.
- Accentuated energy efficiency.
- Machines as chief consumers.
- Tools for wireless communication powered by AI.
- Experience with a customized connection.

5.3.1 Multiband ultrafast speed transmission (THz waves)

The amount of wireless data traffic and associated objects in 6G might significantly grow if there were more than 100 devices in a given cubic yard. Accordingly, THz waves might offer solutions for industries where 5G technology was unable to meet the required high data throughput or hyper delay sensitive regimes (Elmeadawy and Shubair, 2020). As a result, it is anticipated that service providers will use THz waves for communications in locations with a lot of devices or a lot of data. Basic wave equations are used to identify the THz waves as a tiny requirement. The requirement for M2M and human-to-machine communications will be met by 6G, especially as

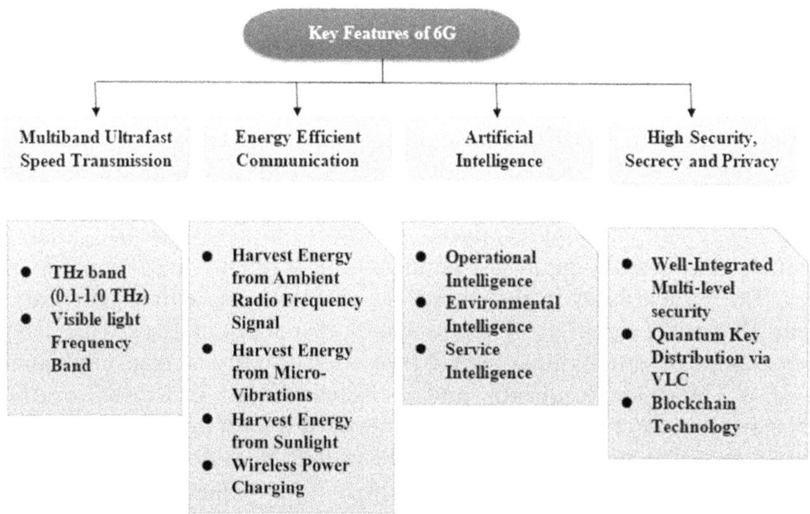

Figure 5.2 Features of 6G.

conscious, autonomous unmanned drone technologies emerge all the time. The idea of the Internet of Everything incorporates this technology (IoE). What can be inferred is that 6G will provide a mega connectivity highly depending on external features with the ability to expertly absorb dissimilar approaches to instantaneously satisfy the needs of a number of applications. (Yang et al., 2019). In order to support Tbps high-speed communications, Rappaport et al. (2019) and Akyildiz et al. (2018) made an in-depth analysis of the THz band between 0.1 and 10 THz. There are various difficulties that must be overcome while using that spectrum. -To make the most of and reuse the THz spectrum of frequencies for communications and assistance for a variety of other activities, it can be organized based on the permeation and perception characteristics of the sub-bands. Extending harmonic qualities, for example, should be enabled in situations supporting a variety of divergent applications. This requirement can be satisfied by careful frequency planning. In the THz areas, it is quite challenging to detect weak signals as sensitivity declines. Because of ambient air quality increases beyond 500 GHz that are located in the middle of the radio spectrum, the THz radio spectrum can be confidently classified into these frequency bands (Oyeleke et al., 2020). When one approaches the THz zone, the extra free area degradation becomes negligible starting at 30 GHz. The narrowed beam, which has an impact on signal gathering and beam tracking in mobile applications and the growing parallelism and complexity of RF hardware, in addition to the extra free area degradation, are other challenges in the higher frequencies that need to be addressed (Saxena et al., 2020).

Researchers in wireless communication concur that a multicellular topology must be used in order to further utilize the elevated spectrum in order to provide better mobile networks (Alsharif et al., 2017). The first approach is in line with the general significant departure from cellular RF range toward mmWave, THz radio spectrum, and recognizable frequency band. Electronic components like drivers must be easily accessible in order to properly utilize the high-frequency spectrum domains. It is encouraging to see how THz digital, optoelectronic, and mixed telecommunication devices are emerging as a result of recent developments in electromagnetic communications. A lower-frequency hybrid that is employed in mixed digital transponders is most likely to appear in the 6G era. The optical laser in this setup has the ability to either generate or send an optical signal at THz frequencies (Sengupta et al., 2018). A few advantages of employing a hybrid link are its ability to transmit signals over a THz–wide frequency and its resistance to inclement weather (Nagatsuma, 2019). Since there is no need for a path link, THz transmission also serves a critical purpose in the uplink. Because it adds a reliable communication method that can withstand ambient light, a hybrid device-to-device system ultimately improves the sensor ratio of the VLC system. The driving force behind 6G will also be ultra-effective, short-range connection solutions, an area where higher-frequency bands will be crucial in the future. The path loss is greatly impacted by particle assimilation, especially at higher wavelengths (1 to 10 dB/km at energies up to 400 GHz). Despite the fact that the element size is two orders of magnitude greater than the equivalent needed for the THz band application, a photonic solution is a viable choice for providing high data speeds with negligible transmission errors. Lallas (2019) focused on the plasmon's possible uses in wireless THz nano-applications and THz band wireless communication after carefully examining the enormous reference material of the technologies and the existing fundamental applications and also hybrid combinations in them.

5.3.2 Efficiency broadcasting

6G will specifically meet and fulfill several needs, including the delivery of a superior efficiency, from the perspective of widespread IoT utilization and with an environment of countless tiny sensors. In keeping with the premise that as energy usage rises, so do capacities and skills to handle extensive multimedia signal analysis, improving the battery life of smartphones is another issue that needs to be addressed (Alsharif et al., 2017). In order to overcome the daily recharging challenges with the majority of radio equipment and improve communication needs, low energy use and long battery life length are two study problems in 6G. Therefore, a comprehensive energy-efficient wireless communication strategy is required for 6G. One of the primary goals of 6G communication is to be rechargeable whenever and whenever it is possible in order to reach 1 pico-joule per bit communication efficiency (Strinati et al., 2019). Elevated THz waves, which enable gadgets to deliver electric beams in a desired direction, are an advantage of 6G communication in addition to directional beam transmission using MIMO multiple antennas. The network's covered devices might be able to get enough energy this way. According to the 6G vision and recommendations given in David and Berndt (2018), research effort should be more concentrated on 6G's battery life and service classes rather than its connection speed and delay.

In order to extend the battery life of 6G devices, a variety of different energy-harvesting techniques must be combined (Niyato et al., 2017; Luo et al., 2019). These techniques must extract energy from micro-vibrations, sunlight, and ambient RF waves in addition to other sources. Long-range wireless power charging offers the potential to increase battery life, which is appealing (Ulukus et al., 2015; Alsharif et al., 2019). Another approach that might potentially offer electricity to mobile gadget sensors over a particular distance is distributed laser charging (Strinati et al., 2019). The four major forms of wireless charging that are compared in Table 5.1 are magnetic pairing, ultrasound imaging pairing, radio frequency, and scattered beam feeding.

Table 5.1 Wireless charging techniques

Technique	*Added benefits*	*Drawbacks*	*Optimal charging range*	*Implementations*
Coupling with magnetic induction	• Easy to implement. • Human-safe.	• Close charging range. • Overheating consequence.	Ranges in size from a few mm to a few centimeters.	• Mobile devices • RFID tags with no contact.
Resonant magnetic coupling	• Alignment is loose. • Excellent charging efficiency.	• Short charging range. • Difficult implementation.	One or two millimeters to one or two meters.	• Electronics on the go. • Home equipment. • Charging of electric vehicles.
Microwave emissions	• Appropriate for mobile apps. • Long effective charging range.	• Ineffective charging and charging by line of sight.	Commonly within a few tens of meters to a few kilometers. suitable for applications on	• RFID cards. • Wireless sensors with embedded medical technology. • Bright LED/
Dispersed laser charge	• Safe and powerful. • Charging of several rx. • Small in size.	• Sightlines are necessary • Poor charging efficiency.	Up to 10 m.	• Mobile gadgets. • Electronics for consumers.

5.3.3 Artificial intelligence

Technology is revolutionized by ML, AI, and Deep Neural Network, which open up new research opportunities in a variety of fields, including IoT and 6G communications. AI and ML are being used to develop 6G communication solutions at the link and system levels. Context awareness, conscience, pooling, and impulsive setup are just a few of the capabilities that AI–powered 6G is anticipated to expand. These materials are referred to as MIMO 2.0 and are extensively discussed in (Nadeem et.al 2019a). ML techniques for designing heterogeneous networks, huge MIMP optimization, and device-to-device (D2D) communication are all summarized in Jiang et al. (2016). Furthermore, a brand-new network design for mobile communication for assessing big data analytics can help with physical layer optimization. An index modulation could help improve the effectiveness of the 6G network. Notably, advanced ML and high intelligence approaches may alter the setup and architecture of 6G networks in addition to improving performance.

5.3.4 Operational intelligence (OI)

Due to 6G's diversity, density, and adaptability, this technology uses inter-algorithms that can function in its extremely complex and dynamic nature. It does this rather than employing conventional techniques to effectively distribute resources to ensure dependable network performance. The performance of such optimization that prioritizes many objectives is often NP—hard to measure in real time. Deep reinforcement learning (DRL), one of the ML and AI technologies that has advanced recently, can help decision makers iteratively improve and ultimately optimize their decisions by leveraging the feedback produced by a loop set up by DRL, which can be referred to as AI's main application in 6G. Such learning algorithms can be used to allocate resources effectively by assisting in optimization.

5.3.5 Environmental intelligence (EI)

The development of intelligent wireless environments and materials has the potential to promote widespread intelligence in many technological fields, including contemporary communications environments. Data centers, IoT gadgets like remotely driven and ground vehicles, and self-organizing and regeneration engines are a few application scenarios that the intelligence-based services could realize. The reliability of D2D users for a 6G network could be enhanced by employing intelligent frameworks. One of the most recent events is the use of reconfigurable intelligent surfaces in adaptive approaches based on sensing for radio wave–tailored modifications. These innovations laid the groundwork for specialized hardware suitable for EI. The periphery and/or internet devices receive the original characteristics that were collected for DNNs. The variability of the devices' capacities for computation and communication makes it difficult to process these features.

5.3.6 Service intelligence (SI)

The key deployment platforms for 6G intelligence include communication services such as digital health, location (both indoors and outdoors), control of various devices, information search, and security. SI can assist in expanding all of these medical applications in a smart and individualized way to promote client happiness, for example by utilizing deep learning algorithms that can improve position accuracy, especially indoors (Belmonte-Hernández et al., 2019). Similar to how the intelligent IoT may assist in the personalization of e-health, data collecting via a multi-model architecture can be helpful. High performance core networks built on top of 6G can be used to increase SI.

5.3.7 High security, secrecy, and privacy

The recent research initiatives have emphasized real-world practical implementations, such as multiple access techniques, physical air interfaces, and 6G data centers. In Giordani et al. (2019), the deployment of cell-less typologies, distributed redistribution of wealth, and 3D amazingly are three features that are highly expected in 6G network architectures. The outcomes of vertically specialized wireless networks are reviewed in Mahmood et al. (2019) along with machine-type communications (MTCs). According to the research, 6G might hasten the development of the first ever ground-breaking standard that would entirely replace the present industry-specific communication protocols and offer an integrated solution permitting wireless integration of all requirements in industry verticals.

5.4 ENVISION OF 6G APPLICATION

5.4.1 One hundred gigabits per second (Gbps)

The majority of individuals believe that using a mobile broadband network only entitles them to stream movies and television shows on the go and swiftly download large files. Speed is required of them (Kulkarni & Kulkarni, 2021). We keep sending out (video) data through the Internet in greater and greater quantities. YouTube video views are estimated to reach one billion hours per month and beyond. Mobile devices are used as an access medium by 75% of YouTube viewers (Kulkarni & Kulkarni, 2021). People hunger for the flexibility to stream and share videos with others in the finest quality, at any time, from any location. Mobile broadband can be seen here as a solution. A 100 Gbps download speed is anticipated as 6G networks continue to proliferate (Akhtar et al., 2020). The download speed is ten times faster on a 5G network and 300 times faster on a 4G network.

5.4.2 Ten million linked devices per km^2

IoT power is determined by the number of connected sensors and devices. Here, an impressive growth is anticipated. From 2025, there will be close to 31 billion IoT devices, up from 12 to 13 billion at the moment, according to research by market research firm Statistic. As the number of devices rises steadily, it becomes difficult to connect them to the Internet per square meters. This graph is referred to as the link density. Currently, there are about 100,000 devices connected to a 4 Gbps per square kilometer. In terms of performance, 5G outperforms 4G with a million connected devices per square kilometer. With the introduction of 6G, an incredible 10 million devices are now linked together every square kilometer (M. W. Akhtar et al., 2020).

5.4.3 A maximum energy consumption of one nanojoule per bit

Greater radio frequency is used by 6G networks to fulfill the demand for larger bandwidths. One issue is that, despite the fact that efficient electricity utilization is a significant issue in the telecommute industry, chip technology cannot work in certain bandwidths. Despite this, the telecommute industry is active. Orange, a telecommuter company, claims that due to the introduction of new gear and software, by 2025 5G network energy usage might be 10 times less than 4G (per transferred gigabit). This number may have increased in the year 2030. On the other hand, the large amount of newly generated data that must be treated poses a risk to recent efforts to increase the energy efficiency of mobile networks. Storage facilities have been engaged in a similar conflict with fiber optic links for years, trying to increase energy efficiency while managing a massive volume of data. Currently, fiber optics operate in test configurations with several hundred femtojoules per bit. (One femtojoule equals 10 to 15 joules). Researchers working on 6G want to reduce energy usage per bit to less than 1 nanojoule (10^{-9} joules).

5.5 FUNDAMENTAL ENABLING TECHNOLOGIES OF 6G NETWORKS

6G enables communications on land, in the air, and at sea with a dependable, rapid network that can support a huge number of devices with ultra-low latency. Researchers from all over the world have identified blockchain, terahertz and millimeter wave communication, tactile internet, non-orthogonal multiple access (NOMA), cellular systems communication, cloud cover computing, AI, ML, computational connectivity and computational algorithms, non-orthogonal multiple input multiple output, and other technologies as key technologies in the realization of the Web of Things. A 5G network is comparable to a 6G network in architecture and incorporates the advantages of 5G, according to the history of mobile phone networks (Figure 5.3). New technologies were also included, and 6G will update some of the 5G technology. So, the 6G system will be powered by a range of technologies. The following article discusses a few of the most significant 6G technologies.

Artificial Intelligence: The core feature of 6G autonomy systems is automatic intelligence. Consequently, AI technology is incorporated into 6G network technologies (Niknam et al., 2020). Future 5G telecommunications will support just a very small amount of AI. The transition from intellectual to intelligent media will be made easier by AI–enabled 6G, allowing radio broadcasts to utilize their full potential. Machine learning advancements are enabling the development of sophisticated intelligent networks for real-time 6G communication. AI alters the real-time data transfer rate.

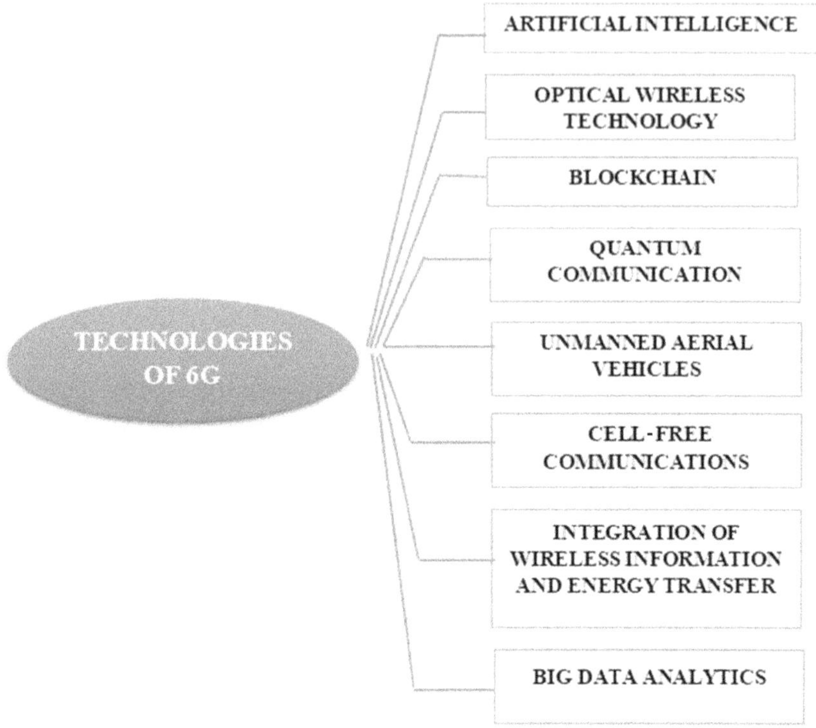

Figure 5.3 Technologies of 6G.

Optical Wireless Technology: Optical wireless communications (OWC) improvements are made to provide 6G connectivity on any usable device, enabling access to networks as well as the use of those networks for backhaul or fronthaul connectivity of the network. Since the creation of 4G communication networks, OWC techniques are now in use. But it's anticipated that it will be used more widely to meet the requirements of 6G communication networks although there are many others, V2X communication, and interior mobile robot locating. Research is still being done to address this technology's drawbacks in order to improve performance when it is used in a particular application. Extremely fast data rates, minimal latency, and secure connectivity can all be provided by wireless optical communications. Due to its optical band, LiDAR emerges as a crucial technology for 6G communications, enabling resolution-rich 3D mapping.

5.5.1 Blockchain

In the near future, communication will manage a large amount of data thanks to blockchain technology. Distributed ledger technology includes blockchain as a subset (Aste et al., 2017). A distributed ledger is a database that's split up among several machines. Every node makes a copy of the ledger, and the exact replication is saved. Using peer-to-peer networks, block chains are administered. It is not required for its existence to be governed by a centralized server or by a centralized authority. By fostering trust among networked programmes, block chain eliminates the need for reliable middlemen. Decentralized tamper-resistance and secrecy, which are intrinsic properties of block chain, allow it to be perfect for a huge range of 6G models. It makes it possible to manage spectrum in a secure and genuine way by enabling transparency, verifiable transactions, and the prevention of unauthorized access. Block chain has possibilities for the future not found in previous systems since it combines network formation with a distributed nature, a consensus protocol, and elevated encryption.

5.5.2 Quantum communication

Ultra-accurate clocks, medical imaging, and quantum computers are all made possible by the properties of quantum mechanics, which include the interaction of molecules, atoms, and even photons and electrons. The entire potential has not yet been realized, though. A variety of cybersecurity-related assaults were defended against using high-level quantum communication technology (Dang et al., 2019). Potentially significant participants in 6G wireless networks are QC and QML. QC provides 6G networks solutions, such as new multiple access technologies like NOMA and RSMA, which consume a significant amount of energy when used for SIC processing, to enhance bandwidth utilization. The same is true for 6G channel estimates, channel coding (quantum turbo codes), localization, load balancing, routing, and multiuser transmissions, all of which rely heavily on QC and QML. QC and QML are able to resolve difficult challenges in ad hoc sensor networks and Cloud IoT by offering the fastest and best path to data streams in the communications infrastructure core.

5.5.3 Unmanned aerial vehicles (UAV)

UAV, or drones, play a crucial role in 6G models. In many situations, UAV innovation provides wireless access at extremely high data rates. Cellular connection for UAVs is provided by the BS entities. A UAV's features, including fast deployment, strong row connectivity, and freedom levels with controlled mobility, are all lacking in fixed BS networks. The most important advancement in 6G communication is therefore considered to be UAV technology. Backscatter communication is employed to support a range of communication tasks in a study by Zeng et al. (2016). An air-interface appropriate for 6G wireless networks can be created using semi detection methods and UAV payloads. Deep reinforcement learning-based durable allocation of resources for accomplishing UAV to enable 6G Intel is a terrific option.

5.5.4 Free Cell Networking

The seamless connectivity of various communication technologies along with their frequencies is essential for 6G systems and ultimately results in a smooth transition for the user from one network to another without requiring manual device settings. Fibroblast and quasi connectivity will replace traditional cellular and orthogonal concepts in the sixth generation of mobile technology. The most effective network is chosen automatically among the available communication technologies. These will all be eliminated by 6G cell-free communications, which will provide higher QoS.

5.5.5 Wireless Data and Energy Transfer (WIET)

A ground-breaking 6G networking technology is WIET. WIET and wireless communication technologies use the same working environment. Autonomous power transfer is utilized during communication to refuel sensors and mobile devices. WIET provides the potential to extend the battery life of wireless charging systems. Therefore, 6G networks can connect gadgets with no batteries.

5.5.6 Big Data Analytic (BDA)

Massive data analysis, or "big data," is a challenging approach that is performed by BDA. This method identifies information such as patterns that are unexpected correlations, hidden, and customer preferences in order to ensure thorough data management. Video, social media, photographs, sensors, and other sources of enormous amounts of data all contribute to this. In 6G networks, this technology is frequently used to manage enormous volumes of data. Automating processes and self-improvement are anticipated in 6G networks to increase their capacity for utilizing vast volumes of data. Big data analytics can be applied, as demonstrated by the E2E latency reduction. By using predictive analytics to determine a user's data proper path, E2E delay can be minimized in 6G systems using ML and BDA.

5.6 APPLICATION OF 6G

Intelligent and sophisticated machines have an impact on contemporary culture. These devices are capable of speaking both to people and to one another. They can be used to help people in a wide range of contexts, including the fields of health care, e-health, transportation, the food service industry, agriculture, education, and so forth (Figure 5.4). Following is a list of the primary 6G utilization scenarios in several fields.

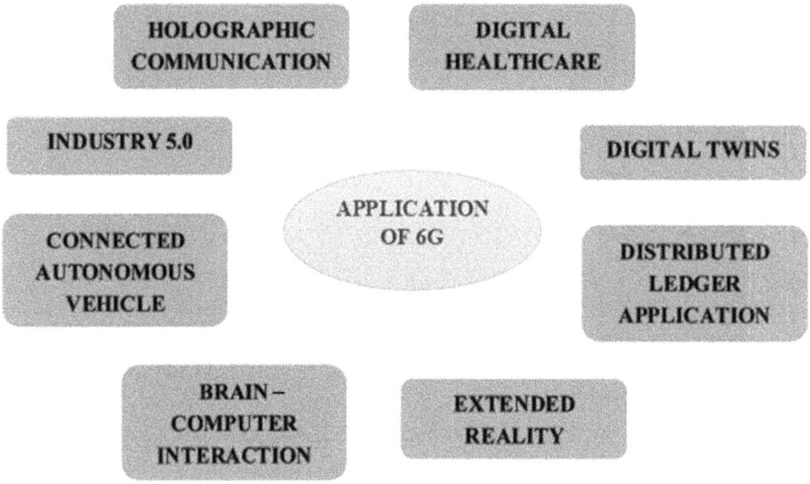

Figure 5.4 Applications of 6G.

5.6.1 Holographic communication

In a wide range of industries, including amusement, medicine, teaching, and business, hologram interactions will be heavily used. Due to the simultaneous operation of hundreds of separate data streams in holographic applications, which involve multidimensional interactions, wireless networks must be able to manage extremely high throughput. When data are destroyed during remote micro surgical or photonic targeted curative operations, restoration of the information is necessary. The security and dependability of communication networks must be considerably increased. With the use of holographic conferencing, users can see distant people and objects in three dimensions in live time with a degree of accuracy that is at minimum as genuine as their actual existence.

5.6.2 Extended reality

By boosting bandwidth and reducing latency, 5G networks have improved the experience of AR and VR. Unfortunately, there are still a number of obstacles in the 6G network that prevent the release of VRs on 5G networks. For instance, while the VR/AR cloud scan now offers customers cutting-edge features, the latency is a significant problem, and the resulting uncertainty creates further challenges. Although 5G bandwidths allow for decent image compression, the cloud-based deployment of VR/AR offers more streamlined and accessible services. For real-time transmission of large uncompressed photo or video files, we must switch to 6G networks. VR and AR experiences are significantly improved with

6G networks. Networks of sensors are used to collect data from the senses and provide feedback to users.

5.6.3 Connected autonomous vehicles

In order to deliver connectivity everywhere, 6G will be broad, offer the best experience for users, and be relevant in a wide variety of situations. We should be most interested in the 6G network's architecture, with a focus on the access and core networks. Sealed modeling and management are used by the twin network to manage the network infrastructure. Given the considerable impact of changes to network infrastructure, it is imperative to embrace new technical components and incorporate them into existing networks. The Security Framework, according to the National Institute of Standards and Technology, should provide cybersecurity, data protection, and security. Protecting 6G users' privacy is essential, especially when they use public transportation like buses, planes, and aircraft. As a result, security frameworks must incorporate the concepts of secrecy by construction, physiological and digital security, and security unification.

5.6.4 Industry 5.0

Personal cooperation with automation and intelligent systems has been termed Business 5.0's next major innovation. The growth of an automated industrial environment relies on 6G. Due to serious security risks, Industry 5.0 apps need to adhere to fundamental security requirements like integrity, availability, authentication, and auditing. Lower operational expenses, a wider variety of devices, and better scalability must all be taken into account while developing Industry 5.0 security approaches. Since 6G connections will be used to transmit command signals and tracking information, they will be in charge of preserving data integrity and security in Economy 5.0. Additionally, the 6G era offers highly automated, scalable techniques and systems for controlling access to confidential resources like Industry 5.0–related property rights.

5.6.5 Digital health-care

New trends in digital health care are emerging. AI–powered intelligent health care will advance through a variety of cutting-edge methods during the coming years. Additionally, the aging population may lead to a greater emphasis on digital health than previously thought. Individualized management and health monitoring are advanced by body area networks (BANs) with smart autonomous technologies. These specialized BANs are capable of gathering health data from a range of sensors, dynamically sharing it, and interacting with network services. 6G is most likely to be the main form of communication for intelligent future health-care services. Thus, security problems in the future of 6G will include equipment, encrypted communications, and security systems for billions of small health devices. As future medical systems' primary communication backbone, 6G networks should protect health information, data confidentiality, and confidentiality.

5.6.6 Digital twins

As a form of communication, things and procedures will be digitally replicated and AI technologies will progress. In the digital environment, intelligent mapping of interactions between people and things will take place. The digital world uses sophisticated algorithm models to mimic and anticipate quantum effects and objects are verified and under control. The flexible solution to issues in the real world should then be given. The digital twin era officially begins with the 6G era. Medical systems in the health-care industry may employ digital twin data to help in diagnostic and therapy choices. Digital product design optimization may contribute to lower costs and higher productivity in the industrial sector. The digital representation, an efficient and reducing industrial control system, has been identified as a key solution for 6G technology.

5.6.7 Brain computer interaction (BCI)

The brain should be connected to technology; this is the basic premise of BCI. The gadgets could be external or inside the person. The four stages of the BCI process are sign capture, feature extraction, feature translation, and final reporting. BCI's main uses in the health-care industry revolve around enabling people with disabilities to control assertive technology. The use of BCI in health care applications occasionally puts patients' lives in danger because of several forms of attacks that restrict the applications' usefulness. The three types of BCI assaults are data processing, data gathering, and brain signal production.

Brain signaling attacks are referred to as adversarial and deceptive stimulus attacks. In an antagonistic onslaught, stunning inputs are given to an ML system with the goal of hindering its proper functioning and output. Conversely, deceptive stimuli try to trick users by presenting them with incorrect sensory information in order to cause a specific brain response. Battery drain, data conversion, and injection attacks are examples of information processor offences. Battery risks drain the battery of a gadget, lowering performance or making it useless. Additionally, because input validation is not performed, injection attacks offer interpretations of input that contain particular features that can change how the inputs are assessed. Attacks on data capture and stimulation such as sniffing, replaying, and spoofing pose a danger to the security of BCI.

5.6.8 Distributed ledger application

Cryptocurrency technology is expected to be employed to share data and spectrum, boosting the reliability of 6G networks. Blockchain technology exchanges information with all parties involved. The three types of attacks include susceptible assaults, privacy preserving assaults, and double assaults. They also offered blockchain-based solutions for 6G networks, including incentive programmers and cryptography algorithms.

5.7 SECURITY AND PRIVACY

The Internet will be accessible to any edge device in 6G, and these devices will frequently deploy AI apps. Since most AI systems rely on data, privacy and security of the data generated are issues that need to be addressed (Sun et al., 2020). The health condition of customers, for instance, may be collected daily by wearable technology, yet this information is sensitive and confidential. If the information is released, clients' privacy may be in jeopardy. Additionally, adversarial attacks on deep learning models are looked into and their real-time execution is demonstrated. To improve the edge computing and AI–based security systems, Lovén et al. (2019) explored edge AI–related issues using security techniques. 6G will require a new authentication and password. Moving to 6G, because 6G is only a theoretical concept at the moment, there are no standards for it. Yet, compared to 5G, it is anticipated to bring about significant gains in terms of speed, latency, security, privacy, and network coverage. It is anticipated that the network would implement even more sophisticated encryption and authentication procedures to guarantee the confidentiality and integrity of data transmission when it comes to the privacy and security of 6G data. This could include employing quantum encryption techniques, which are thought to be virtually impenetrable. Keystream reuse in two successive conversations is a security issue with the VoLTE protocol. An attacker can intercept phone calls by using this flaw to decipher the contents of an encrypted VoLTE call. Many studies predict that 6G will lead to new applications, including mixed reality and autonomous driving, which will be vulnerable to DoS and malicious attack.

5.8 ADVANTAGES, DISADVANTAGES OF 6G TECHNOLOGIES

5.8.1 Advantages

One terabyte of data per second is anticipated to be possible with sixth generation technology. 5G applications and advancements in connectivity, cognition, sensing, and imaging using wireless networks will be made possible by this higher level of capacity and latency. The following are a few benefits of 6G technology to take into account.

Support Higher Number of Mobile Connection

The ability to accommodate more mobile connections than 5G capacity is one of the key advantages of 6G technology. Less device interference results in better service because there will be less of it.

Supports Higher Data Rate

The 6G network enables larger data rates, which is an additional advantage. It's important to keep in mind that, like 5G, this kind of connection will only be possible at frequencies that use the mmWave spectrum. Such high-frequency waves are still incompatible with the current technology.

Revolutionize Health-care Sector

Real-time operations and simulations help medical interns and students learn more effectively. The primary benefit of 6G technology is how it will transform the health-care industry for both patients and medical professionals. Imagine a world where you could learn about your health immediately, rather than needing to wait weeks.

Independent Frequencies

For the 6G standard, control channels are assigned a frequency range of 8 to 12 GHz. Its independent frequencies will have a frequency bandwidth of up to 3.5 kHz. Channels are not overlapping; hence this indicates that data speeds can be increased by giving distinct transmissions room.

5.8.2 Disadvantages

The difficulties with sixth generation technologies that have been noted are not major ones. It would be premature to declare any 6G technology drawbacks at this time until they have been tested in a real-world system because the technology is still in its infancy.

Difficult to Use

The fact that 6G technology is too challenging to utilize is one of its key drawbacks. Many people find technology to be bewildering. Additionally, learning is not simple. For average people, learning this new technology can be quite time and patience consuming.

Expensive

The expensive cost of 6G technology is an additional drawback. The cost of this type of technology is anticipated to be one that consumers are prepared to pay, even though it may seem high in comparison to what they have been accustomed to.

Privacy

With this kind of technology, privacy issues exist. It has been questioned how privately people may be utilizing this technology, and the government or anyone else with access to network monitoring tools can simply access it.

Compatibility Issues

Many users face a dilemma since they want to use this new network but are unable to do so because their device does not support the sixth-generation technology due to compatibility difficulties with older devices. Additionally, others think that this could lead to issues when these technologies spread and are utilized more frequently in the future.

5.9 CONCLUSION

In the way people live nowadays, technology has a significant impact. Businesses, people's quality of living, infrastructure, and numerous other facets of human life have all undergone significant transformation as a result of wireless technologies. People will always struggle to discover effective solutions to a variety of issues, and they must also look for new approaches to advance the field of problem-solving. The technological advancement from 1G to 5G was made possible by the ambition of mankind, and it is continually advancing. Researchers from all over the world are working hard to develop 6G technology, which will be deployed globally by the start of 2030. The main purpose of 5G is communications, just like it was with all prior "Gs," according to Saad. The support of autonomous systems will require these functions. The technique improves spectrum sharing, lowers latency, and increases capacity by making greater use of the terahertz (THz) spectrum and distributed radio access network (RAN).

REFERENCES

Akhtar, M. W.; Hassan, S. A.; Ghaffar, R.; Jung, H.; Garg, S.; Hossain, M. S. The shift to 6G communications: Vision and requirements. *Hum.-centric Comput. Inf. Sci.* 2020 10(1), 1–27. https://doi.org/10.1186/S13673-020-00258-2

Akyildiz, I. F.; Han, C.; Nie, S. Combating the distance problem in the millimeter wave and terahertz frequency bands. *IEEE Commun. Mag.* 2018, 56, 102–108.

Alsharif, M. H.; Kim, J.; Kim, J. H. Green and sustainable cellular base stations: An overview and future research directions. *Energies* 2017, 10, 587. [CrossRef]

Alsharif, M. H.; Kim, S.; Kuruoğlu, N. Energy harvesting techniques for wireless sensor networks/radio-frequency identification: A review. *Symmetry* 2019, 11, 865.

Aste, T.; Tasca, P.; Di Matteo, T. Blockchain technologies: The foreseeable impact on society and industry. *Computer* 2017 50(9), 18–28. https://doi.org/10.1109/MC.2017.3571064

Belmonte-Hernández, A.; Hernández-Peñaloza, G.; Gutiérrez, D. M.; Álvarez, F. SWiBluX: Multi-sensor deep learning fingerprint for precise real-time indoor tracking. *IEEE Sens. J.* 2019, 19, 3473–3486.

Dang, S.; Amin, O.; Shihada, B.; Alouini, M.-S. What should 6G be? *Nat. Electron.* 2019 3(1), 20–29. https://doi.org/10.1038/s41928-019-0355-6

David, K.; Berndt, H. 6G vision and requirements: Is there any need for beyond 5G? *IEEE Veh. Technol. Mag.* 2018, 13, 72–80.

Elmeadawy, S.; Shubair, R. M. Enabling technologies for 6G future wireless communications: Opportunities and challenges. arXiv 2020, arXiv:2002.06068.
Giordani, M.; Polese, M.; Mezzavilla, M.; Rangan, S.; Zorzi, M. Towards 6G networks: Use cases and technologies. arXiv 2019, arXiv:1903.12216.
Jiang, C.; Zhang, H.; Ren, Y.; Han, Z.; Chen, K.-C.; Hanzo, L. Machine learning paradigms for next-generation wireless networks. *IEEE Wirel. Commun.* 2016, 24, 98–105.
Kulkarni, R.; Kulkarni, S. How 6G has an influence on smart cities an overview. *Int. J. Eng. Res. Technol.* 2021, 10(5), 258–275.
Lallas, E. Key roles of plasmonics in wireless THz nanocommunications—A survey. *Appl. Sci.* 2019, 9, 5488.
Lovén, L.; Leppänen, T.; Peltonen, E.; Partala, J.; Harjula, E.; Porambage, P.; Ylianttila, M.; Riekki, J. EdgeAI: A vision for distributed, edge-native artificial intelligence in future 6G networks. *The 1st 6G Wireless Summit*, 2019, 1–2.
Luo, K.; Dang, S.; Zhang, C.; Shihada, B.; Alouini, M.-S. An empirical analysis of the progress in wireless communication generations. In Proceedings of the EAI MobiQuitous, New York, NY, USA, November 2020, 425–434.
Luo, Y.; Pu, L.; Wang, G.; Zhao, Y. RF energy harvesting wireless communications: RF environment, device hardware and practical issues. *Sensors* 2019, 19, 3010.
Mahmood, N. H.; Alves, H.; López, O. A.; Shehab, M.; Osorio, D. P. M.; Latva-aho, M. Six key enablers for machine type communication in 6G. arXiv 2019, arXiv:1903.05406.
Nadeem, Q.-U.-A.; Kammoun, A.; Chaaban, A.; Debbah, M.; Alouini, M.-S. Large intelligent surface assisted MIMO communications. arXiv 2019a, arXiv: 1903.08127.
Nagatsuma, T. Advances in Terahertz Communications Accelerated by Photonics Technologies. In *Proceedings of the 2019 24th International Conference on Opto-Electronics and Communications Conference (OECC)*, Fukuoka, Japan, 7–11 July 2019; pp. 1–3.
Niknam, S.; Dhillon, H. S.; Reed, J. H. Federated learning for wireless communications: Motivation, opportunities, and challenges. *IEEE Commun. Mag.* 2020 58(6), 46–51.
Niyato, D.; Kim, D. I.; Maso, M.; Han, Z. Wireless powered communication networks: Research directions and technological approaches. *IEEE Wirel. Commun.* 2017, 24, 88–97.
Oyeleke, O. D.; Thomas, S.; Idowu-Bismark, O.; Nzerem, P.; Muhammad, I. Absorption, diffraction and free space path losses modeling for the Terahertz band. *Int. J. Eng. Manuf.* 2020, 10, 54.
Rappaport, T. S.; Xing, Y.; Kanhere, O.; Ju, S.; Madanayake, A.; Mandal, S.; Alkhateeb, A.; Trichopoulos, G. C. Wireless communications and applications above 100 GHz: Opportunities and challenges for 6G and beyond. *IEEE Access* 2019, 7, 78729–78757.
Saxena, S.; Manur, D. S.; Mansoor, N.; Ganguly, A. Scalable and energy efficient wireless inter chip interconnection fabrics using THz-band antennas. *J. Parallel Distrib. Comput.* 2020, 139, 148–160.
Sengupta, K.; Nagatsuma, T.; Mittleman, D. M. Terahertz integrated electronic and hybrid electronic–photonic systems. *Nat. Electron.* 2018, 1, 622–635.
Strinati, E. C.; Barbarossa, S.; Gonzalez-Jimenez, J. L.; Kténas, D.; Cassiau, N.; Dehos, C. 6G: The next frontier: From holographic messaging to artificial intelligence

using subterahertz and visible light communication. *IEEE Veh. Technol. Mag.* 2019, 14, 42–50.

Sun, Y.; Liu, J.; Wang, J.; Cao, Y.; Kato, N. When machine learning meets privacy in 6G: A survey. *IEEE Commun. Surv. Tutor.* 2020 22(4), 2694–2724. https://doi.org/10.1109/COMST.2020.3011561

Ulukus, S.; Yener, A.; Erkip, E.; Simeone, O.; Zorzi, M.; Grover, P.; Huang, K. Energy harvesting wireless communications: A review of recent advances. *IEEE J. Sel. Areas Commun.* 2015, 33, 360–381.

Yang, P.; Xiao, Y.; Xiao, M.; Li, S. 6G wireless communications: Vision and potential techniques. *IEEE Netw.* 2019, 33, 70–75.

Zeng, Y.; Zhang, R.; Lim, T. J. Wireless communications with unmanned aerial vehicles: Opportunities and challenges. *IEEE Commun. Mag.* 2016 54(5), 36–42. https://doi.org/10.1109/MCOM.2016.7470933

Chapter 6

Energy efficient wireless sensors architecture with LSTM based on Machine Learning Technique

P. Jayadharshini, T. Abirami, S. Santhiya, N. Bhavatharini, and G. Rithanya

Kongu Engineering College, Erode, India

6.1 INTRODUCTION

A wireless sensor network (WSN), which tracks system, physical, and environmental parameters, is a wireless network with no underlying infrastructure. It is set up ad hoc with a lot of wireless sensors [1]. A WSN uses sensing nodes with an embedded CPU to regulate and keep an eye on the surroundings of a specific area. They are linked to the base station, a processing unit containing part of the WSN system. Base stations in a WSN system are connected through the Internet for the purpose of sharing data.

The "nodes" that make up a WSN might number anywhere from a few to many and each one is linked to additional sensors [2]. Each of these nodes consists of a microprocessor, an electric circuit for effective sensor communication, transceivers with such integrated circuit chips or a link to a transceiver, a source of energy such as a battery or an integrated form of energy harvesting, and a transceiver (Figure 6.1). A sensor node could be as small as a shoebox or a particle of dust, however microscopic sizes have not yet been realised. The cost of a sensor node can range from a few dollars to hundreds of dollars, depending on node sophistication. Resources, such as energy, memory, processing speed, and communications bandwidth, are restricted by size and cost constraints. A WSN's architecture might be anything from a simple star network to an intricate multi-hop wireless mesh network. Two propagation techniques are flooding and routing.

6.2 METHODOLOGY

6.2.1 Protocols

WSN's primary function is to detect and send data using the least number of resources possible. To select a path with the least cost in delay, lifetime, energy, or any other characteristic that is more important to the application, the network layer requires a reliable routing protocol.

DOI: 10.1201/9781003474524-6

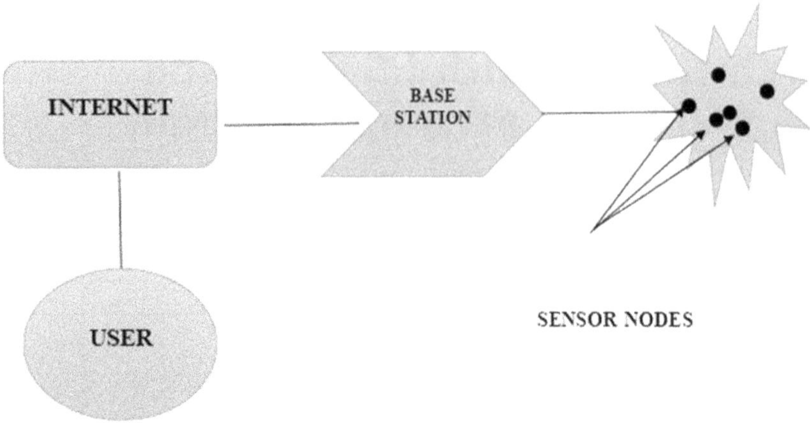

Figure 6.1 Working implication of WSN.

6.2.2 LEACH protocol

LEACH is a TDMA–based MAC technique used in WSN that combines clustering and a simple routing method. By using less energy to maintain clusters, LEACH aims to increase the longevity of wireless sensor systems [3]. A series of protocols called LEACH allows most nodes to connect with cluster leaders, who then assemble and send the information to the sink. Each node utilises a stochastic approach to predict whether it will become a cluster leader in this phase at the beginning of each cycle. The assumption used by LEACH is that each node seems to have a radio capable of making a direct connection to the base station or the closest cluster head, but that using this radio continuously at its full capacity would render it energy inefficient.

For P rounds, P is the targeted proportion of cluster heads; nodes that have already been cluster heads are not permitted to become cluster heads again. After that, there is a 1/P chance that each node will once again serve as the cluster leader. Each node that isn't a cluster head chooses the nearest one at the conclusion of each cycle and decides whether or not to join that cluster [4]. The cluster head then specifies a deadline for each node in its cluster to transmit its data. Only the cluster head and only in accordance with the cluster head's established schedules can initiate TDMA connections with all other nodes.

The LEACH protocol has two stages:

- Set-up phase
- Steady phase

LEACH protocol is split into cycles, with each cycle consisting of two stages. It is a popular hierarchical routing protocol. It is also self-adaptive and self-organising. The LEACH protocol employs rounds as its fundamental unit to save unnecessary energy costs. There are two stages in each round: one for cluster setup and the other for steady-state storage.

6.2.2.1 *Set-Up Phase*

The setup phase is to form clusters and designate the sensor node with the highest energy as the cluster head for each cluster. In the setup phase, there are three crucial steps:

1. Advertisement for cluster heads
2. Cluster configuration
3. Development of a transmission calendar

The establishment of a transmission schedule for each chosen cluster head's member nodes is the third stage. The TDMA schedule is built based on the number of nodes in the cluster. Then, each node broadcasts its data according to the scheduled timing.

6.2.2.2 *Steady Phase*

When the cluster is in steady phase, cluster nodes send data to the cluster head. Only a single-hop gearbox is capable of connecting the sensors in each cluster with the cluster head [5]. The cluster head collects all of the collected data and delivers it to the base station either directly or via another cluster head, along with the static route provided in the source code. After a specific period of time, the network goes back to the setup stage.

6.2.3 EESR protocol

In order to decrease power consumption and data latency while facilitating scaling in the WSN, a flat routing technique called EESR (energy efficient sector-based routing protocol) was especially recommended. Manager nodes, sensor nodes, gateway, and base station are its main components. Following are their duties: Messages from manager nodes are sent through the gateway to the base station, which has more requirements than standard sensor nodes. The gateway receives messages from manager nodes and it connects with sensor nodes. In order to communicate information with sensor nodes, it also interacts with them. Manager nodes and sensor nodes collect information from the environment, which they subsequently transmit one hop distance to one another until they reach the base station.

6.3 MACHINE LEARNING–BASED APPROACHES

Wireless sensor networks keep an eye on environments that are dynamic and change quickly over time [5]. The system designers themselves or external variables are to blame for this dynamic behaviour. Sensor network software uses machine learning methods to adapt to these situations and avoid needless redesign. A lot of useful ideas that enhance resource efficiency and extend the life of the network are also inspired by machine learning.

6.3.1 Supervised machine learning approach

The system model is built using a labelled training set (i.e., pre-specified inputs and known outputs) [6]. This model represents the learned connection between the input, output, and system parameters. Actually, it is frequently used to address a variety of issues in WSNs, including media access control, privacy and intrusion detection, segmentation and device focusing, activity recognition, and query processing as shown in Figure 6.2.

6.3.1.1 K-nearest neighbour (KNN)

The labelling (outcome values) of a nearby data bits are used by this supervised learning technique to categorise a data sample, which is referred to as a query point. For instance, it may be possible to estimate the missing values of a sensor node by using the average readings of surrounding sensors that fall within a certain range of diameters. The closest group of nodes can be found using a number of different functions [7]. Utilising the Euclidean distance

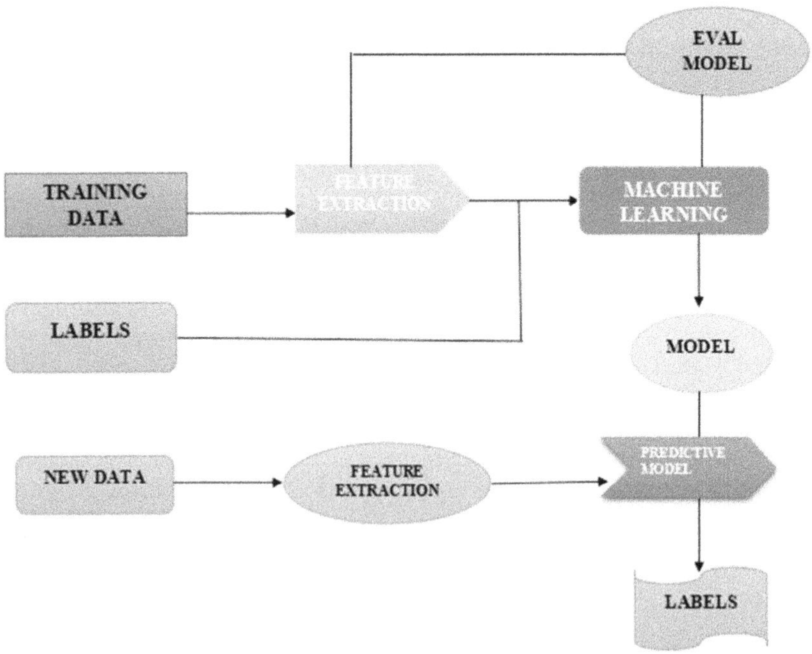

Figure 6.2 WSN working with supervised machine learning.

measure between various sensors is a straightforward way. K-nearest neighbour does not demand a lot of computer resources because the functionality is calculated in reference to adjacent locations. (Specifically, the "k-nearest points," where "k" is a tiny, real number.) The query processing component in WSNs is where the k-nearest neighbour technique is most widely used.

6.3.1.2 Decision tree (D-tree or DT)

Decision tree is a classification algorithm that predicts the data labels by repeatedly passing the independent data. The qualities are similar in respect to the decision circumstances to arrive at a certain category [8]. The DT method, which identifies a few essential elements such the fatality rate, corrupt rate, the time to failure, and the average time to backup, offers a straightforward yet effective method to evaluate link dependability in WSNs. With DT, only single label data may be used, and the best learning trees must be created using nondeterministic polynomial–complete algorithms.

6.3.1.3 Physical networks

It is possible to build this learning algorithm by sliding networks of decision units that are used to identify complex and non-linear functions [9]. Due to the significant computational costs associated with discovering the network weights and the considerable administrative overhead, the use of distributed neural networks in WSNs is still relatively uncommon. However, in centralised solutions, neural networks are able to pick up numerous outcomes and decision borders at once, making them appropriate for employing a single model to address a variety of network problems. For node localisation, one can use the propagation angle and process of checking of the signals that come from anchor nodes [10]. After supervised training, a neural network produces vector-valued coordinates in three-dimensional space representing an estimated node location. Self-organising maps are related approaches to neural networks. One of the crucial uses of neural networks, in addition to function estimation, is the tuning and dimensionality reduction of big data.

6.3.1.4 Support vector machines (SVM)

This method uses labelled training samples to categorise data points as it learns. One technique for determining whether a node is engaging in malicious activities is to use SVM to examine the temporal and spatial linkages of data [11]. SVM splits the space into sections to show supplied WSN data as pointers in the feature space. The separation intervals between these components are as broad as feasible, and fresh reading will be categorised according to which side of the gaps it falls. The SVM approach, which includes evaluating a simultaneous equation with mathematical optimisation programming limitations, provides an alternate to the number of co-neural networks with its quasi and unbounded optimisation problem.

6.3.1.5 Bayesian statistics

Contrary to the majority of machine learning techniques, Bayesian inference only needs a modest number of training data [12]. Without overfitting, Bayesian algorithms quickly learn ambiguous concepts by adapting probability distribution. The key is to upgrade prior beliefs into posterior beliefs using current knowledge (for example, collected data abbreviated as D), where $p(|D)$ is the prior sampling rate of the factor given by the information D and $p(D|)$ is the possibility of the event given by the parameter. Examination of incident uniformity (D) with an unfinished set of information (D) by looking into prior environmental knowledge is one use of Bayesian inference in WSNs. However, this demand for statistical knowledge restricts the widespread use of Bayesian algorithms in WSNs. Gaussian process regression is a statistical learning approach connected to this.

6.3.2 Unsupervised machine learning approach

An unsupervised learning algorithm's main objective is to divide the sample set into various groups by examining their similarities (Figure 6.3). As might be expected, node grouping and data aggregating problems frequently employ this topic of learning algorithms [13].

6.3.2.1 K-means clustering

To classify data into different groups, the k-means algorithm is utilised (known as clusters). Because of its linear complexity and straightforward

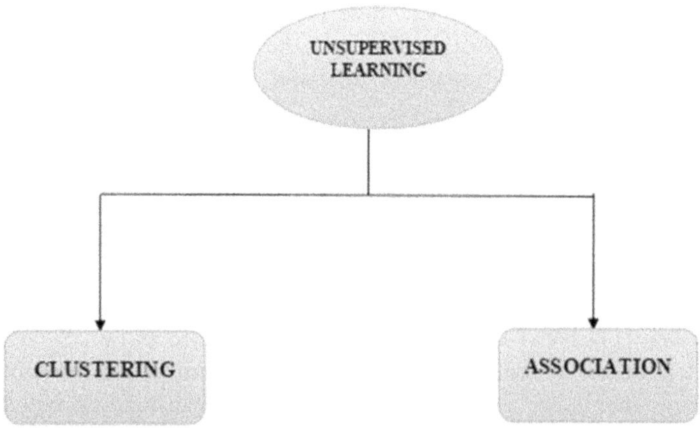

Figure 6.3 Types in unsupervised learning.

implementation, this approach is frequently utilised in research of sensor network segmentation. The steps in k-means to solve this node grouping problem are (a) spontaneously select the first cluster centres; (b) sort each node according to its nearest node using a distance metric; (c) the centroids should be recalculated by using source vertex connections; and (d) if the converging scenario holds true, cease.

6.3.2.2 Principal component analysis (PCA)

PCA is a high-dimensional data compression and dimensional denoising technique that looks for the data's most crucial pieces of data and displays it as a set of new principal components, which are orthogonal variables [14]. The principal components are arranged so that the first component, and subsequent components, correlate to the direction of the data with the highest variance. Consequently, as they have the least information content, the lowest components can be ignored [15]. For instance, PCA finds a few collections of uncorrelated straight relationships of the initial measurements to decrease the quantity of transferred data among sensor nodes. Additionally, the PCA approach provides the issue a simpler solution with taking only a small number of conditions into account in issues with a relatively large number of variables.

6.3.3 Reinforcement learning

With reinforcement machine learning, a sensing node can adapt by communicating with its atmosphere. Using its own expertise, the agent will discover the optimum course of action to maximise its long-term benefits. Q-learning is the most popular reinforcement learning method. Based on the actions conducted at a certain stage, an agent frequently changes its earned rewards.

6.4 OPERATIONAL CHALLENGES

6.4.1 Routing issues in WSN

Although WSNs vary from wireless infrastructure-less networks in a number of ways, designing routes for them represents a difficult issue [16]. There are several kinds of networking difficulties in wireless sensor networks. These significant difficulties are listed as follows:

- Allocating a global IDs system for a significant number of sensor nodes is almost impossible. Therefore, wireless sensor motes are incapable of utilising traditional Internet protocol methods.
- The majority of the time, the produced traffic volume has a lot of duplication, because so many sensor devices might produce identical information at the same time.

When sensor networks are placed in the actual world, a typical issue that might result in inaccurate wearable sensors is node mortality because of energy exhaustion, which is either brought on by regular battery drain or by short circuits [17]. Additionally, sink nodes serve as gateways, storing and transmitting the acquired data. Therefore, issues that influence sink nodes should be found to reduce data loss. The deployment of sensor networks causes congestion issues since several sensor nodes seek to transmit data at the same time. Multiple network floods or improper MAC layer architecture might result in simultaneous communication efforts. The physical length of a connection is another problem.

- Wireless sensor networks require a variety of advanced technologies either because they have many orders of magnitude more connections than ad hoc (when devices connect and interact with one another, an ad hoc network is built on the spot) networks.
- In contrast to MANET nodes, which migrate sporadically, sensory nodes are commonly installed just once throughout their lifespan and are mostly fixed, with a few exceptions.
- In the WSN, elegance reigns supreme. The communication and processing programmes in the sensor's networks should be smaller and more energy-efficient than the typical software used for the same function since the sensor networks are tiny and there are restrictions on energy utilisation.

The following are the main factors that have an impact on a wireless sensor network's design and functionality: wi-fi radio transmission characteristics, media access control schemes, implementation, segmentation, co-ordination, measurement, Internet protocol, physical layers, data integration and propagation, architecture, programming models for sensor networks, development tools, and quality of service, which are some of the topics covered in database-centric and querying hardware, software, and security for wireless sensor networks.

6.4.2 Issues with data collection and clustering

Data Collection: Our method makes use of network nodes operating as transmission agents in the sensing region. Data is sent to an access point as it moves near sensors so that it may be deposited at the target later. To comprehend the important performance indicators such as data transfer, latency to the destination, and power, we provide a mathematical model. Sensory block size, signal creation rate, radio characteristics, and node mobility patterns are among the variables for our approach. Using modelling, we validate our model and demonstrate how much energy can be saved using our strategy as opposed to the conventional ad hoc proposed framework.

The problem in data collection in a sensor network includes many obstacles, including environmental issues. Data collection from sensors and delivery to an architecture wireless router are the goals. Long lengths of time without human intervention are anticipated for these devices (order of months). The main restriction is the sensors' restricted energy budget, which is brought on by their expense and size. The current methods entail creating an ad hoc network among the sensor nodes to transmit data. The following energy-related problems, however, are present. The energy needed to transmit data over one hop in a sparse network is relatively high, to start. This is due to the possibility of far-flung sensors, and the fact that the transmission power needed grows with the fourth power of distance. Second, in an ad hoc network, sensors are required to transfer data for other sensors in addition to their own. Third, the access points on the network are close to hotspots for routing. Nearby sensors have to forward a lot more packets, which quickly depletes their battery power.

Data Clustering: By addressing two significant issues with the size and mobility of the network, clustering techniques for sensor networks increase the network's capacity to grow [18]. The research community has proposed several clustering algorithms. They differ depending on the cluster head (CH) node's features, the overall network architecture, and the node deployment techniques. Depending on the application, a cluster's membership will change. A cluster head is often a node with a lot of energy resources. Cluster heads can occasionally create a second layer in the network. CHs serve as relay nodes in this design

In sensor networks, relay nodes have been proposed to achieve a variety of goals, including data collection, communication range reduction, high availability, etc. [19]. Every node in the WSN is connected to other nodes' sensors, ranging in number from a few to thousands. In order to make WSNs more energy efficient, clustering is a well-known optimisation problem that has been extensively calculated to increase the lifespan of WSNs. There are many different types of clustering. CHs gather information from specific clusters and transmit the combined information to base station.

A collection of uniform sensor nodes makes up a flat network [20]. Any node that transmits data does so either directly or indirectly through any other nodes that are located between it and the CH. Since each node communicates directly with the CH, this architecture's fundamental flaw is that nodes will quickly run out of energy. The bulk of nodes only need to communicate over shorter distances, which significantly reduces the energy consumption of those nodes. For information routing, local data aggregation, distributed management control, and local decision-making, clustering offers an effective architecture. The CH must do more operations in order to aggregate and transport data from the cluster's nodes to the CH, which requires more energy. Rotating the CH's duty between several nodes is one method of saving its energy. The following are the first three clustering phases:

i) Initialisation of the clustering
ii) Cluster development
iii) Cluster upkeep
   1. Therefore, conserving energy, each node only needs to broadcast to their CH.
   2. To again save individual nodes' energy, data collection and aggregation are tasks that are exclusively the responsibility of the CH.
   3. In the event of HWSNs, CHs have greater resources and can thus handle the extra energy demand.
   4. A cluster's nodes are near one another and frequently share data in common. Aggregation is, in general, more efficient because the local CH collect this data.

6.4.3 Issues with event recognition and query processing

Event: For applications of wireless sensor networks, event detection is a significant problem [21]. A sensor network must determine which application-specific incident has happened based on the raw data amassed by individual sensor nodes in order to detect an event. An incident in this context might be anything from a problem with the equipment being tracked to a breach of a secured space. The objective is to enhance network lifetime by providing high-accuracy event detection at low energy cost. Many WSN solutions use event detection as a key component [22]. The topic of event description hasn't, however, gotten enough attention. The vast majority of current methods for event description and detection rely on the use of precise numbers to define event thresholds. However, we think that sharp values can't well manage the frequently inaccurate sensor readings. In this study, we show that the accuracy of event detection is much increased when fuzzy values are used in place of crisp ones. Many WSNs use event detection as one of their primary building blocks. WSNs are used in military applications to spot the entry of hostile troops, while sensor networks for health monitoring and fire detection are used to alert authorities if a fire breaks out anywhere in the monitored region. There hasn't been a lot of study done on developing techniques for event description and detection in WSNs that can enable data reliance and group decision-making.

There are three ways to deal with event recognition:

- Centralised evaluation.
- Decentralised evaluation.
- Distributed evaluation.

Centralised Evaluation: Most frequently used can be the centralised evaluation architecture: A WSN's central base station, which has far greater computer power and energy resources than the sensor nodes, is the only point of communication for all nodes [23]. The semantics of the gathered data are unknown to individual sensor nodes. At the base station, all of the data are collected and analysed. This approach has several benefits, including the high detection accuracy made possible by a thorough centralised overview. The network's performance is bad for a significant number of nodes, which is the fundamental drawback of this method.

Decentralised Evaluation: Within the WSN, smaller clusters are formed using the decentralised assessment technique. A unique node serves as the cluster head in each cluster and carries out this function [24]. Data classification and base station communication are the responsibilities of this node. The fact that this design is fault tolerant against the failure of a single node or the loss of an entire cluster is one benefit. All of the clusters continue to work even if one fails. Because it is possible to place clusters into an energy-efficient idle state when they are not required or activated by an event, this design also provides benefits with regard to energy awareness.

Distributed Evaluation: The decentralised evaluation method and the distributed evaluation method differ in two significant aspects. First, just an event's signalling is sent to the base station; all data processing and event detection take place within the network. A copy of the outcomes is sent to each node that detected the event, and each node then contrasts the results with its own. The absence of any cluster heads in the network is the second issue. Instead, each node is capable of processing data on its own. As opposed to the previously described methods, a leader node is dynamically chosen whenever an event occurs; as a result, a cluster head is not required.

Query Processing: A sensor network that keeps track of environmental factors including temperature, precipitation, pollution, humidity, wind, and so on [25]. These data are transmitted to a database server where they will subsequently be processed. All of the information gathered by the sensors will be stored in a database, which client applications like the supervisor or a third-party application may then access. A view is a hypothetical table that displays the outcomes of a query. A materialised view, as the name indicates, duplicates the data in contrast to a conventional view. According to the value of one or more fields, an index established on a table will provide easy access to records. Additionally, it makes it possible to speed up and simplify search procedures.

- Wireless sensor networks (WSNs) have gained popularity as methods for capturing the real world because of the rapid advancement of sensing and wireless communication technologies. One or more sensing components, a radio component, and constrained computational capabilities make up a sensor.
- It measures the environment physically, including its humidity, light, and other factors like temperature. To the extent of its radio coverage, each node has direct radio communication with the others. The base station, also known as the data sink, has a radio component that

enables communication with neighbouring sensor nodes so that it may gather their data.

- Data collected by sensors placed on the outside of a monitoring area may need to pass through several hops (sensors) before reaching the sink, depending on the size of the monitoring area. Users can query the sensing data at the base station, which then provides the query results. WSNs were initially used in military and academic endeavours. As sensor costs fall and their capabilities rise, applications for WSNs are expanding. WSNs have piqued the interest of both the network and database communities in recent years.

6.4.4 Localisation and object targeting issues

Localisaton: The process of localisation involves locating the sensor nodes [26]. The sensor nodes' position is estimated using localisation techniques and algorithms because their starting coordinates are unknown. It is possible with the use of inter-sensor measurements and the absolute positions of a few sensors. Beacon nodes or anchor nodes are sensors with known coordinate information. By installing the anchors or beacons at known coordinate points or using GPS, their locations may be determined.

Localisation technique includes

- Efficient Energy Consumption in WSNs
- Localisation
- 3D WSNs
- Mobile Wireless Sensor Networks
  1. The development of more dispersed localisation algorithms,
  2. The reduction of location latency while preserving position precision, and
  3. The expansion of mobile sensing to places where data cannot be securely gathered are all priorities.

As more and more sensor networks are deployed, and as the applications become more and more diverse, cooperative localisation research will continue to expand. This demands that localisation algorithms be created in a way that ensures the least amount of bias and variation. They must also be capable of scaling to very high network sizes without significantly increasing energy use or processing workloads [27].

Object Targeting: The primary function that all WNS programmes have in common is object tracking. A runaway, a border trespasser, or an assailant at a military installation might all be represented by this item. Object interception, localisation, and ongoing location reporting to the base station are only a few of the many subtasks involved in tracking these mobile objects. Different signal types may be present on tracked objects that can be detected. Thus, the object targeting can be classified into two types:

1. Network architecture used
2. The approach adopted

Challenges in Object Targeting

- Scalability
- Stability
- Node Deployment
- Computation and Communication Costs
- Energy Constraints
- Data Aggregation
- Sensor Technology and Localisation Techniques
- Tracking Accuracy
- Reporting Frequency
- Localisation Precision
- Sampling Frequency

Solution for the Challenges

- The Naïve Architecture
- Tree-Based Architecture
- Cluster-Based Architecture
- Static Clustering
- Dynamic Clustering
- Hybrid Architecture

Naïve Architecture: The naïve architecture is the most basic and conventional WSN design, in which all sensors are continuously active in an effort to recognise and track things that are within their sensing range and relay data to a single, central sink node. Sensor nodes are able to decide when to sleep based on the information their neighbours supply. As a result, nearby nodes can stay active while distant nodes can go to sleep. The original naïve strategy previously mentioned performs better in simulations than this upgraded approach.

Tree=Based Architecture: This architecture entails making a tree. Some scientists agreed that the entire set of nodes' tree-like subgraph, which represents the collection of nodes, is valid. The sensor nearest to the object is represented as the root of this tree, and when the object travels, additional sensors are added or deleted. Optimised organisation and communication are one such excellent example. The objective is to develop an algorithm, which then modifies its structure in response to the problem put forth and continues doing so in order to consume less energy.

Cluster-Based Architecture: For applications involving dozens, hundreds, or even thousands of sensor nodes, clustering could be a scalable solution. To increase network longevity, scalability calls for effective resource management and load balancing. Effective resource utilisation is achieved by reducing the communication load, reducing the likelihood of data flow bottlenecks, and improving survivability because there is no longer a single point of failure. Load balancing is achieved by reducing the processing load on individual sensor nodes.

The step includes:

- Cluster creation in terms of geography.
- The choice of a few highly capable but sparsely placed sensors to serve as cluster leaders.
- Data aggregation.
- Data transmission.

Static Clustering: Each cluster has a fixed cluster size, coverage area, sensor members, and cluster leader. In other words, the sensor nodes are linked to the same cluster head for the duration of the network. Numerous issues with clustering include the inability to share information, incapacity to handle dynamic situations, and lack of fault tolerance (battery depression).

Dynamic Clustering: A dynamic cluster's members may at any given time be a part of many clusters depending on the movements of the objects. Additionally, because there is only one cluster operating close to the object with a high likelihood, redundant data are suppressed, which enhances tracking accuracy. Energy consumption is also reduced since only one cluster is active at once in response to object movement.

Hybrid Architecture: A combination of one of the aforementioned systems with a prediction mechanism is referred to as a "hybrid architecture." As a result of the unavoidable prediction errors, these methods offer recovery techniques to accommodate for incorrect object localisation. Regrettably, these algorithms are frequently too complex to be used on sensor nodes with limited resources. The cluster head attempts to foretell the target's potential position through each iteration of the tracking algorithm. This prediction indicates that the remaining nodes remain inactive and that only the nodes situated inside the expected zone are switched on. Regarding network longevity, energy efficiency, and tracking precision, the proposed method has demonstrated its viability.

- Use an efficient energy-saving technique to decrease the number of sensor nodes that are constantly active.
- Tracking algorithms are concerned with removing (or decreasing) correlated, inconsistent, or redundant data in order to reduce not only the number of packets delivered but also the number of collisions and interference in the shared medium.

6.4.4.1 Issues in medium access control

We look at a static wireless sensor network architecture that allows for concurrent broadcasting and receiving at the same node. However, a transmitter or a receiver must be employed by each node. The core of interference is nevertheless represented by allowing these circular zones to extensively overlap, despite the fact that this is an oversimplified and unrealistic assumption. Every node is considered to be in the transmission range of at least one other node. We take into consideration random huge (or renewable) energy sources in addition to constant traffic load distributions among transmitters. A topology-based, greedy method is provided to gradually pick which distinct receiver groups to activate in a time-division mechanism.

- As the first stage of receiver activation, a random node is chosen to start the first receiver group. Either a rigid priority-based standard is used, or the choice is made at random.
- When this occurs, all other nodes within a specific range are recognised as transmitter nodes by the active receiver node. Distributed control information can make this possible.
- The experiment is performed numerous times using a transmitter node that has already been developed.
- As a consequence, we create a variety of receiver activation groups until each node performs both transmitter and receiver tasks at least once over an entire cycle of activation periods.

There is a suggestion for a two-layered time-division mechanism based on group TDMA and receiver activation as a way to schedule links and allocate resources with suboptimal yet polynomial time solutions. In order to increase the lifespan of nodes, a topology-based greedy algorithm is utilised to select which receiver groups should be activated in which discrete time intervals and to select temporal allocations based on the total battery energies remaining available at the transmitter groups. We quantify the impact of group TDMA and energy-efficient receiver activation on performance using numerical examples.

Although it falls short of providing a comprehensive solution to the MAC problem in sensor networks, it does offer a useful method to handle the coordination of transmissions and receptions in such a network. The focus is on the fundamental problem rather than on its immediate implementation, despite the fact that this is just the first step in a process to explore how sensor networks work.

6.5 RESEARCH-BASED MACHINE LEARNING–BASED WSN

According to the developers of sensor networks [27], machine learning is a collection of methods and software used to create forecasting models. On the other hand, machine learning experts view it as a wonderfully rich field with massive themes and patterns. Understanding such concepts is advantageous for those who wish to use machine learning with WSNs. When employed in various applications, algorithms provide several advantages for flexibility.

Supervised Learning: The system model is constructed via supervised learning from a labelled training set. This model is intended to illustrate the discovered relationship between the system parameters (input, output, and parameters). In the context of WSNs, the main supervised learning algorithms are addressed in this class. In actuality, supervised machine learning techniques are extensively employed to address a number of problems in WSNs, including localisation and object targeting.

Unsupervised Learning: There are no classifications required for unsupervised learning. An unsupervised learning algorithm's basic goal is to separate the test sample into various groups by assessing their similarity.

Reinforcement Learning: An agent can learn through interaction with its surroundings thanks to reinforcement learning, for example a sensor node. By drawing upon its own experience, the agent will discover the appropriate course of action to pursue in order to optimise its long-term benefits. Q-learning, which uses reinforcement learning, is the most used method. With a distributed design like that seen in WSNs, in which each node wants to select behaviours that were anticipated to maximise its lengthy benefits, this method is simple to implement. The broad and effective usage of Q-learning in the WSN routing problem is something that must be noted.

6.6 PRACTICAL APPLICATIONS OF WIRELESS SENSOR NETWORKS

Cataclysmic Relief Operation: If an area is believed to have been impacted by a calamity, such as a wildfire, drop the sensor nodes on the fire from an aeroplane. Track the data from each node and produce a temperature map in order to establish suitable plans and techniques to tackle the issue.

Moral Applications: WSNs are especially useful in military operations for spotting and tracking the movements of allies and enemies since they may be instantly deployed and self-organise. The sensor nodes can be used to monitor the battlefield in case it needs additional equipment, manpower, or ammo. Biological, nuclear, and chemical attacks may potentially be detected by the sensor nodes.

Authentication for the Environment: There are a plethora of applications for these sensor networks in the environment. These tools can be used to watch and record the movement of animals and birds. It is possible to monitor the earth, the soil, the atmosphere, irrigation, and precision farming with these sensors. They can also be used to detect other natural calamities including earthquakes, fires, floods, chemical and biological epidemics, and others.

Medical Applications: Utilising WSNs, a patient can be continuously monitored in health applications. Animal behaviour may be seen, as well as its internal workings. One can do diagnostics. Additionally, they support monitoring both patients and doctors as well as keeping an eye on how drugs are administered in hospitals.

Other Applications Keeping an eye on and controlling the air quality.

- Monitoring and controlling traffic conditions.
- Meteorological conditions are monitored and managed.
- Characteristics of an environmental monitoring system.

Autonomy: The weather stations must be able to run on batteries for the duration of their deployment.

Relability: To avoid unplanned breakdowns, the network must carry out straightforward and predictable tasks.

Robustness: Numerous issues, including poor radio connection (for instance, in the event of snowfall) or hardware breakdowns, must be taken into consideration by the network.

Flexibility: Depending on the demands of the applications, one must be able to swiftly add, relocate, or delete stations at any moment.

6.7 RESEARCH AND FUTURE DEVELOPMENTS

Compact Sensing and Sparse Coding: There are small changes between these two methods, but they both deal with obtaining a sparse representation. With fewer data points than the original signal, compressed sensing is primarily concerned with the issue of solving an underdetermined system of linear equations.

Resource Management Using Machine Learning: Strategic alignment with the purpose is more akin to human resource management (HRM). AI (artificial intelligence)–based methods are being used more and more in HRM operations. Businesses are focusing on more sensible answers. Machine learning (ML) algorithms are currently improving a number of human resource management processes. This report provides a summary of key HRM operational tasks where ML–based solutions, in particular, can be used to increase the accuracy of HRM operational processes.

Dispersed and Accomodative ML Techniques in WSN'S: Devices with restricted resources, like WSNs, benefit from distributed ML algorithms. Dispersed learning algorithms use less computing power and memory space than centralised learning algorithms. The nodes can quickly change their predictions of future behaviour to the conditions of the environment thanks to decentralised learning mechanisms. Due to these factors, dispersed and accommodative learning algorithms are suitable in data sorting inside networks without taxing nodes with heavy computing workloads.

6.8 CONCLUSION

It is necessary to employ tools and protocols that can deal with specific issues and limitations because wireless sensor networks are different from traditional networks in a number of ways. Because of this, WSNs require cutting-edge solutions for real-time and energy-conscious networking, encryption, organisation, location, node aggregation, information retrieval, damage diagnosis, and information security. A wire-free sensor network's ability to adapt to its surroundings may be enhanced using a range of techniques given by machine learning. The development of lightweight and distributed message passing techniques, online learning algorithms, hierarchical clustering patterns, and the application of machine learning to the resource management issue in wireless sensor networks are other issues that have not yet been solved and need further study.

REFERENCES

[1] Khan, AA. (2016). A survey of routing protocol in wireless sensor networks [thesis]. Lahore: GCU. Available from: http://library.gcu.edu.pk/theses.htm

[2] Bakr, BA, Lilien, L (2011). LEACH-SM: A Protocol for Extending Wireless Sensor Network Lifetime by Management of Spare Nodes. In *International Symposium on Modelling and Optimization in Mobile, Ad Hoc and Wireless Networks (WiOpt)*. New Jersey, Princeton: IEEE.

[3] Nallusamy, R, Duraiswamy, K (2011). Feedforward Networks Based Straightforward Hierarchical Routing in Solar Powered Wireless Sensor Networks. *WSEAS Transactions on Communications*, 10(1), 24–33.

[4] Tian, S, Zhang, X, Wang, X, Sun, P, Zhang, H (2007–11). A Selective Anchor Node Localization Algorithm for Wireless Sensor Networks. *IEEE International Conference on Convergence Information Technology*, Korea, pp. 21–23.

[5] Saud Althobaiti, Ahlam, Abdullah, Manal. (2015). Medium Access Control Protocols for Wireless Sensor Networks Classifications and Cross-Layering. *Procedia Computer Science*, 65, 4–16. doi:10.1016/j.procs.2015.09.070

[6] FrancescoMario, Di, DasSajal, K, Giuseppe, A (2011-08-01). Data Collection in Wireless Sensor Networks with Mobile Elements. *ACM Transactions on Sensor Networks*, 8(1), 1–31, TOSN. 8. 7. doi:10.1145/1993042.1993049

[7] Xia, Feng, Tian, Yu-Chu, Li, Yanjun, Sun, Youxian. (2007-10-09). Wireless Sensor/Actuator Network Design for Mobile Control Applications. *Sensors*, 7(10), 2157–2173. doi:10.3390/s7102157

[8] Akyildiz, IF, Su, W, Sankarasubramaniam, Y, Cayirci, E (2002). A Survey on Sensor Networks. *IEEE Communications Magazine*, 40(8):102–114.

[9] Dargie, W, Poellabauer, C (2010). *Fundamentals of Wireless Sensor Networks: Theory and Practice*. John Wiley and Sons. doi:10.1002/9780470666388

[10] Hac, A (2003). *Wireless Sensor Network Designs*. John Wiley & Sons Ltd.

[11] Raghavendra, C, Sivalingam, K, Znati, T (2004). *Wireless Sensor Networks*. Springer.

[12] O'Donovan, Tony, O'Donoghue, John, Sreenan, Cormac, Sammon, David, O'Reilly, Philip, O'Connor, Kieran A (2009). A Context Aware Wireless Body Area Network (BAN) (PDF). *Pervasive Computing Technologies for Healthcare*, 2009.
[13] Heinzelman, W, Chandrakasan, A, Balakrishnan, H Energy-Efficient Communication Protocols for Wireless Microsensor Networks. *Proceedings of the 33rd Hawaaian International Conference on Systems Science (HICSS)*, January 2000. Paper.
[14] Varshney, Shweta, Kuma, Rakesh. (2018). Variants of LEACH Routing Protocol in WSN: A Comparative Analysis. *2018 8th International Conference on Cloud Computing, Data Science & Engineering (Confluence)*. IEEE, 199–204.
[15] Worlu, C, Jamal, AJ, Mahiddin, NA (2019). Wireless Sensor Networks, Internet of Things, and Their Challenges. *International Journal of Innovative Technology and Exploring Engineering*, Blue Eyes Intelligence Engineering & Sciences Publication. ISSN: 2278-3075, Volume-8 Issue-12S2. doi:10.35940/ijitee.L1102.10812S219
[16] Roy, Nihar Ranjan, Chandra, Pravin (2018). A Note on Optimum Cluster Estimation in LEACH Protocol. *IEEE Access*. 6(21), 65690–65696. doi: 10.1109/ACCESS.2018.2877704
[17] Russell, Stuart J, Norvig, Peter (2010). *Artificial Intelligence: A Modern Approach* (Third ed.). Prentice Hall.
[18] Mohri, Mehryar, Rostamizadeh, Afshin, Talwalkar, Ameet (2012). *Foundations of Machine Learning*. The MIT Press.
[19] Geman, S, Bienenstock, E, Doursat, R (1992). Neural Networks and the Bias/Variance Dilemma. *Neural Computation*, 4. pp. 1–58. doi:10.1162/neco.1992.4.1.1
[20] Maity, A (2016). Supervised Classification of RADARSAT-2 Polarimetric Data for Different Land Features. arXiv preprint arXiv:1608.00501. doi:10.5281/zenodo.832427
[21] Buhmann, J, Kuhnel, H (1992). Unsupervised and Supervised Data Clustering with Competitive Neural Networks. *[Proceedings 1992] IJCNN International Joint Conference on Neural Networks*. Vol. 4. IEEE. pp. 796–801.
[22] Roman, Victor. (2019-04-21). Unsupervised Machine Learning: Clustering Analysis. *Medium*. Retrieved 2019-10-01.
[23] van Otterlo, M, Wiering, M (2012). Reinforcement Learning and Markov Decision Processes. *Reinforcement Learning*. Adaptation, Learning, and Optimization. vol 12. Springer, Berlin, Heidelberg. doi:10.1007/978-3-642-27645-3_1
[24] Russell, Stuart J, Norvig, Peter (2010). *Artificial Intelligence: A Modern Approach. Pearson education*, (Third ed., pp. 830, 831), 978-0-13-604259-4.
[25] Riveret, Regis, Gao, Yang (2019). A Probabilistic Argumentation Framework for Reinforcement Learning Agents. *Autonomous Agents and Multi-Agent Systems*, 33 (1–2), 216–274.
[26] Zang, Z, Qi, JD, Cao, YJ (2010). A Robust Routing Protocol in Wireless Sensor Networks. In *IET International Conference on Wireless Sensor Network*. China: IET pp. 276–279.
[27] Sohraby, K, Minoli, D, Znati, T (2007). *Wireless Sensor Networks: Technology, Protocols, and Applications*. John Wiley and Sons, doi:10.1002/047011276X

Chapter 7

Healthcare 4.0

Blockchain technology application in healthcare ecosystem

Jeevesh Sharma and Swati Jain
Manipal University Jaipur, Jaipur, India

7.1 INTRODUCTION

In today's dynamic world, which is expanding at an exceptionally rapid pace, humans cannot separate themselves from the existence of the internet and various industries. Industry 4.0 is strongly related to the concepts of crowdsourcing, circular economy (sharing economy), green economy, and bioeconomy and the ever-changing environment, where the technology has become a part of everyone's lifestyle. The healthcare industry is not an exception. New-generation administrators, economists, and physicians in the healthcare industry must thrive in this technology-driven environment to cope with the current advancements in the medical industry, which has adopted the digitalization era. Authentication, security, immutability, distributed ledgers, and decentralized storage are just a few of the built-in features that blockchain technology offers. The excitement has given way to practical use cases in sectors like healthcare. Researchers from academia and business have begun to investigate healthcare-focused blockchain applications, developing on the technology that are already available. Some of these purposes include identity authentication, smart contracts, and fraud recognition. Owing to additional regulatory obligations to protect patient's health records and information, the healthcare industry has needs in terms of safety and secrecy. Cloud space and the proliferation of mobile health appliances have made record and data sharing more common in the age of the Internet, but they have also increased the risk of hostile assaults and the possibility that private information may be compromised when it is shared. The sharing and privacy of this information are issues since it is getting easier to get health information via smart devices and because individuals are visiting several providers.

Cloud-based solutions, which are intended to reinforce corporate operations and push more customer interactions online, have become increasingly prevalent because of COVID-19. Due to this trend, the major electronic health records (EHR) suppliers are being compelled to move their offerings to the cloud and form alliances with companies that enterprise software solutions like customer relationship management (CRM) and enterprise

 DOI: 10.1201/9781003474524-7

resource planning (ERP). Emerging transformative technologies like artificial intelligence (AI), machine learning, telehealth, blockchain, robotic surgery, process automation, image interpretation, joint replacements, and monitoring devices like sensors, wearables, and ingestibles are supplying real-time and continuous data about our health. Additional adoptions in seemingly unimportant fields like medical billing give more advantages. The way that healthcare is provided in the future is being redefined by this. As a result of their expanded use of cloud computing, hospitals are already prominent targets for ransomware attacks and other forms of online criminality. Blockchain technology might be used to protect health data from similar threats. Data security, cost, reluctance to change, and other factors are the key obstacles appearing in this new healthcare ecosystem.

Numerous healthcare applications can benefit from the application of blockchain technology (Tanwar et al., 2020). The implementation of blockchain in EHR can maintain the confidentiality of patient's sensitive data and guarantee that only authorized parties should have access to the information (Hathaliya et al., 2019). Due to technological advancements and the current scenario, there exists a need to think about Healthcare 4.0 among the academic and corporate communities, as it is one of the most significant areas of interest in the modern world. The main objective of this study is to elaborate the blockchain technology's current features and uses in the healthcare business and potential for the future. The chapter presents the status of blockchain technology usage in the Indian context and compares it with the traditional healthcare ecosystem. The chapter also analyzes various techniques provided by blockchain technology in the healthcare ecosystem. Along with this, it also discusses the challenges being faced by the healthcare sector in the current environment.

Further, the chapter is divided into sections. The development and future of Healthcare 4.0 are covered in Section 7.2. Section 7.3 includes the introduction of blockchain technology and outlines its necessity in the healthcare industry. Section 7.4 discusses the numerous benefits and applications of blockchain technology in the healthcare sector. The use of blockchain in the Indian healthcare sector and the legal framework are discussed and covered in Section 7.5. Section 7.6 presents the issues and challenges related to blockchain technology and healthcare and Section 7.7 summarizes and concludes the study. Lastly, Section 7.8 provides future directions.

7.2 HEALTHCARE 4.0

The emergence of Industry 4.0 has led to exponential changes in business operations and in this regard the healthcare industry is not an exception (Bongomin et al., 2020). Industry 4.0 deals with digital transformation and usage of various technologies in various industries. Industry 4.0 has created a new paradigm for the healthcare sector also. Orthopedics, dentistry, and other medical specialties have all been impacted by the disruption and revolutionary wave of Industry 4.0, which is massively upgrading multiple business sectors (Rehman et al., 2019; Javaid & Haleem, 2020). The approach of healthcare is about to undergo a fundamental transformation toward the so-called "Healthcare 4.0," which is an era of smart, connected healthcare (Aceto et al., 2020). This recent notion of Healthcare 4.0 has emerged because of the significant disruption to the healthcare sector. There is a tremendous change in the traditional and contemporary healthcare system. The Healthcare 4.0 setup makes it possible for patient care to advance in exciting new ways. Doctors and patients may get more information more quickly in Healthcare 4.0 (Javaid & Haleem, 2019). The traditional healthcare industry has never seen the level of connectivity that Healthcare 4.0 has provided. Resultant to this people can therefore make better decisions and lead better lives (Jayaraman et al., 2020).

The healthcare ecosystem has experienced the transition of healthcare services from Healthcare 1.0 to the current arena. Smart healthcare tools and applications, application of blockchain technology, mobile healthcare services, digitization of the healthcare system, and functioning of the e-healthcare system are the key characteristics of Healthcare 4.0, which shows a drastic transformation in the traditional healthcare industry, that is, Healthcare 1.0 (Aceto et al., 2020). Figure 7.1 summarizes the evolution of Healthcare 4.0 from Healthcare 1.0:

The figure points out the key features of each healthcare scenario. Following are the medical industry's most promising developments brought by Healthcare 4.0:

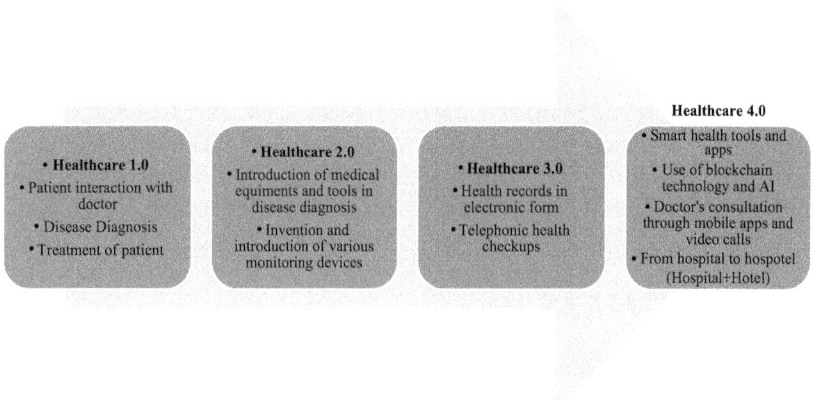

Figure 7.1 Evolution of healthcare 4.0.

Source: Author's compilation.

Using Connected Devices to Make Healthcare More Responsive: A network of sensors, monitoring devices, and other healthcare devices is the foundation of Healthcare 4.0. The well-known term Internet of Things (IoT) also applies to the healthcare industry, although experts refer to the "Internet of Medical Things" (or IoMT) in the context of hospitals. The development of a network of interconnected medical devices enables new developments in medical operations. For instance, employing a set of precise instruments like the da Vinci Surgical System enables doctors to conduct procedures remotely. This guarantees that surgical procedures are always of the highest caliber among surgeons. Although storing and transferring patient data is a crucial function of wearable devices, their contribution doesn't end there. They can also be helpful in ensuring worker safety. Businesses like Smart-Cap offer monitoring tools to assess employee attentiveness and stop workers from dozing off at work. It is simple to understand how this could help healthcare professionals, even though it hasn't yet been implemented in hospitals.

Virtual Reality (VR): VR must be mentioned while discussing new and developing technology in healthcare. Patients can temporarily escape the boring hospital environment by put on a VR headset and travelling to different worlds. It has been observed that there are other ways in which VR technology and healthcare are connected. For some patients, like burn victims, VR has been demonstrated to have pain-relieving abilities that go beyond simple escape. While learning how to manage chronic pain, this could be a helpful strategy to enhance patient results. VR headsets are now portable enough to take with you when you leave the hospital, providing patients with a new method of pain management. VR can be helpful for teaching medical professionals new surgical techniques. With the use of this technology, doctors can acquire new knowledge without risk. After all, in virtual surgery, mistakes have no repercussions.

Using AI to Provide Fresh Resources: As AI develops, it can contribute more and more to the well-being and safety of patients. Without compromising the needs of the patients, this enables the person to focus on the more technical aspects of providing high-quality patient care. The chatbot is one instance of AI that has already affected patient care. Chatbots offer new patients a consistently great experience by assisting them in politely and quickly responding to screening questions. In some cases, chatbots may even be used to deliver therapy for mental health issues (Javaid & Haleem, 2019). Chatbots can assist users in determining the type of assistance they require, although they do not take the place of expert assistance. Social companion robots could play a significant role in patient's lives in the future. They can take on some of the responsibilities of hospital staff or even work as hospital receptionists.

Improving Hospital Eating: One thing unites almost everyone who goes to the hospital; they all need to eat. But it is frustrating that so many patients dread mealtimes because they hate eating at the hospital. But hospitals might soon be able to improve the kind of food that patients eat and how they eat it thanks to current technologies. In order to prepare healthy meals for their patients, many hospitals have started to grow their own vegetables, placing a heavy emphasis on using plants instead of animals. IoT technologies are extremely beneficial in agriculture, despite the logistical challenges that may arise. For instance, hospitals can automate greenhouse operations using IoT equipment. They would be able to enjoy fresh vegetables without having to pay the heavy cost of hiring a full gardening crew. By allowing interactive "room service" dining for patients, technology can also be helpful. Hospitals might use this technology as well to increase efficiency, like how "smart hotels" let customers order food through touchscreens.

Rethinking the Actual Hospital Setting: Doctors are now able to obtain patient information remotely because of wearables and other IoMT devices. As a result, typical operations like routine checkups may no longer require physical areas in healthcare institutions. As a result, the hospital construction and operation process might be rethought by the industry. This has already generated a few intriguing concepts. Mercy Virtual Care Center, the first telehealth-only facility in the world, is noteworthy for its method of patient care. There aren't any patients despite the facility's size. Instead, the facility acts as a sizable remote dispatch center, linking medical professionals with patients in need of assistance who are located far away. Healthcare company Higi and its sizable network of health-monitoring kiosks demonstrate how smart kiosks can significantly contribute to streamlining healthcare. Kiosks can act as convenient health-monitoring stations since they can fit a lot of healthcare-monitoring equipment in a small space. For people who cannot afford personal health monitoring gadgets, this is especially crucial. Therefore, emergence of Healthcare 4.0 has facilitated individuals and provided a better version of healthcare services. The advanced customer services provided by the hospotel (Hospital plus hotel) has become the key ingredient of the present healthcare system. Section 7.3 introduces blockchain technology and Section 7.4 describes the role and status of blockchain technology in the present healthcare industry.

7.3 BLOCKCHAIN TECHNOLOGY

In 1991, the first version of blockchain technology was created to store and protect digital data. Blockchain is a common and public ledger, which can be accessed by various people at the same time. One of its key benefits is that the data is recorded and hard to alter without agreement from all parties

(Tanwar et al., 2020). In blockchain technology, every new record turns into a block with an exclusive hash for recognition, according to IBM. By linking the blocks all together into a chain of documents, a blockchain is produced (Nakamoto, 2008).

Blockchain technology's core concept is to offer a basis for assistance between unidentified and unreliable things without the need for a centralized safekeeping and verification authority, as in the current cloud computing architectures, while also correlating the distributed features of mobile (smart health) devices (Dinh et al., 2018). Blockchain enables immediate data access and does away with centralized control. This structure's blocks are scattered around the computer node and are all interconnected with one another. This makes it more complicated for hackers to modify the information. Furthermore, it also offers safe transactions, lower compliance expenses, and expedited data transfer processing.

A further benefit is that using blockchain can potentially address security issues. Decentralized networks, transparency, and immutability are the three characteristics that best describe blockchain technology. It performs the role of a distributed ledger and keeps track of previous transactions. The network's nodes, which are made up of the numerous participating systems, distribute this transaction history. Each transaction is put into data blocks and encrypted to keep them safe, creating a mess that acts as a unique signature. The blocks cannot be changed since any attempt by a hacker to modify the information will lead to a new hash value, or signature, which will invalidate the transaction. All users who are participating will be made aware of any record tampering (Wenhua et al., 2023). To get secured access to the information, users will have to decide between public, private, and permissioned blockchains. A public blockchain is open and transparent, enabling any user to contribute to system activities, in contrast to a private blockchain that restricts data sharing to specific parties. Based on this fact it is anticipated that 10% of worldwide GDP will be deposited on blockchains by the year 2025, as per the World Economic Forum. When it comes to EHR around all hospital systems, blockchain will be able to give stakeholders immediate access from anywhere. As a result of this retrieving a patient's medical history will take less time, and repeated medical procedures won't cost as much money.

Therefore, introduction of blockchain technology in the healthcare industry will benefit the service providers as well as the customers, that is, the patients. As doctors can access the real-time data of their respective patients, the accessibility of medical services has increased for the patients and overall costs will be reduced for all stakeholders. Section 7.4 elaborates the application of blockchain technology in healthcare.

7.3.1 Blockchain technology in healthcare

Distributed ledgers, immutable technology, and cryptographically secure technology are also part of Industry 4.0's new Internet of Things applications

(Bongomin et al., 2020). This technique makes use of multiple transaction lists that are duplicated and shared by numerous groups or parties. The healthcare sector is one in which blockchain technology has immense potential due to the more patient-centric approach to the healthcare system, the capability of blockchain technology to connect scattered platforms, and its ability to increase the accuracy of electronic health data. The major challenge of the healthcare industry is to secure the privacy of patients' medical data. In the digital era, record sharing and data sharing are becoming increasingly widespread because of cloud storage and the use of mobile health devices (McGhin et al., 2019; Wenhua et al., 2023). The previous healthcare system's "volume-based care" system is being completely replaced by a newer model called "value-based care" system. Value-based healthcare puts emphasis more on individualized, patient-centric services than the conventional model, in which charges for patients are based on the customer services rendered regardless of their medical importance. Using big data to improve healthcare services and make treatment more affordable are two significant revolutionary advances of blockchain technology (Tanwar et al., 2020). Also, big data is essential to the fourth industrial revolution as well as to the healthcare system, which includes drug testing, clinical trials, patient information management, patient involvement, remote patient intensive care, and so on.

The healthcare industry is one of those that deserves attention for several reasons. First, due to the vast amount of data that needs to be recorded, kept, and shared every minute due to the increasing worldwide population and COVID-19, health service providers are under a tremendous amount of pressure (Gupta et al., 2022). A repository of vaccination data is now necessary due to the ongoing global campaign to vaccinate against COVID-19 and travel restrictions related to travelers' level of immunization. Second, the spread of fake medications is a major issue. Third, with the inception of COVID-19, medical supply chain management has taken on a position of significant importance. In India's second wave earlier this year, there was a severe scarcity and black-market sale of medical equipment (Fiore et al., 2023). Health informatics (HI), a subfield of medical informatics, is concerned with using technology to deliver healthcare in the clinical setting. Healthcare informatics is developing as a result of changes in technology and healthcare operations. With new technological advances, HI offers essential, indivisible knowledge bases to healthcare providers and organizations, enabling them to offer patients higher-quality services. The process of collecting data for clinical research, selecting an appropriate diagnostic technique, explaining test results, and acquiring and preserving medical and patient data are all examples of information processing and communication in the healthcare sector.

In this healthcare ecosystem, which is enabled by blockchain technology, patient records will have a unique identity, and every time the new information will be added to the previous data, and this system is well connected to the EHR (Elangovan et al., 2022). The patient's information will only be accessible by the appropriate stakeholders with the owner's consent, giving 360° controls over their data.[1] In a manner similar to this, IBM created the private Health Utility Network, a blockchain-enabled ecosystem that serves as an association among rivals and enables healthcare organizations to develop, share, and implement customer-centric payment processing solutions, facilitating safe and efficient healthcare information exchanges. In this direction, Aetna, Anthem, Health Care Service Company, PNC Bank, Cigna, and Sentara are a few of the healthcare companies that have teamed up. Figure 7.2 explains the implication and process of blockchain in the healthcare ecosystem. With the help of blockchain, the detailed information of patients and doctors are recorded in a digital manner, which enhances the overall medical treatment system.

7.3.2 Implications of blockchain in healthcare

Most medical professionals are seeking to replace traditional health services with e- Health with innovative technologies like the IoT, VR, and AR (Aceto et al., 2020). Using cutting-edge information and communication technologies, eHealth's primary goal is to transform the administration of health information and enhance the healthcare system. The healthcare industry has special security and privacy considerations since there are more statutory regulations to secure patients' medical information (Sookhak et al., 2021). Cloud computing along with the adoption of digital healthcare devices have

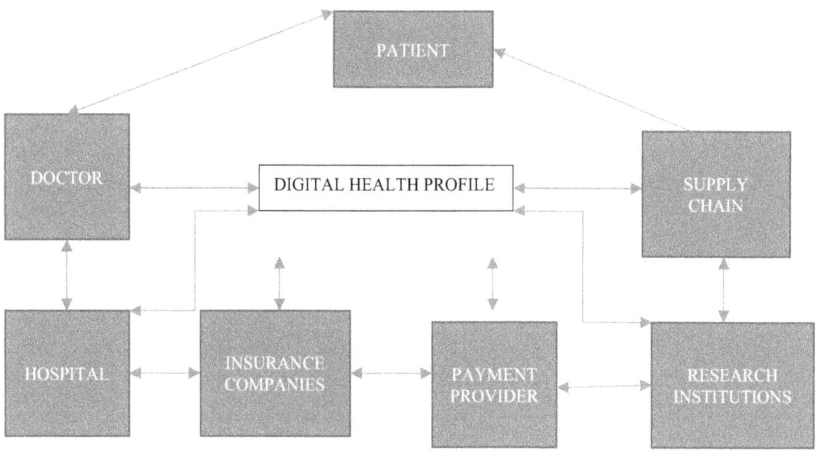

Figure 7.2 Blockchain in healthcare ecosystem.

Source: Kirill. https://www.dataart.com/blog/blockchain-in-healthcare-complex-challenges-overshadowed-by-the-hype-need-to-be-overcome

made data interchange more common in the digital age, but they have also increased the risk of hostile attacks and the compromising of personal information. The issue of sharing and privacy of this information is increased when patients visit several providers and have access to health information via smart devices (McGhin et al., 2019). Therefore, this section explains how the healthcare sector is executing blockchain technology to make system more transparent and easier to access. There are many reasons behind approaching blockchain technology in the healthcare sector. Technology has made many things easygoing and resolved many challenges also. Figure 7.3 presents some points related to execution of blockchain technology in healthcare, followed by a description of it.

Management of Medical Records: Medical records are very crucial for any patient and hospital, so keeping records up to date in a safe place is very important. So that patients and doctors can access records very easily and timely, but the lack of proper management of records is very common and hence arise security and privacy issues in the management of medical records. Even the already accessible data is scattered and restricted, the lack

Figure 7.3 Execution of blockchain technology in healthcare.

Source: Author's compilation.

of a medical data management system might have serious consequences (Hathaliya et al., 2019). Blockchain technology has enormous potential in managing medical records. The integrity of the data throughout the system is first and foremost guaranteed by the decentralized network that distinguishes blockchain. The system is far more difficult to hack because there isn't a single, centralized server. Second, the likelihood of data preservation in its original form is higher since the data is resistant to modification. This electronic health record system will give patients control over their own data and make data sharing across healthcare stakeholders easy, visible, and reliable. Because of its decentralized nature, immutability, data provenance, dependability, resilience, smart contracts, security, and privacy, it can be used to administer and preserve patient EHRs.

Logging Data: Getting and recording information over time is the process of logging. It makes it possible to keep tabs on all kinds of activities, such as the archiving, accessing, or updating of information, files, or software programmes inside of a system. For ensuring data access, privacy, and control hospitals must use NEM multi-signature blockchain contracts with the blockchain-based platform idea (Fan et al., 2018). With this concept, patients can monitor personal records and may choose who can access it.

Tracking Pharmaceutical Products and Medical Equipment: India is not an exception when it comes to the problem of false pharmaceutical products. These could cause deaths if used. Therefore, a modern system for monitoring the chain of custody from the manufacturer to the patient becomes crucial. Blockchain technology enables a lower chance of third parties interfering with the supply chain of real items. The distribution of drugs or pharmaceuticals is one use case for blockchain technology in the healthcare sector, notably in the context of health-related supply chain management (Fiore et al., 2023). The adoption of blockchain technology in this industry has created a safe and secure platform and addressed some of the most common problems that the pharmaceutical industry has, like the supply of poor medications that could trouble a patient.

Medical Research: A blockchain that holds information about trial subjects' permission for clinical research is another possible area. The consent cannot be altered until the ledger is properly updated. Additionally, this network will serve as a resource for a ready supply of test volunteers for researchers as needed. Blockchain smart contracts boost data and medical trial confidence by preventing data manipulation and underreporting of undesired results. Blockchain technology was used in clinical studies to improve the trustworthiness, dependability, and transparency of data administration (Nugent et al., 2016). Blockchain's tamper-proof and cryptographic features boost data logging for complicated clinical trial data administration by preventing all types of manipulation.

Transplantation and Organ Donation: A well-maintained and coordinated system of storage and transportation of the organ to the recipient is required in this sector, as well as coordination and real-time data inputs from numerous stakeholders, including the donor, the practitioner, or the hospital where the donation is to take place. The system is now plagued by improper data, which results in an illegal organ supply and ineffective delivery. The availability, status, and location of an organ can be time-stamped on a blockchain that connects all the stakeholders, making the system more effective and eradicating the problems while maintaining adequate security of the clients' data (Miriam et al., 2023).

Remote patients monitoring: Another blockchain application in the healthcare industry was remote patient monitoring. Biological data from various mobile applications must frequently be acquired to provide remote patient monitoring outside of conventional healthcare contexts, such as hospitals. In order to monitor diabetes patients, Saravanan et al. (2017) recommended an end-to-end secured system, a new approach to healthcare (i.e., Secured Mobile-Enabled Helping Device for diabetics), and formalized data access.

7.4 INDUSTRY REPRESENTATIVES TO IMPLEMENT HEALTHCARE FACILITIES THROUGH BLOCKCHAIN

Blockchain is a novel technology that is still being developed. It can be used in innovative manners in the healthcare sector. The discovery of economical remedies and advanced therapies for many diseases relies on all the well-known participants and healthcare providers being able to share and distribute information in a smooth, effective way. There are several industrial blockchain professionals who can effectively treat healthcare viewpoints and overall growth. Many affiliated industrial or medical care sponsors or providers have assisted with investigations and studies to use blockchain techniques in healthcare. Table 7.1 illustrates some agencies that are practicing blockchain technology at the ground level as its core domain.

Therefore, a digital medical record long-term, nationwide blockchain network will boost efficiency and support outcomes that are better for patient health. Blockchain, specifically, is a digital book that stores a shared, permanent record of transactions conducted from linked transaction blocks.

7.5 BLOCKCHAIN IN COVID-19

One such area that has been hardest damaged by the current pandemic is the healthcare system. The biggest difficulty for most states and international organizations was to develop a precise system that could analyze the newly identified instances of the pandemic in progress and gauge the likelihood of its spread. To combat the COVID-19 emergency, a novel approach is required. By managing data relevant to patient records, immunization

Table 7.1 Industry representatives in enabling blockchain implementation healthcare services

Healthcare aspects	*Blockchain practices by service providers*
Recording of patient information	BURSTIQ FACTOM MEDIAL CHAIN GUARDTIME
Updating treatment & prevent costly mistakes	SIMPLY VITAL CORAL HEALTH RESEARCH ROBOMED
Breakthrough in Genomics	NEBULA GENOMICS ENCRYPGEN DCO.AI
Medical Supply Chain Management and Drug Traceability & Care	CHRONICLED BLOCKPHARMA TIERION

Source: Haleem et al., 2021

records, and the delivery of medicines from the manufacturer to the patient, blockchain technology has the potential to greatly improve medical healthcare during this pandemic crisis. A patient's medical history can be processed more securely through a peer-to-peer network. The prior illnesses of patients who are experiencing COVID-19 symptoms can also be traced in the records by using blockchain technology (Gupta et al., 2022). Only the concerned authorities can access this secure data for monitoring and future action. During the COVID-19 epidemic, the government has made vaccination mandatory for all citizens. A vaccination system can be made safer and more secure using blockchain technology. Through this study, immunization records can be connected via a data storage infrastructure. The programme uses a blockchain to limit illegal access (Srivastava et al., 2022).

Among other things, the adoption of blockchain can improve clinical data management and enable communication between several supply chain participants. The epidemic has sped up the diffusion of false information, triggering impulsive conduct and anxiety among the populace. All the data that the general public and the government get may be thoroughly evaluated and verified thanks to the blockchain tracking system. By creating a tracing method concerned with records on COVID-19 patients from diverse sources, the project investigates the possibility of blockchain to end the illness. The number of new and recovered cases as well as fatalities is tracked using Ethereum's blockchain technology, which is based on smart contracts and oracles. Figure 7.4 shows various blockchain applications used during COVID-19:

Therefore, the application of blockchain technology in the healthcare industry has helped in the enhancement of the healthcare ecosystem. The

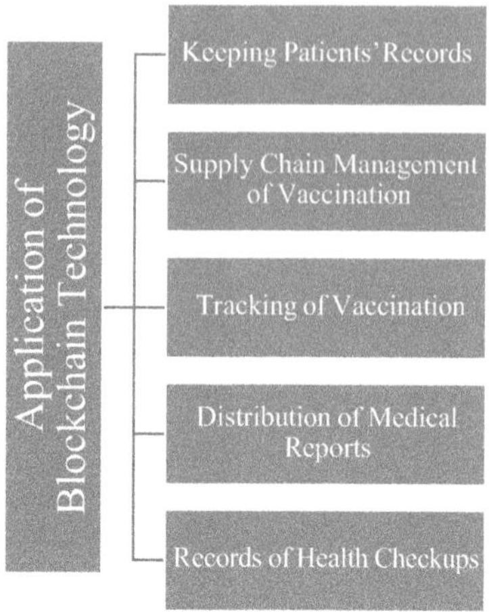

Figure 7.4 Application of blockchain technology during COVID-19.

Source: Author's compilation.

technology has enabled a close interaction of medical practitioners and patients at a lower cost; it also provided the implementation of hi-tech equipment for providing advanced medical services.

7.6 BLOCKCHAIN TECHNOLOGY: INDIAN HEALTHCARE SYSTEM

In this direction, India has already achieved some progress: Thynk-Blynk has partnered with a Hyderabad-based Call-Health venture that offers healthcare platforms to use Chain-Trail, its own blockchain technology, to enable multiple medical facilities enablers to integrate and safely disseminate their information on the network.[2] Health coverage for the poor and deprived society is still underrated in developing countries, even though there have been substantial gains made in the supply of healthcare facilities. Healthcare inequities are widespread in major developing nations like India. Disparities based on gender, region, and socioeconomic position persist despite recent improvements in access to healthcare over the previous decade or so. Lack of integration and fragmented health data are two of the main factors for inconsistent healthcare service delivery. Owing to these challenges, a blockchain-based integrated healthcare framework was created with a focus on providing timely and appropriate healthcare to all the country's citizens. Patients' secure, comprehensive, and unchangeable medical records from any treatment center in the country are easily accessible because of this technology. The safety and confidentiality of medical information is also ensured when patients choose biometric authentication to allow healthcare professionals to access their information (Dhagarra et al., 2019).

Due to several characteristics, including the size of the population it serves, the existence of both organized and unstructured healthcare services, economic disparities, and governance structure, among others, the healthcare system is a complex and distinctive organization. To persuade the Indian healthcare system to adopt a modified and more effective structure, the current body of research strongly emphasizes factors like accountability, backup and recovery, decentralization, standardization of procedures and practices, traceability, and data integrity. The recent development of blockchain technology gives host systems many of the features. It's crucial to comprehend the significance of blockchain technology in the Indian healthcare sector and community. It's also critical to assess the requirements of the key stakeholders for the implementation of blockchain technology in the Indian healthcare system with relation to any issues with attribute prioritization, customization, or design (Shukla et al., 2021).

Blockchain enables an effective and corruption-resistant implementation of national health insurance plans while maintaining a safe and transparent system of cohesive healthcare benefits that puts patients first (Pandey & Litoriya, 2020). Healthcare fraud may be reduced as blockchain technology becomes more widely used in these services. By making healthcare more patient-centric, the application of blockchain technology possesses the ability to completely alter the industry. It promotes the secure exchange of data in a secure mode among the many parties. Many agencies and service providers have access to medical information. The amount of data that hospitals must manage each day increases along with the number of patients and the complexity of their illnesses. Data is produced by many different types of devices, including IoT devices and EHR. The complete security of information exchange methods is the most important factor in delivering effective medical care. They enable all other parties, including healthcare practitioners, to confirm the accuracy of the data. Here is where blockchain is useful (Miriam et al., 2023).

The following are the benefits of blockchain in healthcare:

Data reliability: This can be done by examining the timestamp of the data and performing rigorous medical assessments. Since there is no need to rely on other agencies, the audit will be cheaper, and the privacy of the information will be ensured.

Drug traceability: Because blockchain transactions are real time and immutable, fraudulent drug dealers can be quickly identified. The operational data is stored on the blockchain from the manufacturing of a medicine to its delivery to the retailer. All the networks connected can be read at any moment, and the entire route of drug transportation is simply verifiable.

Clinical trial data security: Using blockchain technology, users can validate the validity of medical documents stored in the system.

Maintaining patient records: In a blockchain-enabled system, the patient ID and signatures are generated for each block of patient health information. The entities can obtain pertinent data with the use of the proper application programming interface without disclosing the patients' identities. The patient has the same control over who gets access and to what extent (full or partial).

Increase access to and recording of medical records: It is difficult to maintain an EHR since it can vary from one healthcare provider to another for the same patient. According to the criteria, a block-compatible system must allow the transmission of records from one physician to another.

Saving money and time: Information in the medical field is dispersed across numerous entities. It costs time and money to obtain the patient's medical history from their former healthcare practitioner. Blockchain use in the healthcare sector can lower both. In addition, the doctor's credentials can be checked.

The use of blockchain is gradually expanding in India: The Reserve Bank of India (RBI) became aware of the widespread use of cryptocurrencies in open marketplaces around the world in the year 2013. Users, owners, and traders have been warned against using "virtual currency" for any purpose, but the legality of its use has not been addressed. Most bitcoin exchanges reduced their activities starting in 2017 in response to RBI's opinions. In April 2018, these operations were further hampered when RBI forbade all banks and financial institutions operating under its auspices from engaging in direct transactions with or rendering any services to any entity engaged in cryptocurrency trading. However, RBI recently lifted its ban on cryptocurrency exchanges.

The patient, service provider, payer, pharmaceutical industry, clinical technology, technology providers, and government regulator are the seven main players in the Indian healthcare ecosystem. The type of communication affects the degree of protection. In this regard, various solutions have been put up. In the Indian eHealth system, communication levels have been created based on the amount of sensitivity. Passwords, smart cards, and others are used in the Indian eHealth system.

Healthcare Policy-Legal Framework: In 2017, the National Health Policy (NHP) was announced by the government of India with aim of ensuring the welfare of the Indian population and confirming that everyone has unrestricted access to high-quality medical treatment. NHP plans to establish a digital health technology environment among many other objectives.[3] The safety of client information and confidentiality rights, as well as data gathering, storing, and disseminating, will be essential to its creation. In 2017, the government of India introduced a programme called Digital Information Security in Healthcare Act (DISHA) with the

> motto to ensure the protection of patient confidentiality and the well-being of health data. By regulating the entire data retrieval process, as well as by safeguarding users' rights, this programme is anticipated to reorganize the e-health details of people with easy access in the future. In this regard, the gap between the creation of laws and regulations and their implementation could be filled by a connected promising block-chain network with a health data exchange architecture.

Even though blockchain technology is in a nascent stage in the healthcare industry, India is moving quickly to adopt it. The Indian government has begun developing a national framework to support the wider adoption of blockchain, including property records (a blockchain-aided scheme for managing land records), supply of pharma drugs, Super-Cert (a blockchain solution for educational certificates), the supply chain for immunizations, insurance, and organic farming.

7.7 ISSUES AND CHALLENGES

Despite all the sophisticated capabilities it offers, blockchain still has some restrictions and challenges that are required to be resolved. The absence of expertise is the key challenge in using this advanced technology in healthcare services (Haleem et al., 2021; Pandey & Litoriya, 2020). This section examines the problems and obstacles that blockchain technology in healthcare faces, as well as potential solutions to these problems. The health sector will adopt blockchain technology through various ways of applications and devices because of its exponential growth. Private health information about an individual may be handled by these programmes and tools. Figure 7.5

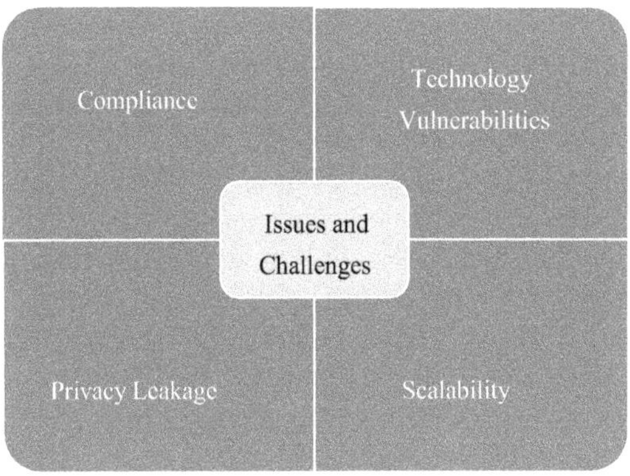

Figure 7.5 Major issues and challenges.

Source: Author's compilation.

presents some major issues and challenges related to the blockchain-based healthcare system.

Compliance: It is crucial to highlight that under data protection rules, health information is typically categorized as sensitive personal information. Given that a blockchain network's inherent structure defies any set form or norm, the adoption of a blockchain system for healthcare could clash with data protection rules. First, in a blockchain-based network where each machine serves many purposes as a node or a miner, it may be challenging to define the role of a data fiduciary or a data controller. As a result of the extensive compliance requirements that data privacy regulations impose on these organizations, this could become a substantial burden. Second, the majority of data protection regulations provide a right to deletion of data or the right to be forgotten. Over an unchangeable blockchain, this becomes very troublesome. Thirdly, these laws also grant a right to amend data, which could be problematic since it would require a majority of nodes to agree in order to identify the block and re-hash it.

Technology vulnerabilities: There are also some weaknesses in the implementation and architecture of the Blockchain system. Block-holding attacks, 51% attacks, double-spending attacks, selfish mining attacks, hashing attacks, collision attacks, difficulty-increasing attacks, and concerns about blockchain anonymity are all examples of blockchain-specific vulnerabilities (Tosh et al., 2017). When malicious actors successfully mine blocks but do not submit those blocks back into the system, this is known as a withholding attack. Instead, the miner merely uploads portions of the block that do not fully satisfy the system's requirements (McGhin et al., 2019).

Privacy leakage: In blockchain technology, data storage is a crucial problem that can be solved on-chain or off-chain. By employing the EHRs to the cloud and giving control over data management, cloud data storage may be able to solve this problem. Cloud computing, however, lacks both security and credibility while inspecting about patient medical records and breach their privacy by disclosing EHRs and other sensitive data.

Scalability: The idea of scalability describes how the suggested approach manages an increasing volume of work without requiring new storage, computing, or communication resources. Scalability, however, is defined differently in the context of blockchain technology. With the increasing number of patients and caretakers in society, the current system ought to be scalable and user-friendly. The technology-based healthcare approaches, sensors, communication devices, and software should be employed concurrently (Jaiswal & Anand, 2021). In a contemporary healthcare system with a collaborative healthcare system involving many healthcare professionals, scalability is known as a crucial component (Nguyen et al., 2019).

Therefore, the use of blockchain in healthcare has a number of requirements and obstacles also, including global interconnectivity, data safety, data reliability, cost efficiency, clarity, and intricacy (Kumar et al., 2018). The expense of storing and processing such large amounts of data is extremely high. Additionally, the improper handling of health records and data breaches have increased worries about data privacy and reduced confidence in the current system of health regulatory processes.

7.8 FUTURE DIRECTIONS

Blockchain technology has been adopted by the healthcare industry. Although technology has made many things easier for hospital management and patients, still there is a major challenge behind technology implications. The absence of trained professionals is the main challenge when utilizing this advanced or progressive technology and techniques in healthcare. Blockchain applications need further technical research. Along with this, the duties of pharmaceutical companies and regulators also need to be updated. Technologies and innovations in the health industry will increase rapidly in future years. Almost certainly, blockchain will expand in the healthcare industry. Blockchain will revolutionize healthcare by enabling patient-level public and private key encryption of medical data.

This technology makes treatment results and progress easier to define, improving healthcare apps. Considering technology, it is anticipated that patient records, infringement prevention, improved interoperability, procedural simplification, drug and prescription control, and supply chain monitoring would all be addressed. As a result, blockchain is expected to do very well in the sector of healthcare.

7.9 CONCLUSION

The present chapter analyzes the uses of blockchain technology in Healthcare 4.0. For this purpose, the study started by explaining the Healthcare 4.0 scenario followed by blockchain technology and its application in healthcare. It can be observed that there may be opportunities for the adoption of blockchain technologies in the healthcare industry. Decentralization of data, data integrity, access control, data versioning, and nonrepudiation are some of the reasons behind the adoption of blockchain technology in healthcare industries. A branch of medical informatics called health informatics (HI) focuses on the clinical field and the application of technology to the delivery of healthcare. Healthcare informatics is developing because of changes in technology and healthcare. One of the major factors behind the adoption of technological advancement is to offer and prove more effective medical facilities along with high security and confidentiality of patient information related to medical history. This information can be accessed by doctors and other related parties in real time.

Blockchain technology is a distributed ledger system in which blocks (records) are linked to one another. It has numerous applications in the healthcare industry. This technology was first developed for the banking and finance industry, but it is currently aimed toward delivering safer methods of data sharing between patients and other related groups in the healthcare systems. Researchers have begun to investigate possible use cases for blockchain in the healthcare industry due to the technology's enormous potential and practical usability. To combat the COVID-19 epidemic, blockchain-enabled solutions are emerging, such as managing personal information or medical history records, immunization records, and tracing and supply of drugs from the manufacturer to the patient. Blockchain technology has the potential to greatly improve medical healthcare during this pandemic crisis. Additionally, blockchain has been utilized or proposed for future vaccines, medical supplies, and drug supply chain management. Hospital information and e-health record techniques are utilized extensively around the globe. The present healthcare information systems have a number of drawbacks, such as inadequate security measures, and are primarily cloud-based, kept by a single data contractor (Elangovan et al., 2022). Numerous data breaches, problems with data veracity, and data sharing have resulted from this, leaving patients vulnerable to financial risks and even social stigma.

NOTES

1 https://www.forbesindia.com/article/great-lakes-institute-of-management/blockchain-in-indian-healthcare-system/57281/1 (accessed on 22 Feb 2023).
2 https://www.forbesindia.com/article/great-lakes-institute-of-management/blockchain-in-indian-healthcare-system/57281/1 (accessed on 22 Feb 2023).
3 https://www.forbesindia.com/article/great-lakes-institute-of-management/blockchain-in-indian-healthcare-system/57281/1 (accessed on 22 Feb 2023).

REFERENCES

Aceto, G., Persico, V., & Pescapé, A. (2020). Industry 4.0 and health: Internet of things, big data, and cloud computing for Healthcare 4.0. *Journal of Industrial Information Integration*, *18*, 100129.

Bongomin, O., Yemane, A., Kembabazi, B., Malanda, C., Mwape, M. C., Mpofu, N. S., & Tigalana, D. (2020). The hype and disruptive technologies of industry 4.0 in major industrial sectors: A state of the art.*Journal of Engineering. 200*, 8090521. Preprints. DOI: 10.1155/2020/8090521

Dhagarra, D., Goswami, M., Sarma, P. R. S., & Choudhury, A. (2019). Big data and blockchain supported conceptual model for enhanced healthcare coverage: The Indian context. *Business Process Management Journal*, *5*(7), 1612–1632.

Dinh, T. T. A., Liu, R., Zhang, M., Chen, G., Ooi, B. C., & Wang, J. (2018). Untangling blockchain: A data processing view of blockchain systems. *IEEE Transactions on Knowledge and Data Engineering*, *30*(7), 1366–1385.

Elangovan, D., Long, C. S., Bakrin, F. S., Tan, C. S., Goh, K. W., Yeoh, S. F., ... & Ming, L. C. (2022). The use of blockchain technology in the health care sector: Systematic review. *JMIR Medical Informatics*, *10*(1), e17278.

Fan, K., Wang, S., Ren, Y., Li, H., & Yang, Y. (2018). Medblock: Efficient and secure medical data sharing via blockchain. *Journal of Medical Systems*, *42*, 1–11.

Fiore, M., Capodici, A., Rucci, P., Bianconi, A., Longo, G., Ricci, M., ... & Golinelli, D. (2023). Blockchain for the healthcare supply chain: A systematic literature review. *Applied Sciences*, *13*(2), 686.

Gupta, B. B., Mehla, R., Alhalabi, W., & Alsharif, H. (2022). Blockchain technology with its application in medical and healthcare systems: A survey. *International Journal of Intelligent Systems*, *37*(11), 9798–9832.

Haleem, A., Javaid, M., Singh, R. P., Suman, R., & Rab, S. (2021). Blockchain technology applications in healthcare: An overview. *International Journal of Intelligent Networks*, 2, 130–139.

Hathaliya, J. J., Tanwar, S., Tyagi, S., & Kumar, N. (2019). Securing electronics healthcare records in healthcare 4.0: A biometric-based approach. *Computers & Electrical Engineering*, 76, 398–410.

Jaiswal, K., & Anand, V. (2021). A survey on IoT-based healthcare system: potential applications, issues, and challenges. In *Advances in Biomedical Engineering and Technology: Select Proceedings of ICBEST 2018* (pp. 459–471). Springer Singapore.

Javaid, M., & Haleem, A. (2019). Industry 4.0 applications in medical field: A brief review. *Current Medicine Research and Practice*, *9*(3), 102–109.

Javaid, M., & Haleem, A. (2020). Impact of industry 4.0 to create advancements in orthopaedics. *Journal of Clinical Orthopaedics and Trauma*, *11*, S491–S499.

Jayaraman, P. P., Forkan, A. R. M., Morshed, A., Haghighi, P. D., & Kang, Y. B. (2020). Healthcare 4.0: A review of frontiers in digital health. *Wiley Interdisciplinary Reviews: Data Mining and Knowledge Discovery*, *10*(2), e1350.

Kumar, T., Ramani, V., Ahmad, I., Braeken, A., Harjula, E., & Ylianttila, M. (2018, September). Blockchain utilization in healthcare: Key requirements and challenges. In *2018 IEEE 20th International Conference on e-Health Networking, Applications and Services (Healthcom)* (pp. 1–7). IEEE.

McGhin, T., Choo, K. K. R., Liu, C. Z., & He, D. (2019). Blockchain in healthcare applications: Research challenges and opportunities. *Journal of Network and Computer Applications*, *135*, 62–75.

Miriam, H., Doreen, D., Dahiya, D., & Rene Robin, C. R. (2023). Secured cyber security algorithm for healthcare system using blockchain technology. *Intelligent Automation & Soft Computing*, *35*(2), 1889.

Nakamoto, S. (2008). Bitcoin: A peer-to-peer electronic cash system. *Decentralized Business Review*, 21260.

Nguyen, D.C., Pathirana, P.N., Ding, M., Seneviratne, A. (2019). Blockchain for secure EHRs sharing of mobile cloud based E-health systems. *IEEE Access*, *7*, 66792–66806.

Nugent, T., Upton, D., & Cimpoesu, M. (2016). Improving data transparency in clinical trials using blockchain smart contracts. *F1000Research*, *5*, 2541.

Pandey, P., & Litoriya, R. (2020). Implementing healthcare services on a large scale: Challenges and remedies based on blockchain technology. *Health Policy and Technology*, *9*(1), 69–78.

Rehman, M. U., Andargoli, A. E., & Pousti, H. (2019). Healthcare 4.0: Trends, challenges and benefits. In *Australasian Conference on Information Systems* (pp. 556–564).

Saravanan, M., Shubha, R., Marks, A. M., & Iyer, V. (2017, December). SMEAD: A secured mobile enabled assisting device for diabetics monitoring. In *2017*

IEEE International Conference on Advanced Networks and Telecommunications Systems (ANTS) (pp. 1–6). IEEE.

Shukla, R. G., Agarwal, A., & Shekhar, V. (2021). Leveraging blockchain technology for Indian healthcare system: An assessment using value-focused thinking approach. *The Journal of High Technology Management Research*, *32*(2), 100415.

Sookhak, M., Jabbarpour, M. R., Safa, N. S., & Yu, F. R. (2021). Blockchain and smart contract for access control in healthcare: A survey, issues and challenges, and open issues. *Journal of Network and Computer Applications*, *178*, 102950.

Srivastava, S., Pant, M., Jauhar, S. K., & Nagar, A. K. (2022). Analyzing the prospects of blockchain in healthcare industry. *Computational and Mathematical Methods in Medicine*, *2022*, pp. 1–24.

Tanwar, S., Parekh, K., & Evans, R. (2020). Blockchain-based electronic healthcare record system for Healthcare 4.0 applications. *Journal of Information Security and Applications*, *50*, 102407.

Tosh, D. K., Shetty, S., Liang, X., Kamhoua, C. A., Kwiat, K. A., & Njilla, L. (2017, May). Security implications of blockchain cloud with analysis of block withholding attack. In *2017 17th IEEE/ACM International Symposium on Cluster, Cloud and Grid Computing (CCGRID)* (pp. 458–467). IEEE.

Wenhua, Z., Qamar, F., Abdali, T. A. N., Hassan, R., Jafri, S. T. A., & Nguyen, Q. N. (2023). Blockchain technology: Security issues, healthcare applications, challenges and future trends. *Electronics*, *12*(3), 546.

Chapter 8

Inventory tracking via IoT in the pharmaceutical industry

Galiveeti Poornima
Presidency University, Bangalore, India

J. Vinay
Cloud Operations, SECURONIX

P. Karthikeyan
National Chung Cheng University, Chiayi, Taiwan

V. N. Jinesh
School of CSE & IS, Presidency University, Bangalore, India

8.1 INTRODUCTION

The amount spent globally on health is increasing. In 2017, the figures were $7.8 trillion and $1080 million per person[1]. The annual cost of pharmaceuticals, representing approximately 10% of healthcare spending, was estimated to be $1.25 trillion in 2019. This represents a significant part of the healthcare budget [1]. Since pharmaceuticals account for up to 40% of the healthcare budget in developing countries, this number is higher [2].

Despite their high cost, pharmaceuticals have many benefits [2]. Despite a substantial investment in funds, they frequently experience deficits, losses, and waste due to damage, spoilage, contamination, overproduction, expiration, and improper use in health facilities [2, 3]. The cost of inventory management and waste disposal increases when money is spent on unnecessary or substandard drugs [4]. An appropriate inventory management system and qualified personnel must address these issues and implement adequate pharmaceutical supply [2, 3].

Due to its impact on both financial and medical outcomes, pharmaceutical inventory management is a challenging but linear time-invariant in the healthcare sector [3, 5, 6]. Since a pharmacy is an inventory, this term refers to the procedures used to establish and maintain an appropriate supply of pharmaceuticals while keeping waste and financial losses to a minimum [7, 8]. Standard practice typically involves both a buffer stock and an essential supply. Stocks are divided into "basic," which is enough to meet typical demand, and "safety," which allows for unforeseen fluctuations in sales.[2]

DOI: 10.1201/9781003474524-8

According to the report released by the World Health Organization (WHO), the main objective of managing pharmaceutical inventory is to ensure consistent provision of pharmaceuticals to operational units and patients while simultaneously reducing the expenses associated with inventory and procurement [2]. This can be accomplished by keeping expenses related to inventory and procurement to a minimum. Overstocking essential medications can lead to poorer patient care quality, drug waste, and financial loss, so it is essential for pharmaceutical inventory management to strike a balance between stock on hand and new orders [8]. Furthermore, it functions as a safeguard mechanism during critical points in the supply chain, mitigating the impact of variations in customer demand and the uncertain nature of the requesting cycle [8]. The availability of pharmaceuticals could be interrupted if they are not handled properly. Various methods are employed to evaluate the performance of management systems, including vital, essential, and non-essential (VEN) analysis; ABC-VEN matrix analysis; XYZ (high, medium, and low value) analysis; fast, slow, and nonmoving (FSN) analysis; and FSN-XYZ matrix analysis.

Several researchers have conducted studies on pharmaceutical products and inventory management in hospitals and pharmaceutical companies, using conventional inventory control techniques such as ABC, VEN, FSN, and matrix analysis, which are commonly used [9]. The ABC analysis model gives a clear and fair picture of how much money was spent on inventory from the budget. The VEN analysis facilitates the selection of multiple drugs for employment in a drug supply system and for procurement purposes [10].

As defined by the WHO and the International Pharmaceutical Federation, one of the eight most important jobs of a pharmacist is managing money, materials, people, time, and information. It is essential for professional success both at the individual and organizational levels [11].

The inventory of any industry constitutes the majority of its assets and working capital. Each organization must maintain a specific type of inventory to function properly. Stock management guarantees material availability and reduces investment when necessary. Inventory decisions that involve a lot of risk and directly affect the organization's bottom line include funding, promotion, sourcing, and procurement management.

Inventory management affects pricing and production in supply chain management (SCM). Best customer service is possible by keeping inventory costs as low as possible, a possible goal of inventory management. This is done by setting the right policies to replenish the inventory.

8.1.1 Classical inventory models in brief

The efficient functioning of the supply chain is heavily dependent on effective stock management, which includes raw materials, work-in-progress components, and finished products. Inventory-related costs constitute a substantial fraction of the total expenses accrued in the supply chain. The act of devising and overseeing stock levels aids in optimizing inventory management's utilization of available items. This results in a reduction in expenses and an increase in customer satisfaction. Inventory management typically involves two distinct types of inventory reviews, categorized according to their mode of execution: continuous reviews and periodic reviews. Under the continuous inventory review system, it is imperative to monitor stock levels regularly. Initiating an order when the inventory level drops below a predetermined threshold is recommended. The amount that has already been decided is called the ordering point. The expected demand, the cost of holding, the cost of ordering, and other factors will determine how much to order. The economic order quantity (EOQ) and economic production quality (EPQ) models are two fundamental classical models that involve a consistent stock review and are characterized by their simplicity. These models are widely used in inventory management. The EOQ model postulates the immediate reception of the ordered quantity upon order placement. In contrast, the EPQ framework posits that ordered goods are obtained incrementally during production.

8.1.2 Industy 4.0

Two predominant perspectives exist regarding the conceptualization and application of the Internet of Things (IoT) and Industry 4.0, with respect to their respective terminologies. One way of thinking is that the terms "Industry 4.0" and "IoT" are the same and can be used interchangeably. According to a different perspective, IoT constitutes a constituent element of Industry 4.0, enabling the realization of the concept of Industry 4.0 [12].

The German government initially introduced the concept of Industry 4.0, also known as the Fourth Industrial Revolution. Implementing diverse technologies, including IoT, RFID tags, iOS, cloud computing, big data, and cyber-physical systems (CPS), offers a new strategy for improving organizational performance. The introduction of the steam engine to industry marked the beginning of the first Industrial Revolution. The shift from traditional to mass production posed novel challenges and complexities for the industrial domain. The Second Industrial Revolution started when factories started using electricity more widely. The Third Industrial Revolution occurred as the electronic world and information technology developed. Every link in a supply chain has new perspectives thanks to Industry 4.0. Companies could reduce waste, become more responsive, and make decisions in real time with Industry 4.0. Natural and artificial systems in physical space, known as CPS, are closely integrated with communication, computation, and control systems (cyberspace). The amalgamation of detectors, data acquisition systems, communication networks, and cloud computing has led to the rise of CPS, which has become an essential infrastructure element in various industries.

In contrast, the sector has produced essential data due to the extensive implementation of sensors and control systems [13]. Such a large amount of data management requires careful consideration [14]. For this, cloud storage is used. The efficient operation of the production line can be achieved with minimal human intervention and real-time data analysis. This can be facilitated by enabling self–decision-making methodologies for the machines in CPS, thereby reducing interaction errors. As a result of the use of these new technologies, occupational psychology has become a major problem and an essential part of the workforce. When smart manufacturing is used to automate a CPS system, the system will become more self-sufficient in making decisions and connecting machines. This will be the case because smart manufacturing utilizes artificial intelligence.

The implementation of Industry 4.0 has facilitated advances in various aspects of the supply chain, including, but not limited to, manufacturing, distribution, and transportation. Technologies from the "Industry 4.0" movement impact several industries, including aerospace, agriculture, construction, food and beverage, pharmaceuticals, and services. Within the framework of Industry 4.0, there are many prospects for implementing sustainable manufacturing practices. The use of digital cyber networks facilitates the transmission of data within the closed-loop supply chain, thus promoting sustainable manufacturing practices. To ensure the environmental sustainability of products and processes, they must incorporate reuse and remanufacturing as integral components of their closed-loop life cycles.

Organizations need specialized infrastructures to introduce cutting-edge business models to implement Industry 4.0 technology. To provide intelligent products and services in a wholly digitalized environment, novel business models, particularly disruptive ones, are imperative. According to Lin et al. [15], the implementation of automated virtual metrology has the potential to achieve the objective of zero defects in automated metrology and progress towards Industry 4.1 as the subsequent phase, thereby allowing the realization of Industry 4.0. Despite the considerable attention that Industry 4.0 has garnered in various fields, its practical implementation in everyday life has not yet been widely and effectively realized [16]. Academics and businesspeople have not done much research on IoT–based supply chain solutions.

8.1.3 IoT

The terminology "Internet of Things" was initially introduced by an executive director associated with the Auto-ID center. The concept of network connectivity has enabled the notion that machines can function as a unified system without human intervention, consequently minimizing the probability of errors or inefficiencies. This approach to machine creation paints a picture of manufacturing and production processes such as the smart factory. One of the Industrial Revolution 4.0 modules in this theory is receiving a lot of attention: the IoT driver. Integrating the digital and physical realms is a powerful means of communication that finds application in diverse domains. Its implementation improves the intelligence of processes, operations, and activities within the value chain, offering novel solutions to transform these activities. Wireless technology based on the Internet serves as a way in which various gadgets are interconnected, enabling them to communicate with one another and improving their operational efficiency. Sensors play a crucial role in improving situational awareness in various systems, providing a significant amount of real-time data. According to Sun et al. [17], the utilization of IoT technology can considerably influence production processes through the improvement of resource utilization efficiency, the enhancement of transparency and accessibility throughout the supply chain, the facilitation of real-time SCM, the optimization of supply chain operations, and the enhancement of supply chain responsiveness.

8.1.4 IoT enabled by RFID technology

Radio frequency identification (RFID) technology can improve the overall efficiency of the supply chain, from warehousing to transportation, by enabling real-time connectivity and information exchange. By improving the visibility and traceability of products, RFID helps improve inventory flow. According to He et al. [18], implementing RFID technology can mitigate loss of inventory and misplacement and minimize transactional errors and supply chain discrepancies. RFID vocabulary tags were initially defined using IoT [19]. Using RFID readers with an Internet interface enables global detection, tracking, and real-time monitoring of tagged products. RFID is a prerequisite for the implementation of the IoT. RFID systems are made up of various components that facilitate the transmission of data. These components are available in various dimensions, designs, and configurations. The applications of each of them exhibit slight variations from each other. The fundamental constituents of readers and tags have mainly remained unchanged. An RFID system can consist of one or more readers and tags.

The objects' distinct IDs are stored on the tags that are attached to them. Readers emit a reconnaissance signal to detect and retrieve the identification information of RFID tags within their vicinity. Creating an interactive map representing physical objects and the subsequent transformation of the physical world into a digital model suggests a practical solution that can offer benefits in shipping, electronic health, and security. From a physical perspective, RFID tags are similar to adhesive labels. Passive tags are activated through the reader's signal, energizing the tag's antenna without a power supply. The energy is subsequently utilized to furnish the RFID tag microchip, which is responsible for transmitting the stored identification information. Semi-passive and active tags are distinct categories of RFID technology that operate on independent power sources. A battery in semi-passive tags powers the microchip that houses the ID. Semi-passive tags employ heat emitted by the reader to facilitate data transmission. At the same time, active tags require battery power to transmit information to related readers. Despite their higher cost, these two types of reader have the potential to provide better coverage, as stated in the research by Atzori et al. on IoT [20].

8.1.5 IoT application development

Considering what the term Internet of Things means, it is possible to envision a great deal of software being used in areas such as environmental protection, public welfare, and personal protection. Furthermore, applications of the IoT, such as smart homes, smart cities, self-driving cars, IoT retail stores, agriculture, wearables, telehealth care, hospitality, and intelligent SCM, are everywhere.

8.2 INVENTORY MODELS INCLUDE DATA FROM INDUSTRY 4.0

Suppliers previously influenced production strategies. Today, customers have demands. Adjust the manufacturing plans accordingly. Firms that want to be competitive need data to plan production and make the best decisions. Demand dictates production rates. Production line machines should work on demand. Demand should affect production proportionally. To smooth production, warehouse inventories must cover changes in demand. Therefore, inventory replenishment, evaluation, and order quantities should reflect changing demand.

Using Industry 4.0 technologies, particularly IoT, which enable connected equipment to function collaboratively and in unison due to changes in demand, inventory management should take on a more significant share of the responsibility for adjusting operations related to inventory in response to these shifts. In the Industry 4.0 environment, the idea of an intelligent factory and the fact that inventory management is an essential part of SCM mean that the rules for replenishing inventory need to be looked at and new rules need to be made for how to use technologies from Industry

4.0. As a result, it appears doubtful that traditional methods could supply the necessary materials for production or assembly lines with sufficient lead time and without shortage. We review the relevant literature after taking all of these factors into account.

8.3 IOT FOR PHARMACEUTICAL

The pharmaceutical industry faces a significant logistics challenge, which can be defined as the inability to transfer information in a timely, precise, and consistent manner while pharmaceutical products are in transit. Attaining uninterrupted surveillance and real-time visibility of goods throughout the supply chain poses a formidable obstacle that requires resolution. An emerging Internet-based technology called IoT was launched to help solve this problem. The Internet of Things is an international infrastructure that links the cloud for the Internet to physical and digital objects in the real world. During transit, the commodities equipped with sensor devices generate live data through this linkage, which then transmits updates to the site concerned.

The IoT provides a unique object identification system in addition to the object's location at any given point in time within the global network. IoT as a value-creation system is thought to be enabled by wireless sensing devices and RFID. In the foreseeable future, RFID tags are expected to be equipped with detectors and additional value-added features that will enable them to establish communication with their network, monitor their environment, and capture geographic information. These cutting-edge tags with a distinct IP address or addressable number are connected to pharmaceutical products. Figure 8.1 depicts the three layers of the IoT architecture. These are the perception, network, and service layers [21].

The data extraction layer, the perception layer, constitutes a crucial component of the IoT. A range of technologies, including sensors, RFID tags and reader-writers, wireless sensor networks (WSN), intelligent terminals, cameras, global positioning systems (GPS), electronic data interfaces (EDI), and mobile devices, are employed to gather various types of data and information from other physical equipment. Information is collected at the perception layer and transmitted downward through the transport layer to the

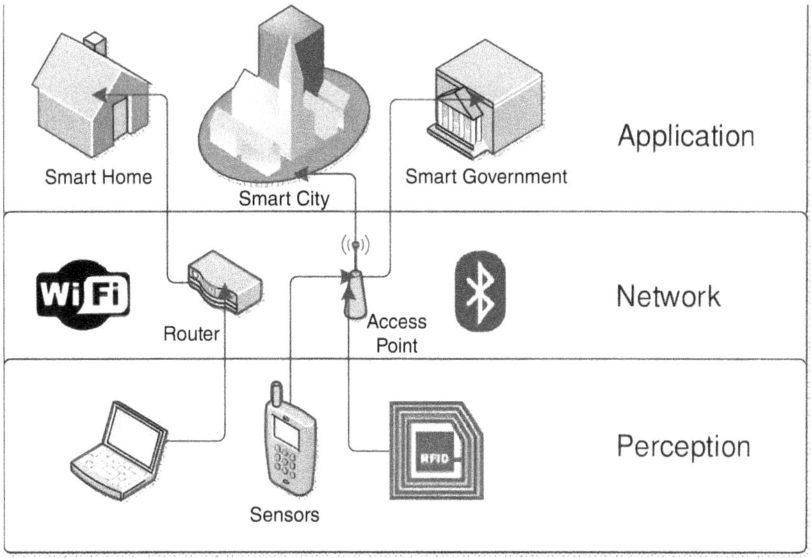

Figure 8.1 Architecture of IoT.

network layer. The network layer is responsible for the transmission of data. The network layer improves a variety of technologies, including WSN, mobile communication networks, radio access networks, and other communication devices. These technologies include the global system for mobile communications (GSM), general packet radio service (GPRS), wireless fidelity (Wi-Fi), worldwide interoperability for microwave access (WiMax), Ethernet, and many others. This stratum offers a reliable, robust, and credible framework for the network infrastructure of enterprises that operate on a grand scale. The information transmitted from the network to the service layer, also known as the application layer, is stored and managed in this location. This specific stratum encompasses a range of underlying strata, with the data management stratum and the application service stratum being the most prominent. Screening, reorganizing, cleaning, and combining are the methods used by the data management sub-layer to process complex data and uncertain information. It also offers directory services like market-to-market (M2M) services, quality of service (QoS), geometrics, and other comparable services. Cloud computing technologies, service-oriented architecture (SOA), and other techniques are used to achieve this. The application service sub-layer provides an effective user interface for business applications and end users. Examples of enterprise applications include logistics and supply, disaster warnings, environmental monitoring, agriculture, production, and other related industries. Additionally, it reads the data as complete information based on how it was collected.

Good Distribution Practices (GDP) Working Document QAS/04.068 was just released by the WHO. This document applies to "manufacturers, suppliers, distributors, brokers, wholesalers, traders, transport companies, processors, etc." who are involved in the distribution of pharmaceutical products. Each activity involved in the distribution of pharmaceutical products should be carried out by the principles of good manufacturing practices (GMP), good storage practices (GSP), and GDP, according to the description of pharmaceutical logistics found in the working document. Businesses are putting significant effort into this direction to ensure adequate security and secure distribution without compromising the product. Enhancing a company's technological infrastructure should be a priority for businesses looking to implement intelligent logistics. For example, a pharmaceutical company can track pharmaceutical products through the supply chain using the IoT.

8.4 REVIEW OF LITERATURE

8.4.1 Pharmaceutical sector

The geriatric patient demographic presents a significant challenge to the pharmaceutical sector regarding providing adequate care. In particular, it must produce superior results quickly and affordably. It is projected that the demographic of individuals aged 65 years and older will exceed that of individuals under the age of 18 by 2035. The Baby Boomer cohort experiences a significant rise in the mean number of different prescriptions per customer from 3.15 to 8.85 upon reaching the age of 65. Furthermore, the demand for effective drug delivery methods is increasing tremendously [22]. According to statistical data from the US Census Bureau and other government-affiliated organizations, the current proportion of individuals aged 65 and above constitutes more than 16.9% of the total population. Future projections suggest an increase to approximately 22% within the next 30 years[3]. Figure 8.2 illustrates the ascending nominal pharmaceutical expenditure in the United States between 2002 and 2021. To sustain industrial expansion amidst this trend, it is imperative to prioritize acquiring reliable and up-to-date data that can be transmitted back to the manufacturer.

The pharmaceutical industry can be categorized and segmented based on each drug's brand name and the kind of substance being manufactured. Singh [23] explains that pharmaceuticals are manufactured as either trademarked or generic medications and can be synthesized through various biologic methods, including the creation of molecules with differing sizes. Generic medications are available over the counter and do not require a prescription, whereas branded medications are often more specialized and do. Disparities in packaging specifications among branded and generic pharmaceuticals can

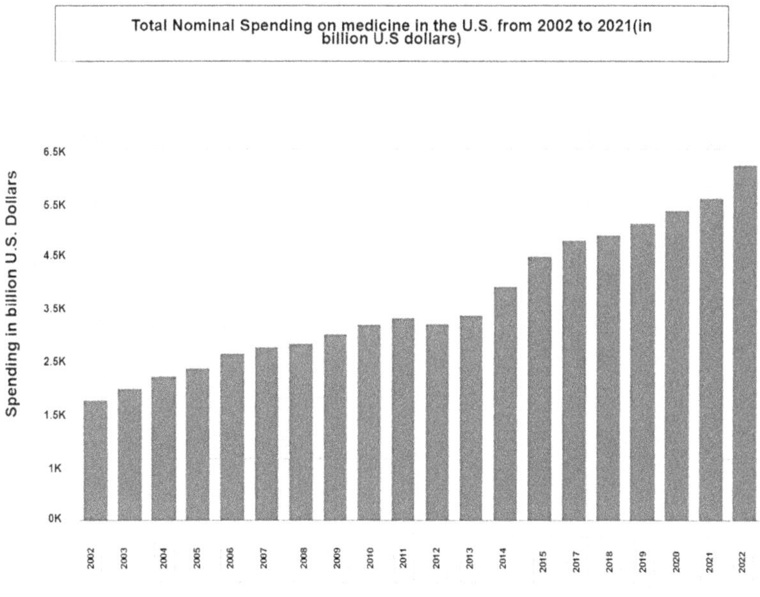

Figure 8.2 The aggregate nominal expenditure on pharmaceuticals in the United States.

affect the physical dimensions and surface area accessible for an IoT device solution designed for monitoring objectives.

Figure 8.3 shows the switch from brand-name to generic dietary supplements in prescriptions and customer behavior. Price is a significant element in the change in customer preference from brand-name to generic medications, as generics provide a more affordable option than brand-name medications [24]. Pharmaceutical companies mark up their branded medications during the exclusivity period to recoup the expenditures of R&D invested in creating the new drug. Although a higher percentage of prescriptions are filled with generic drugs, their share of sales remains mostly stable.

For information to be integrated from the consumer through the distributor, wholesaler, and producer, there must be a drive for backward transparency in the PSC [25]. Drug compounds are produced using molecular or biological methods and have varying lead times, packaging, and delivery requirements. Enhanced visibility across the supply chain could furnish pertinent information to the manufacturing process, thereby enabling the fulfillment of specific requirements. Pharmaceutical producers can keep up with the constantly changing demands of consumers because of their quick time to market and sound inventory management practices. The availability of generics over the counter without a prescription has the potential to alter the pharmaceutical sales volume to reflect the transition in consumer inclination from branded to generic products. Pharmaceutical firms, like the

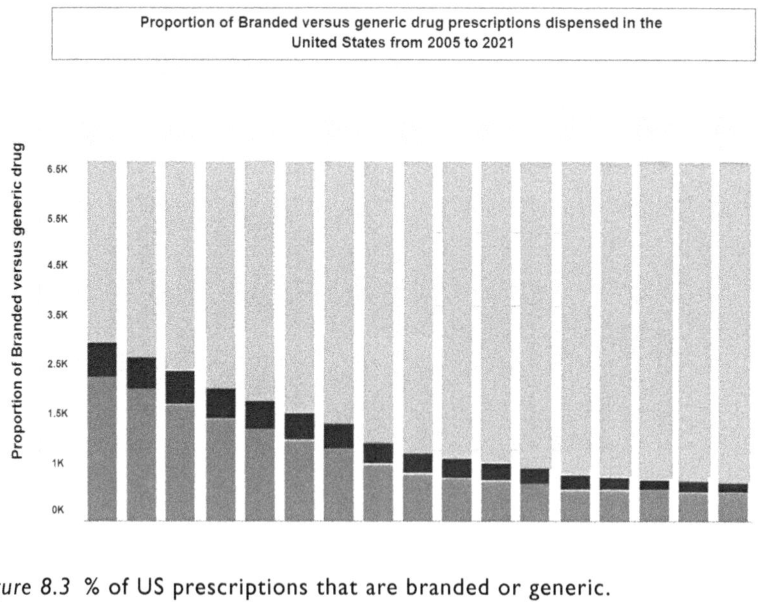

Figure 8.3 % of US prescriptions that are branded or generic.

sponsor, will be able to adjust and be agile in the market if they have precise, real-time inventory management.

8.4.2 PSC organization

The PSC comprises various entities, such as manufacturers, storage facilities, transporters, distributors of wholesale goods, retail pharmacies, medical professionals, and end users or subjects, all of which possess a significant stake in the process. According to Shah [26], pharmaceutical drug manufacturing is occasionally conducted in batches or campaigns to augment inventories over a continuous time frame. Manufacturing in batches does not provide flexibility in response to changing customer or market demand. Additionally, it can result in waste along the entire supply chain, with inventory being the most common place. Strategic inventory positioning is a key strategy to reduce inventory expenses. Inventory maintenance by pharmaceutical companies is of interest due to its potential to secure operating capital that could be allocated to other areas within the organization [27].

The partnership between the manufacturer and the pharmacy will determine the multichannel distribution strategy [28]. The figure presented in Figure 8.4 portrays a PSC arrangement, commencing from the preliminary phases of research and development and culminating with the final consumers while disregarding the intermediate function of wholesale distributors in the dissemination procedure.

However, the producer often sends the finished product to a wholesale distributor via their distribution network. Accordingly, the wholesale distributor assumes responsibility for storing, gathering and delivering merchandise to the ultimate location. According to Beier's study on management, the majority of pharmaceutical purchases made by pharmacies, specifically 86.6%, occur through the wholesale distribution channel [29]. Supply chains must be designed to work in harmony with the key business procedures of end users. The relationships between pharmacy management and wholesale distributors have always been the foundation of the network [23].

Knowing the traits of medicinal drugs, such as end-of-life thresholds and their uncertainty in the demand cycle, should determine the best inventory management strategies. Pharmaceutical companies are more vulnerable to hazards and counterfeiting as the complexity and levels of supply chain networks become more varied and their spread increases [30]. The PSC networks require enhanced visibility due to the dynamic consumer behavior in the industry, which poses potential risks in terms of consumption and usage. The dynamic behavior of consumers in the industry gives rise to potential risks associated with consumption and use that may not conform to established procedures for purchasing and inventorying. Singh [23] posits that the feasibility of incorporating reverse logistics with buybacks during the termination phase of a product's life cycle can be attributed to the contradiction

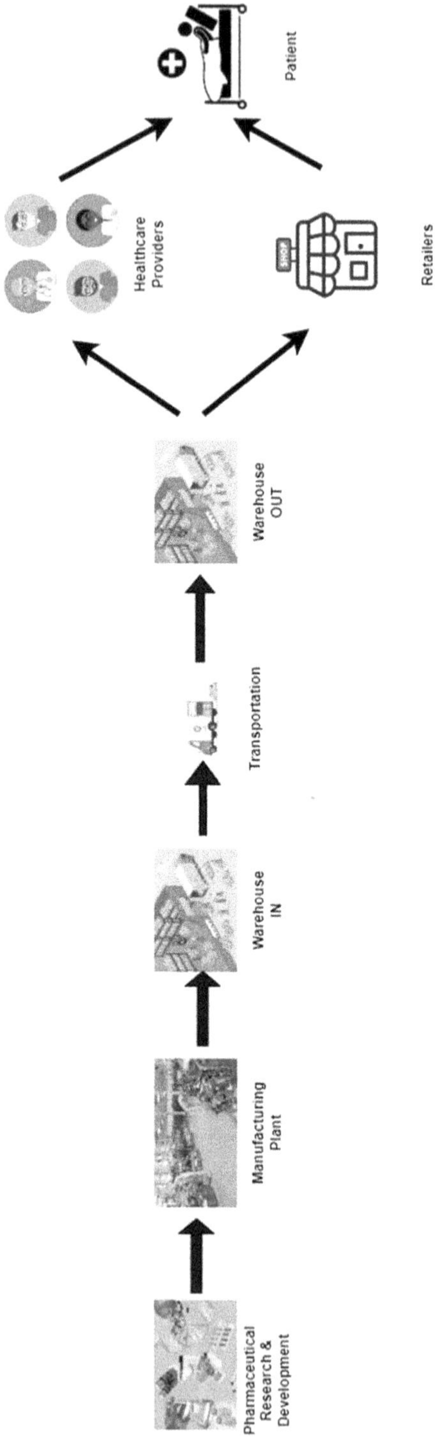

Figure 8.4 Design of the PSC.

between pharmacy procurement trends and the data accessible to the ultimate consumer during the point of purchase.

8.4.3 PSC wholesalers

Pharmaceuticals are delivered to pharmacies around the world by wholesale distributors. The aggregation of demand and inventory, along with typical warehouse tasks, is the main duty at this point in the supply chain. The pharmaceutical sector is subject to stringent regulations, which challenge the distribution process. Wholesalers are in charge of managing adequate inventory levels to satisfy all of their customers, while also informing their suppliers of the inventory levels to prompt the production and delivery of manufacturers [26]. As per Beier's [29] study on metrics of quality for leading wholesale distributors, the three most prominent qualities were identified as follows:

- The order will be fulfilled within a time frame of 24 hours from the moment it is placed.
- Digital ordering.
- Regular delivery schedules.

Wholesalers endeavor to anticipate pricing fluctuations that impact demand, as two out of the three primary metrics prioritize delivery. As a result, manufacturers issue anticipatory requests for large orders [26]. According to Iacocca et al. [28], the investigation of inventory policies for wholesale distributors indicates that combining inventory at the distributor's level, rather than dispersing it across various pharmacy locations, can improve the system's effectiveness in terms of resources and flexibility.

Pharmaceutical companies often lose track of their inventory as the goods shift ownership from manufacturer to wholesaler at the point of exchange. From a control perspective, the inventory ceases to be under the entity's purview despite being yet to be consumed. The exchange of data regarding product inflows and outflows between wholesale distributors and manufacturers is limited in scope, as it fails to furnish manufacturing firms with a comprehensive understanding of consumer demand and utilization patterns. The industry as a whole could experience manufacturing and inventorying inefficiencies as a result of this lack of transparency.

8.4.4 Pharma supply chain reverse logistics

Pharmaceutical recalls and inventory buy-backs are the result of drug expiration. [23]. This reverse logistics requirement can be solved by proper inventory management and inventory turn procedures.

Any return requires a logistics route, financial impact, regulated drug disposal, legal implications, and network inventory reconciliation [23]. Reverse logistics programs are primarily implemented due to pharmaceutical waste disposal, as stated by Campos et al. [31]. Pharmaceutical companies need positive customer perception to compete. Organizations gain credibility by reducing their environmental impact. Brand recognition and speed-to-market keep companies relevant. Creating a novel product through research and development in the manufacturing sector is time-consuming and spans several years.

The implementation of appropriate inventory principles by pharmaceutical companies is imperative in light of the legal and regulatory consequences associated with drugs that have reached their end-of-life or end-of-use, necessitating the monitoring of inventory expiration patterns. Pharmacists must adhere to the first-in-first-out (FIFO) approach in managing their inventory to ensure optimal utilization of drugs within their respective life cycles, despite limited visibility into product location and consumption.

8.5 HEALTHCARE PHARMACEUTICAL INVENTORIES

Healthcare inventory systems have incorporated operational management principles while emphasizing stakeholder management and relationships to establish internal processes. Certain companies utilize inventory policies in SCM to ascertain optimal inventory levels and ordering procedures.

According to research, political and experience-based managers primarily influenced inventory management and ordering policies, as opposed to data analysis and policy [32]. In a conventional supply chain, inventory management is characterized by a limited number of stakeholders participating in the decision-making process. Nurses, doctors, pharmacists, and financial analysts have different views on quality service and inventory levels in healthcare [32]. Given the dire consequences of stock shortages for patient care, healthcare professionals feel compelled to maintain excess inventories.

8.5.1 Hospital pharmacy inventories

De et al. [32] state that hospital pharmacies experience greater demand volatility than retail pharmacies. Additionally, they face significant internal stakeholder pressure to uphold inventory levels due to patient movement and employee participation. Patients with life-threatening diseases suffer more from hospital inventory shortages. 43.8% of *Journal of Business Logistics* respondents set safety stock levels using judgment or experience [29]. Due to loyalty to the product and the previous success of treatment, physicians and providers are more likely to switch from branded therapies to generic therapies [33]. Figure 8.5 shows how stakeholders and perception affect inventory management. The absence of effective inventory management practices can fail, leading hospitals to lose track of their overall

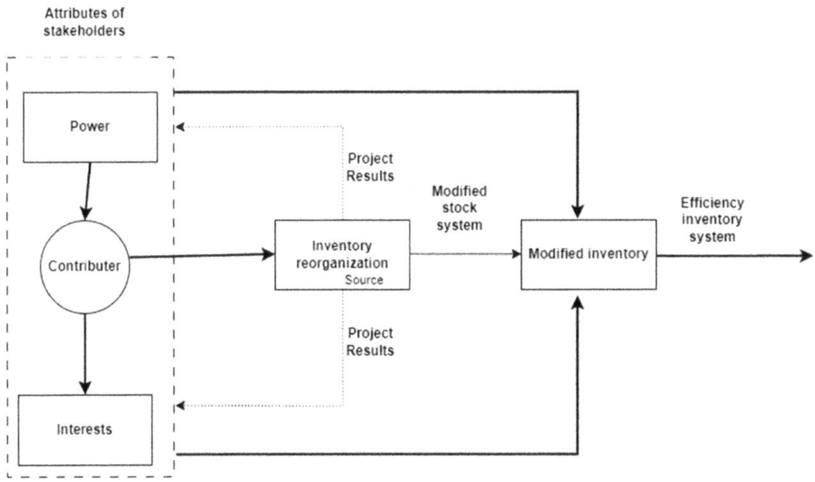

Figure 8.5 Management of hospital inventories.

inventory expenses, encompassing costs associated with ordering, carrying, and holding.

Internal and external distribution networks are present in hospital supply chain networks. Complications arise in hospital pharmacies due to the extensive array of products they handle and their intermediary role between hospitals and patients. The task of adequately assigning the products to the required units is in the hands of the pharmacy [34]. The hospital employs a sophisticated inventory management system that oversees the distribution of pharmaceuticals and the inventory brought in by vendors and the pharmacy. The complexity of inventory management in hospital pharmacies is depicted in Figure 8.6, illustrating the growing number of final internal nodes involved in drug distribution.

Managing hospital inventory systems is faced with various obstacles, including but not limited to demand prediction, procurement, reception, handling, replenishment procedures, distribution, and inventory management. The extent of these limitations exhibits variability contingent upon the hospital ward or affiliated division [35]. Managing inventory policies within hospital networks is a multifaceted undertaking that is typically approached individually, with consideration given to past performance and the level of expertise available locally. This approach is necessary due to the intricate nature of stakeholder requests and the involvement of various processes.

8.5.2 Retail pharmacy inventories

Complex consumer factors affect retail pharmacy demand. The bullwhip effect has been observed in supply chains due to demand volatility and low

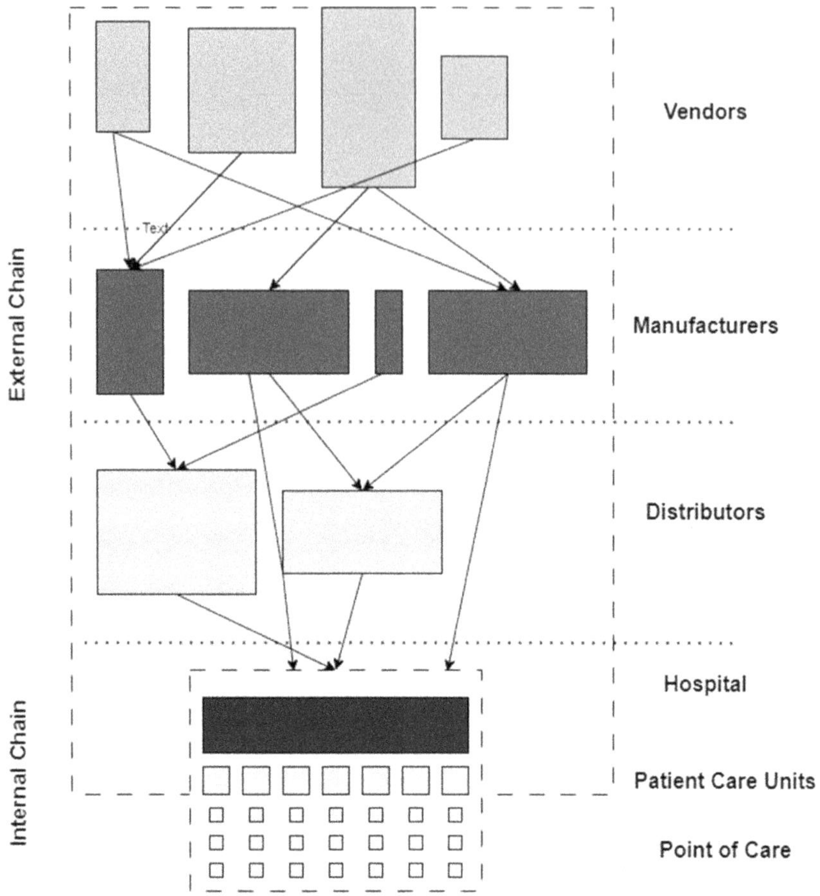

Figure 8.6 Supply chains in hospitals.

inventory planning and software utilization, leading to the spread of volatility throughout the supply chain [36]. Depending on risk and intermediary reaction, minor retail fluctuations can have significant implications for the manufacturer. It is imperative for pharmacies to precisely predict demand and establish a uniform ordering protocol that is grounded on an EOQ principle [32].

8.6 INTERNET OF THINGS

Sensors convert nonelectronic inputs into electronic signals [37, 38]. This electronic signal generates actionable information. The IoT is a wireless sensor network. Price, capability, and size have improved as sensing and computing technology advances, driving sensor deployment [37, 38]. These advances are expanding the technology and applications of IoT sensing.

To efficiently process and analyze the signals and data produced by these sensors, it is necessary to establish interconnectivity among them within a network. Holdowsky explains that connecting sensors to networks involves the utilization of gateways, routers, or similar devices [37, 38]. The author's research supports this assertion. The configuration of a network is influenced by various factors, including but not limited to the data transmission rate, energy consumption, and the scope of graphical representation. Since the year 2002, there has been an increase in data rates. The network data rates experienced a significant increase from 2 Mbps in 2002 to 1 Gbps in 2020. The US Internet transit prices dropped from $120Mbps in 2003 to $0.63Mbps in 2015. [37, 38].

Sensors and wireless networks have enabled massive increases in connected devices. According to Holdowsky's report [37, 38], there has been a substantial rise in interconnected devices worldwide, with an increase from 0.5 billion in 2003 to a projected 42.1 billion in 2019. The current trends in the industry necessitate heightened data visibility and analysis, while technology consumers seek expedited network configurations, thereby contributing to the anticipated surge in the number of devices.

8.6.1 Narrow-band IoT technology

Narrow-band (NB) IoT technology infrastructure is common. NB IoT uses ubiquitous carrier networks for portable devices. Many devices can function within a spectrum range of 200 MHz in conjunction with the existing infrastructure. According to a source, the appeal of NB IoT devices can be attributed to their superior indoor coverage, low power consumption, cost effectiveness, and ample connection capacity[4]. The low cost of NB IoT devices is crucial.

The public and private sectors use the NB IoT most. Huawei (2016) illustrates intelligent utilities, alarms, asset tracking, and smart agriculture. These applications exhibit low maintenance expenses, minimal device involvement, and extensive geographical reach. Huawei says that the China Unicorn marking meter program and Vodafone's 2015 water metering solution have been shown to work. The cost effectiveness, ease of upkeep, and extensive reach of the NB IoT render it an attractive option for enterprises that incorporate technology into their products. However, commercial sensing products that use the technology are often large or nonfunctional. Applications with limited space face challenges due to the absence of advancements in reducing the size of NB IoT technology. Public and private IoT ventures must consider this trade-off early in the design phase.

Wireless network standards predominate. Star topologies entail a configuration in which peripheral devices are exclusively linked to a central device that is an intermediary between the peripheral devices and the secondary cloud network. Wireless mesh networks (WMNs) comprise interconnected devices linked to each other and the cloud network via a particular device. WMNs utilize a mesh topology for radio nodes as opposed to a star topology, as noted by Liu et al. in their publication [39]. Figure 8.7 illustrates the comparison between star and mesh networks.

The IoT of the mesh network has several benefits. Expanding the coverage of the network is easy. Adding a router within the network's range usually increases coverage in a new area [39]. Devices automatically connect to the network once they are powered on. The implementation of flexibility in WMNs can decrease the level of technical proficiency required and minimize the degree of disruption involved in servicing or expanding such networks. According to Liu et al. [39], accessibility can potentially lower the expenses associated with setting up networks in rural regions that lack dependable coverage. WMN systems also have lower energy costs than other IoT networks. According to Liu [39], compared to the star network topology, most devices in wireless networks are connected over shorter distances, thereby eliminating the need for a central node to facilitate continuous connectivity between remote devices—the reduced connection span results in a decrease in the power consumption of the primary device.

WMNs mitigate certain limitations. The WMN requires a lot of hardware. Without a WMN, routers and connection points must be installed everywhere. The end-node IoT devices in WMNs can transfer limited data [39]. WMNs may not be ideal for IoT systems that need a lot of data transfer. According to Liu [39], in contrast to the star network topology, where a central node is required to maintain connections between remote devices, most devices in wireless networks are connected over shorter distances. The central device uses less power because the length of the connection interval is getting shorter.

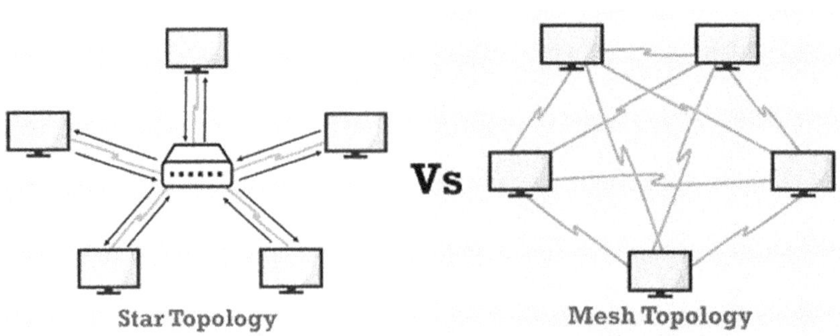

Figure 8.7 Star vs mesh topology.

8.6.2 RFID

RFID is a novel approach to monitoring the movement of goods along the supply chain. RFID systems comprise three components, namely a tag, an interrogator, and a controller, as illustrated in Figure 8.8 [40]. Transponders, or tags, are small chip-based devices with antennas and batteries [40]. A tag is put on the thing to be tracked. The interrogator finds and analyzes the tag's attributes, which is the part of the system that reads and writes. The interrogator transmits and receives electromagnetic waves when the tag is within the designated reading area. Finally, the controller receives and updates the interrogator data. The controller receives updated tag data from the interrogator through the local area network or wireless network, as stated in Hunt's work on RFID technology [40]. Subsequently, the controller, typically a personal computer or workstation, proceeds to refresh the database with the data obtained from the interrogator [40]. Figure 8.8 shows this three-tier process.

Implementing these networks enables the establishment of multiple tags and transponders within a singular system, thereby providing considerable adaptability for diverse applications. Applications for RFID technology can be either active or passive. Active RFID can transmit a signal over distances of up to several miles without the need for a stationary reading device. However, this technology is not being thoroughly considered for this use case due to its current clunky size, short battery life, and high price. On the other hand, there are numerous ways to implement passive RFID. Passive RFID technology facilitates communication between a sensor and a stationary reader, whereby the sensor is not required to possess an internal power source. Instead, the current reader obtains information from the sensor and transmits it upwards via a pre-existing wireless network. According to Kerr [41], there exist four discrete categories of RFID technology networks, namely low frequency (LF), high frequency (HF), ultra-high frequency (UHF), and microwave. When thinking about RFID networks, the three main factors to consider and trade-offs are 1) read range, 2) tag size, and 3) tag cost.

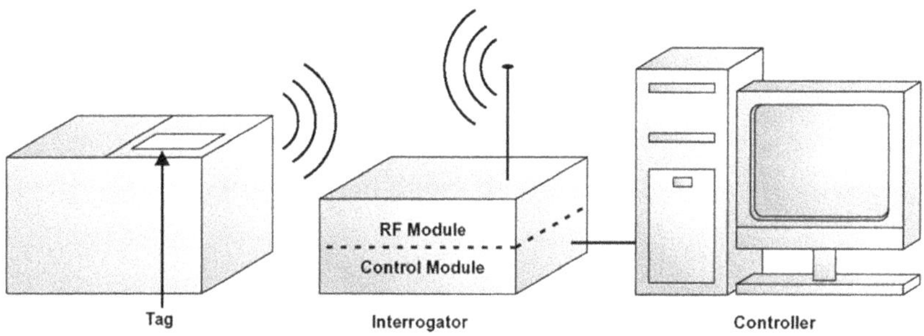

Figure 8.8 RFID system device configuration.

With LF technology, the read range of passive RFID tags is less than 0.5m; with UHF technology, it is about 4–5m [41]. These distances are small compared to other scales of IoT network technology that can connect many miles, even though they are significantly different on a small scale. Although it varies, the size of the tag is generally small. Passive RFID tags, such as hotel key cards that may fit in a wallet or security tags attached to expensive clothing in stores, are examples of small tags that are hard to see. Furthermore, tags are usually less than one US dollar, rendering them economically feasible for most use cases.

As already mentioned, to work with passive RFID technology, stationary readers and mobile tag devices must be in close proximity [41]. It can be difficult to identify the necessary readers and related support networks. Moreover, establishing the requisite commercial and monetary connections to foster these reader communities across diverse entities can be challenging, if not unfeasible. Passive RFID systems are known to have a low financial threshold for implementation. However, their optimal performance is often observed when deployed within a single organizational setting.

8.6.3 Bluetooth network design

Bluetooth technology allows for short-distance connections between devices. To enable wireless communication between equipment and devices, the main objective behind the development of Bluetooth technology was to facilitate rapid communication between wireless devices, including portable cell phones, headphones, and keyboards [42]. Although Bluetooth was first used for wireless audio communication, it has quickly expanded to include a wide range of uses, such as tracking, advertising, and external beacons for smartphones with Bluetooth. Figure 8.9 depicts the gadgets in question.

Using an unlicensed frequency band ranging from 2.4 to 2.483 GHz is among the notable attributes of Bluetooth technology, as reported by Lawrence [42]. This frequency band was chosen because it can be used in the majority nations of the world [42]. Classic Bluetooth and Bluetooth Smart (also known as Bluetooth Low Energy, or BLE) are two Bluetooth

Figure 8.9 Bluetooth devices.

connectivity options for point-to-point, mesh, or other networks[5]. Although Bluetooth Smart uses significantly less power and is generally less expensive than Bluetooth Classic, Bluetooth Classic can transmit at a higher communication range.

In general, using Bluetooth to build an IoT network has many benefits. For example, it has a low entry barrier in terms of development. Apple and Google, two smartphone manufacturers, have developed sophisticated tools to create applications that can access an external Bluetooth device. As a result, IoT device developers can create devices that are immediately compatible with major smartphone manufacturers, eliminating the need for an expensive and time-consuming integration configuration. Furthermore, the simplicity of the development process has led to the widespread adoption of Bluetooth technology in various commercial domains. Finally, devices that are enabled with Bluetooth technology show small form factors and low energy consumption. Bluetooth tags are often available in sizes similar to those of a quarter or a house key.

In addition, it should be noted that Bluetooth device tags exhibit the most economical mean cost among all IoT devices. This low cost barrier gives applications in ambiguous environments a great deal of flexibility. The main drawback of Bluetooth technology is its limited range. In general, the detection range of Bluetooth tags is limited to a few hundred feet from the active devices responsible for their reading, such as cell phones. Furthermore, these tags exhibit passivity and require the involvement of a connection device to initiate data transmission. The proliferation of Bluetooth-enabled devices can lead to congestion and signal interference. Due to signal congestion at nearby frequencies during peak hours caused by Bluetooth usage that is not regulated or licensed, connections may be slow or impossible during these times [41].

8.6.4 Technology comparison

When choosing the most practical course of action of the sponsor, five key factors must be taken into account: battery life, cost, device size, communication range, and beacon specifications. An aggregated view of the general characteristics of the critical variables of each IoT network is shown in Table 8.1.

Table 8.1 A comparative analysis of IoT technologies

Platform	*Battery life*	*Size comparison*	*Avg. cost*	*Read range*	*Local beacon*
Narrow Band	3–10 years	Calculator	$5	<38 km	No
RFID	3–5 years	Coaster	$0.10–$0.15	0.0005–.1 km	Yes
Bluetooth	0.5–3 years	Coin	$1	<.152 km	Yes/No

8.7 INVENTORY

Inadequate management of pharmacy inventory can negatively impact patient safety and result in unfavorable financial results for the business. The results mentioned previously can be attributed to many factors, including but not limited to the availability of lapsed, substandard imitations or deteriorated goods; insufficiency of essential products; unfulfilled prescriptions; and the failure to update the formula. Performing weekly stock reviews to verify amounts and monthly checks to identify expired products is a recommended practice to enhance patient safety. Implementing software systems that can alert the pharmacist when inventory levels are approaching critical thresholds or expiration dates can be used to achieve this objective.

8.7.1 Inventory-associated costs

The pharmacy field involves four distinct expenses associated with inventory management, namely purchasing payments, purchasing expenses, carrying expenses, and shortage costs, as outlined by Carroll and DeYoung in their 1998 publication on accounting practices [43]. The cost incurred by the pharmacy to procure the goods is commonly referred to as the acquisition cost. Expenses associated with purchasing products include the costs of order placement and receipt, shelf stocking, and invoice settlement. Carrying costs, or expenses related to storing goods, entail additional fees that emerge from unanticipated events such as destruction or loss. Stockout costs, also referred to as shortage costs, are monetary expenses incurred as a result of the unavailability of a product.

8.7.2 Methods of inventory management

As per Desselle and Zgarrick's study in 2005 [44], inventory management is a continuous process that involves strategic planning, systematic organization, and efficient inventory control. The main aim of this procedure is to reduce the amount of capital tied up in inventory while maintaining an equilibrium between the availability and requirement of commodities. The process is designed to minimize purchasing and carrying expenses while ensuring a sufficient inventory of products available to meet the demands of both the customer and the prescriber. The proficient administration of medications is an essential facet of the operational framework in diverse pharmacy contexts, encompassing community and healthcare practices. Inadequate inventory management leads to an unnecessary escalation in transport and procurement expenses and a disruption in the equilibrium between supply and demand. Adequate training in management techniques is imperative for pharmacists and pharmacy students.

According to Desselle and Bouldin [44, 45], pharmacies employ three different techniques, namely visual, periodic, and perpetual methods, to effectively manage their inventory. The visual method involves requesting the pharmacist or another designated staff member to conduct a visual comparison between the existing inventory and a predetermined list of the required number of products. The pharmacist initiates a procurement request when the level of the inventory falls below the predetermined threshold. The pharmacist initiates a procurement request when the inventory level falls below the pre-established threshold. Items are purchased promptly upon depletion of the minimum threshold quantity. Perpetual inventory management is the most general and practical approach for handling pharmacy inventory in developed nations. A computerized system is used to constantly and methodically monitor the inventory. This system involves entering the available inventory into the computer software and automatically subtracting the necessary number of products from the list each time a prescription or medication order is filled.

Pharmacists have various tools, including the ability to run a computerized system (continuous method) and to perform a yearly count using visual and periodic procedures. By employing this technique, the pharmacist can contrast the product quantities shown on the computer screen with those that are, in fact, present on the shelves. This approach will evaluate and verify the accuracy of pharmacy financial records and identify and correct any potential variations caused by changes in supply and demand. Furthermore, pharmacists must adhere to the regulations established by their respective pharmacy boards and groups on the management of inventory for specific items, including prohibited substances, immunizations, and biological products.

8.7.3 Assessment of inventory control

Given the substantial resources devoted to the management of pharmacy inventory, it is imperative to assess how well a pharmacy works its inventory. A recommended methodology is to calculate the inventory turnover rate (ITOR) as a pragmatic approach. This can be performed for the entire pharmacy inventory, a specific department (e.g. parenteral products), or a particular product, as suggested by Huffman [46]—an assessment of inventory management. The ITOR is calculated by the ratio of the cost of goods sold (COSG) to the average amount of inventory. The average inventory is computed by taking the values of the initial and final stocks and averaging them over an agreed-upon period.

According to the National Community Pharmacists Association, independent community pharmacies are characterized by an ITOR value of 10 or higher. This implies that within a particular timeframe, the pharmacy has successfully sold its entire inventory ten times [47]. To assess the efficacy of inventory management, it is advisable to analyze the ITOR in conjunction with the ITOR metrics from the previous period, such as the previous year. According to McCaffrey [48], a higher ITOR indicates that the stock was efficiently purchased, sold, and replenished in a specific period. In contrast, lower ITOR values indicate suboptimal inventory management practices within the pharmacy, whereby the merchandise remained stagnant and was not used on the shelves. In this scenario, the pharmacist should consider various possibilities, including overordering the correct product, erroneous ordering of an incorrect product, or potential inventory management oversights, such as inaccurate input of product information or quantities into computerized software without conducting a physical inventory count.

In addition to ITOR, the pharmacist must determine the correlation between net profit and inventory turnover over the specified time frame. The percentage net profit (PNP) calculation, characterized as the percentage relationship between net profit and average inventory, can be employed to achieve this goal. A better PNP value signifies high marks on the merchandise being retailed. An increase in the ITOR value over a period of time, coupled with a decrease in PNP during the same duration, suggests that a greater volume of products was sold, although at a diminished profit margin due to reduced markups. The measurement of net profit percentages is relative. Therefore, it is imperative to juxtapose absolute net profit in aggregate currency denominations, such as the US dollar, over pertinent time frames. To make informed assessments of the efficacy of pharmacy inventory management, it is advisable to evaluate PNPs compared to the previous period, similar to the approach taken with ITOR.

8.7.4 Inventory management determinants

When assessing their pharmacy inventory management, pharmacists are required to take several factors into account. These factors include inventory, product classification (generic or brand), size, policies regarding returned products, unclaimed prescriptions, inventory shrinkage, and formulary utilization.[6] Generic products are associated with lower acquisition costs than their brand-name counterparts, which in turn leads to lower inventory costs. The reduction in inventory investment can be attributed to the relatively smaller inventory size of primary product lines compared to their entire product line counterparts. This decision should be weighed against the needs of both patients and healthcare providers. Most product vendors, including manufacturers and wholesalers, implement protocols for the return of merchandise.

Examples of such policies include giving future orders credit, substituting products, and giving cash back to the pharmacy. Pharmacists can use these benefits before policy expiration due to effective inventory management. In America, community pharmacies fill approximately 1. 5% of the prescriptions that go unclaimed [49]. Pharmacists should monitor these prescriptions and establish a deadline (for example, two weeks) by which products must be returned to the shelves. It is imperative to ensure that patients or caregivers are properly notified of prescriptions that have been filled or refilled, especially in the case of elderly patients and those with chronic diseases exhibiting a growing tendency towards nonadherence to medication.

Due to inventory shrinkage, community pharmacies can lose up to 4.5% of their sales [50]. The term "inventory shrinkage" also described losses caused by theft, shoplifting, and robbery. Regrettably, inventory shrinkage in community pharmacy settings is primarily attributed to employee theft, as reported by Desselle et al. in 2005 [44]. It is imperative that pharmacists not only recruit personnel who exhibit traits of trustworthiness and openness but also establish and enforce robust measures for security, observation, and monitoring. It is imperative to prioritize sufficient security measures for controlled substances when monitoring shrinkage, particularly in light of the increasing difficulty of theft of such chemicals.

In hospital pharmacy settings, formulas are utilized to improve inventory management practices. Because of this, pharmacists can keep one equivalent therapeutic product of each drug in a class on hand, reducing overall inventory costs. The restrictions of using restricted lists and formulas can challenge community pharmacy settings to achieve an equilibrium between the availability and requirement of pharmaceutical products.

8.7.5 Inventory management and information technology

The use of information technology in the management of pharmacy inventory is a widespread practice in developed countries, where computerized systems are widely used in different pharmacy practice environments. Utilizing technology improves the efficiency, accuracy, and precision of inventory management and assessment techniques. The implementation of technology in inventory management can be observed through various practices, such as the use of handheld scanning devices to scan barcodes on product packaging or shelf labels for periodic inventory control. Additionally, electronic submission of purchase orders is facilitated by entering scanned data into a web-based system, such as eProcurement.

Currently, technology is extensively utilized in virtually all aspects of pharmacy operations, encompassing the processes of order placement, product procurement, storage, and financial transactions. To improve pharmacy inventory management, pharmacists should incorporate modern technologies into their daily practice. RFID microchips, commonly called "tags," have emerged as a contemporary technology that facilitates the distribution of products from manufacturers to wholesalers and pharmacies. The objective of these tags is to encapsulate data about the pharmaceutical product, spanning from its inception date to its receipt and subsequent withdrawal from the storage racks of the pharmacy. In addition, computerization in pharmacy inventory management enables pharmacists to dedicate additional time to providing pharmaceutical care and other associated services to patients and customers, as Awaya et al. [49] stated in their research.

The use of information technology in pharmacy operations can improve inventory management and evaluation by effectively reducing procurement costs and mitigating inventory shrinkage caused by theft. Furthermore, the integration of product barcode scanning within pharmacy practice, specifically in hospital pharmacy settings, has reduced the probability of medication errors [50].

8.8 INVENTORY MANAGEMENT

The right kind of material must be available at the right time, in the right quantities, and at competitive prices. This can be determined scientifically by calculating the amount of stock that needs to be kept on hand to meet production demands.

Any production operation will typically take into account the return on investment when purchasing capital equipment, and many appropriation requests will be denied if the rate of return is too low. Commitments for inventories must be taken into account in the same manner, and it goes without saying that buying and holding a product for one month rather than two gives a better return on investment and inventory turnover. This oversimplifies the costs involved in inventory decisions, which include ordering costs, out-of-stock costs, clerical costs, computer costs, and quality control costs, among many others that are too numerous to list here. Inventory can comprise anywhere between 35% and 80% of working capital, and some of these top pharmaceutical companies have global inventories of close to $700 million, according to an examination of their annual reports, showing that they have a higher return on equity. Reducing inventory can free up much-needed cash that the company can use to invest in more lucrative endeavors and pay down debt. A well-managed inventory can exert significant financial leverage. A company's division responsible for inventory planning and control is called PPIC.

Processing Purchase Order: The process starts with identifying the need. The relevant department determines its needs, obtains the department head's approval, and with that approval, a designated person sends a purchase request to the purchasing department to begin a purchase. The user department must create a budget before sending a purchase request for the purchase of property, plant, and equipment. If the proposed budget is approved by the department head or higher authorities, as necessary, a purchase request is sent to the purchasing department. The planning division also chooses the quantity and timing of raw materials for packaging or raw materials. When materials are needed, this department communicates with the purchasing department. The request for a quote or tender was received by the purchasing department. The supplier has been chosen after receiving the bid or quotation. The supplier could be domestic or foreign. The purchase department then issues an order for the purchase if the terms and conditions are favorable to both the company and the chosen supplier. The factory issues a purchase order for any raw materials or packaging. The purchase department keeps a purchase register in which they keep all the necessary details about a consignment.

Receiving Material, Goods and Services: In most cases, the user department that issued the purchase order or, in rare circumstances, the authorized department, is the one who receives the goods and services. The quality assurance department (QAD) of the factory receives the materials. After receiving materials, goods, and services, the receiving department issues a gross revenue retention to the purchasing department in addition to a monthly recurring revenue for materials. The purchase department receives the invoice or bill in the meantime. The product has been inspected by the inspection or QAD, by the user department or by the authorized department before being used by the user department, which is at the time of delivery. On the basis of a sample test, QAD examined the materials and issued a certificate.

8.9 FORMS OF INVENTORY ADMINISTRATION

Four distinct inventory management categories exist, as outlined in a reputable source.[7]

- Just-in-time management (JIT)
- Materials requirement planning (MRP)
- EOQ
- Days sales of inventory (DSI)

8.9.1 JIT

The JIT inventory management approach involves the receipt of goods from suppliers solely at the point when they are required, thus minimizing inventory holding costs. This approach aims primarily to mitigate the expenses associated with inventory holding and enhance the frequency of inventory turnover. Implementing the JIT methodology involves meticulous planning of the entire supply chain and the use of advanced software to facilitate the process until delivery. This approach improves efficiency and minimizes the possibility of errors by closely monitoring each process. Implementing a JIT inventory management system can produce significant important effects.

- **Reduces Inventory Management:** Implementing a JIT strategy mitigates overproduction, whereby the quantity of a given item available for purchase exceeds the level of demand, resulting in a buildup of unsold inventory.
- **Cost of keeping a warehouse open decreases:** The cost of warehousing is a significant expense, and the retention of surplus inventory

can result in a two-fold increase in holding payments. In addition, the implementation of a JIT system results in the reduction of warehouse maintenance expenses. As the order fulfillment process is initiated only upon receipt of a customer's order, the merchandise in question is effectively pre-sold before it arrives at the vendor's premises, thereby preventing the need for prolonged storage. Organizations that adhere to the JIT inventory methodology can reduce the number of goods stored in their inventory or even eliminate the need for warehouse facilities.

- **Gives the manufacturer more control:** The JIT model is characterized by the manufacturer's complete authority over the production process, which operates on the demand-pull principle. The company can address customer demands by promptly adjusting the production levels of high-demand and low-performance products. The flexibility of the JIT model enables it to adapt to dynamic market demands. Toyota follows a JIT inventory system, where raw materials are not procured until a customer order is received. Implementing this strategy has facilitated the organization in maintaining a low inventory level, thus decreasing its expenses and empowering it to respond quickly to fluctuations in demand without having any concerns regarding existing inventory.
- **Local sourcing:** The implementation of the JIT methodology necessitates the initiation of manufacturing activities solely upon receipt of an order. Consequently, procuring raw materials from nearby sources is imperative to ensure timely delivery to the production unit. Additionally, the practice of procuring goods and services locally reduces both the time and the cost associated with transportation. Consequently, this generates a demand for numerous additional companies to operate simultaneously, thus enhancing labor force participation rates within that specific population.
- **Smaller investments:** The JIT model is designed to acquire only the necessary inventory, reducing the working capital required for financing the purchase. Consequently, the enterprise's return on investment would be elevated due to the reduced quantity of stock maintained in the inventory. The utilization of JIT models involves the implementation of the right first-time principle, which entails the execution of tasks with utmost accuracy during the initial attempt, thereby minimizing expenses associated with inspection and rework. This results in a reduced investment for the company, minimized expenditures for error correction, and increased profitability resulting from the sale of a product.

8.9.2 MRP

The material requirements planning (MRP) system determines the materials and components needed to produce a specific product. The process comprises three fundamental stages: conducting a comprehensive assessment of the available materials and components, determining the required supplementary materials, and arranging their production or procurement. The MRP system is a computerized inventory management tool that allows efficient and cost-effective procurement of necessary inventory for production purposes. MRP is predominantly executed through specialized software. MRP improves the efficacy, adaptability, and financial benefits of production activities. Adopting this methodology holds promise in increasing the efficiency of industrial workers, streamlining the caliber of merchandise, and curbing costs related to workforce and resources. Implementing MRP can help manufacturers quickly address heightened demand for their goods while avoiding production setbacks and stock insufficiencies that may lead to customer attrition. This, in turn, can promote revenue expansion and stability.

The use of MRP is prevalent among manufacturers and has indisputably served as a crucial facilitator in the expansion and extensive accessibility of reasonably priced consumer goods, elevating the quality of life in numerous nations. The absence of automated mechanisms for handling intricate computations and data organization in MRP procedures would have hindered the ability of individual manufacturers to expand their operations at the pace they have achieved over the past 50 years since the advent of MRP software.

The MRP system uses data collected from the material bill (BOM), inventory records, and the master production schedule to determine the requisite components and their corresponding temporal allocation for the manufacturing process. The BOM is a systematic inventory comprising all the essential materials, sub-assemblies, and additional components required to produce a given product. The BOM typically depicts a hierarchical arrangement of these elements, with their respective quantities specified in a parent-child association. At the apex of the hierarchy lies the parent, representing the outcome.

The BOM enumerates two distinct classifications of components, specifically those characterized as independent demand and those classified as dependent demand. The terminological expression "independent demand item" refers to the final product occupying the highest hierarchical level. Determining the quantity of a product that manufacturers will produce involves carefully considering confirmed orders, market conditions, past sales, and other relevant indicators. This information is used to create a forecast which is then utilized to make informed decisions about the number of units produced to meet the anticipated demand.

In contrast, the dependent demand items refer to the required components and raw materials of the final product. The demand for each item depends on the quantity required to produce the subsequent higher-level component in the BOM hierarchy.

The majority of companies employ MRP as a means of monitoring and controlling interdependencies, as well as determining the requisite quantity of items within the timeframes outlined in the master production schedule. In other words, MRP is a system utilized for inventory management and control, which involves purchasing and monitoring the necessary components required to produce a given product.

Lead time, which refers to the duration between the placing of an order and the delivery of the corresponding item, is a crucial notion within the context of MRP. Numerous lead-time variations exist. Material and factory or production lead times are two frequently encountered temporal factors in manufacturing. Material lead times refer to the time required to order and receive materials. In contrast, factory or production lead times denote the time taken to manufacture and transport the product after all materials have been obtained. The term "customer lead time" refers to the time that elapses between placing a customer's order and the eventual delivery of the requested product or service. The MRP system is responsible for computing a significant number of lead times, while a subset of them is determined by operations managers and manually entered. The principal aim of MRP is to guarantee the availability of materials and components during production and to ensure that manufacturing occurs according to the predetermined schedule.

Other advantages of MRP include the following benefits:

- shortened customer lead times to boost customer satisfaction;
- less expensive inventory;
- effective inventory management and optimization: businesses can reduce the risk of stock- outs and their detrimental effects on customer satisfaction, sales, and revenue without spending more than necessary on inventory by purchasing or producing the ideal amount and type of inventory;
- increased manufacturing efficiency by accurately planning and scheduling production to use labor and equipment as efficiently as possible;
- increased output at work;
- greater price competition for the goods.

MRP has shortcomings, such as:

- Although MRP is intended to guarantee adequate inventory levels at the necessary times, businesses can be persuaded to hold more inventory than is essential, which increases inventory costs. Additionally, an MRP system anticipates shortages earlier, which can result in overestimating lead times and inventory lot sizes, particularly in the early stages of deployment, before users gain the experience to understand the amounts required.
- Lack of flexibility: MRP's accounting for lead times or details that affect the master production schedule, such as the productivity of factory workers or problems that can delay the delivery of materials, is also somewhat rigid and simplistic.

- Information about critical inputs, particularly demand, inventory, and production, must be accurate for MRP to function effectively. In addition, later stages may experience errors that one or two inaccurate inputs have amplified. Therefore, effective data management and integrity are necessary for the use of the MRP system.

8.9.3 EOQ

The optimum order quantity, or EOQ, is the number of units a business should buy to meet demand while reducing inventory costs such as holding, shortage, and order costs. Ford W. Harris created this production-scheduling model in 1913, which has since been improved. Demand, ordering, and storage costs are considered constant in the EOQ formula.

The formula to calculate the EOQ is

$$Q = \sqrt{\frac{2DS}{H}}$$

where:

Q = EOQ units
D = demand in units (typically on an annual basis)
S = order cost (per purchase order)
H = holding costs (per unit, per year)

The EOQ formula aims to determine the ideal quantity of units of products to order. If accomplished, a business can reduce its purchasing, shipping, and storage unit costs. Companies with extensive supply chains and high variable costs use an algorithm in their software to calculate the EOQ. The EOQ formula can be changed to determine production levels or order intervals.

An essential cash flow tool is the EOQ. The equation can assist a business in managing the amount of cash held in the inventory balance. In addition to human resources, inventory is often a company's biggest asset, so they must keep enough of it on hand to meet customer demand. If EOQ can reduce the amount of inventory, the money saved can be invested or used for other business needs.

The inventory reorder point of a business is determined by the EOQ formula. If the EOQ formula is used in business processes, ordering more units becomes necessary when the stock reaches a certain level. Therefore, the company can continue to fulfill customer orders and prevent running out of inventory by deciding on a reorder point. A shortage cost, or revenue lost because the company lacks enough stock to fill an order, occurs if the company runs out of inventory. A lack of inventory could also result in a client's loss or decreased future charges for the business.

The EOQ formula takes for granted the idea of constant consumer demand. Additionally, the calculation assumes that the ordering and holding costs will not change. Because of this, the formula has a difficult or impossible time considering business events like shifting consumer demand, seasonal variations in inventory costs, lost sales due to inventory shortages, or purchase discounts that a business might realize to buy inventory in larger quantities.

8.9.4 DSI

The days sales of inventory (DSI) financial ratio shows how long it typically takes for a business to convert its stock, which includes goods that are still being manufactured, into sales in days.

The term DSI can be used to refer to a variety of concepts, including the average age of the inventory, days of inventory outstanding (DIO), days of inventory (DII), and days of sales in the list The figure indicates how long a company's current inventory stock will last and the liquidity of the merchandise. Although the average DSI differs from industry to industry, a lower DSI is generally preferred because it indicates a shorter period to clear off the inventory.

The formula for calculating DSI is

$$DSI = \frac{\text{Average Inventory}}{\text{COGS}} \times 365\,\text{days}$$

where:

DSI = days of inventory sales
COGS = cost of goods sold
S = order cost (per purchase order)
H = h
Iiiio4lding costs (per unit, per year)

A business needs raw materials and other resources, which make up its inventory and have a cost, to make a marketable product. Using the stock to create a saleable product is another expense. These expenses include labor costs and payments for utilities like electricity, which are included in the COGS, which is the price paid for purchasing or producing the goods a business sells over a specific period. The average inventory value and the COGS over a particular period or as of a specific date are used to calculate DSI. Mathematically, 365 for a year and 90 for a quarter determine the number of days in the corresponding period. Sometimes, 360 days are used in its place.

The numerator value represents the inventory valuation. The average daily cost the business incurs in producing a marketable product is represented by the denominator (Cost of Sales / Number of Days). The net factor indicates the typical number of days the company needs to sell its inventory.

Depending on accounting procedures, two different DSI formulas can be applied. The average inventory amount is used in the first version as the amount reported at the end of the accounting period, such as the fiscal year ending on June 30. This version displays the DSI value as of the specified date. Another variation uses the average value of the Start Date Inventory and the End Date Inventory; the resulting number represents the DSI value "during" that specific period. Therefore,

$$\text{Average Inventory} = \text{Ending Inventory}$$

or

$$\text{Average Inventory} = \frac{\text{Beginning Inventory} + \text{Ending Inventory}}{2}$$

A lower value of DSI is preferred because it shows how long a company's cash is anchored in its inventory. A lower number suggests that a business is using its stock more effectively and frequently, which results in rapid turnover and the potential for higher profits (assuming that sales are profitable). However, a high DSI value suggests that the company may need help managing its high-volume, obsolete inventory and may have overspent on it. To achieve tall order fulfillment rates, the business should keep much stock in anticipation of brisk holiday season sales.

DSI is a metric to assess how well a company manages its inventory. A sizable portion of a business's operational capital requirements is inventory. This efficiency ratio determines the average time a company's cash is locked in the list by counting the days it keeps the inventory before selling it.

However, this number should be considered cautiously because it frequently lacks context. Depending on several variables, including the product type and business model, DSI tends to vary significantly across industries. As a result, it is crucial to evaluate peers in the same industry. For example, companies in the technology, automotive, and furniture industries can afford to hold onto their inventories for extended periods, but those in the perishable or fast-moving consumer goods (FMCG) industries cannot. As a result, comparisons for the DSI values should be made using specific sectors.

8.10 PRINCIPLES OF INVENTORY MANAGEMENT

The article outlines five fundamental principles of inventory management.[8]

- demand forecasting
- warehouse flow
- inventory turns/stock rotation

- cycle counting
- process auditing

8.10.1 Demand forecasting

Inventory is considered to be among the top five expenses for businesses, depending on the industry. Precise demand prediction holds the most significant potential for cost reduction among the various tenets of inventory management. The presence of either an excess or a shortage of inventory can result in substantial financial implications for a business. Therefore, the precision of the forecast is crucial, regardless of whether it pertains to end-item stocking or essential component sourcing. Implementing effective maximum and minimum inventory management strategies at the individual inventory line level, considering lead times and safety stock levels, is crucial for ensuring the timely availability of required resources. Moreover, this approach prevents the occurrence of expensive overstocks. The presence of idle inventory results in additional expenses associated with the handling and storing of fast-moving items due to the loss of storage space.

8.10.2 Warehouse flow

The old-fashioned notion that warehouses are unsanitary and disordered is no longer tenable and financially burdensome. Lean manufacturing principles, such as the implementation of the 5S methodology, have been incorporated into the warehousing industry. Implementing sorting, ordering, systemic cleaning, standardization, and discipline maintenance measures can prevent financial losses from inefficient processes.

The fundamental tenets of inventory management are analogous to those of other industrial procedures. The lack of organization incurs financial expenses. Therefore, it is imperative to establish a formal and standardized process for every operation, from housekeeping to inventory transactions, to achieve consistently exceptional outcomes.

8.10.3 Inventory turns/stock rotation

Effective inventory management at the lot level is crucial for reducing operational expenses in specific sectors, such as the pharmaceutical, food, and chemical warehousing industries. Therefore, the metric of inventory turns is a crucial factor in assessing the efficacy of one's implementation of inventory management principles.

Establishing a benchmark for the efficacy of stock rotation is of utmost importance in evaluating demand prediction and inventory management within a warehousing context.

8.10.4 Cycle counting

Cycle counting is considered a fundamental approach for ensuring the precision of inventory records. This facilitates the assessment of the efficacy of current procedures and upholds responsibility for potential sources of error. In addition, cycle counting has financial implications. For example, specific industries necessitate regular 100% tallies. These procedures are accomplished either using perpetual inventory count maintenance or by conducting full-building counts.

8.10.5 Process auditing

The process of identifying error sources proactively commences with conducting process audits. One of the fundamental tenets of inventory management is to conduct frequent and timely audits. Therefore, it is recommended that process audits be performed at every stage of the transactional process, encompassing the receipt and shipment of goods and all inventory-related transactions that occur in between.

8.11 METHODS OF INVENTORY CONTROL

There exist two primary categories of inventory control systems.

- periodic system
- perpetual system

The selection of an appropriate inventory control system is contingent upon the nature, magnitude, and composition of the inventory. This section comprehensively analyzes the aforementioned types, including their advantages, disadvantages, and conditions.

8.11.1 Periodic system

Periodic inventory control systems refer to inventory management systems that regularly count and record inventory at predetermined intervals. The overall frequency for conducting inventory counts is every month, although alternative intervals such as quarterly, semiannually, or annually are also viable options.

Periodic inventory control systems offer the benefit of being comparatively uncomplicated and straightforward to execute. However, rare inventory control systems are deemed less precise compared to perpetual inventory control systems, which can result in stockouts.

The periodic inventory control method is deemed most suitable for small-scale enterprises that possess limited inventory and experience infrequent inventory turnover.

8.11.2 Perpetual system

Perpetual inventory control systems are a type of inventory management system that enables the continuous tracking and recording of inventory in real time. The integration of inventory software with the point of sale (POS) system is employed to achieve this objective.

Perpetual inventory control systems possess an advantage over periodic inventory control systems in accuracy. However, endless inventory control systems are associated with the drawback of increased complexity and the potential for higher implementation costs.

The perpetual inventory control system is deemed most suitable for enterprises that operate across multiple locations, maintain substantial inventory levels, and experience high ITOR. This solution is particularly well-suited for enterprises specializing in selling FMCG and requiring stringent inventory management practices.

Various techniques and methods are employed in the domain of inventory control. The four prevalent techniques for managing inventory are as follows:

- **ABC Analysis:** The ABC analysis is a widely used method for categorizing inventory into three groups.
    - The items denoted by the letter 'A' are deemed to be of utmost importance and thus warrant the highest degree of attention.
    - While A items hold greater significance, B items are also considered necessary.
    - The elements denoted by the symbol 'C' are deemed to possess relatively lower significance and, consequently, are accorded comparatively lesser consideration.

  The implementation of ABC analysis has the potential to assist inventory managers in directing their attention toward the items that have the most significant impact on the business.
- **Last In, First Out (LIFO) First In, First Out (FIFO):** The LIFO and FIFO inventory valuation methods are commonly employed in accounting practices. LIFO stands for last in, first out, while FIFO stands for first in, first out. The process under consideration presumes that the sale of inventory items occurs in the order of their most recent manufacture or purchase. In contrast, the FIFO method presupposes that the sale of things happens in the order of their being in stock for the most extended duration.

  Inventory valuation is contingent upon multiple factors, including but not limited to accounting regulations, tax ramifications, and inventory turnover.

- **Batch Tracking:** Batch tracking is an inventory management technique that involves categorizing inventory into distinct 'batches' and allocating a unique identifier to each batch. The identifier may be a barcode, lot number, or serial number.

  Batch tracking enables warehouse managers to monitor the following details:
  - The origin of the items is being queried.
  - The destination of the commodities.
  - The potential expiration of the items.
- **Safety stock:** Safety stock refers to the store of inventory maintained to fulfill customer demand in case of any unexpected disruption in the supply chain. This phenomenon can address stock deficiencies, unanticipated consumer demand surges, and other unforeseen circumstances.

The practice of maintaining safe stock can aid businesses in mitigating stockouts and minimizing lost sales. However, excessive safety stock levels can result in the immobilization of working capital and give rise to challenges in inventory management.

8.12 COMPONENTS OF INVENTORY MANAGEMENT

Professional inventory management involves the delicate equilibrium of satisfying the requisites and desires of the production personnel, the sales department, and the inventory supervisor. Determining the supplier's availability of goods and materials is a crucial factor that must be considered. Effective inventory management is a critical concern for businesses, regardless of whether they manufacture or resale third-party products. It is a complex undertaking that demands careful attention. Even so, any enterprise can employ three fundamental constituents of inventory management to construct a proficient stock management framework.

8.12.1 Inventory forecast analytics

One of the most challenging tasks for inventory managers is comprehending the levels and patterns of customer demand. Gut intuition may prove helpful in certain instances, but it is only sometimes reliable. In the absence of a dedicated forecasting system and established procedures, sales forecasting can devolve into a form of conjecture that has the potential to incur significant costs in terms of both time and finances for your enterprise. A viable resolution entails the utilization of proficient predictive analytics, such as those furnished by an inventory analysis and planning application, that are grounded on mathematical and statistical methodologies. Sales and inventory software, such as DataQlick, offers statistical data and visual representations that eliminate the need for speculation in predicting future sales.

Instead, this software utilizes factual information from previous sales to inform inventory management decisions.

8.12.2 Optimized purchase orders

The number of suppliers has increased significantly; however, this phenomenon must be clarified for generating purchase orders. In addition, the growing intricacy of supply chain models may pose a challenge. Implementing safety stocks is a potential solution to the issue at hand. However, it is essential to note that this approach may result in the immobilization of capital, which can be a significant challenge for small businesses. Inventory analysis and planning applications incorporating an intelligent purchase order feature like DataQlick can facilitate availability and cost management.

8.12.3 Inventory control

Effective inventory control can simplify the laborious task of managing inventory by promoting stock transparency, ultimately optimizing storage space and enhancing the financial aspect of inventory management. However, manual inventory control requires more efficiency and susceptibility to inaccuracies. Therefore, inventory control software is essential for effective inventory management because quick actions and accurate data are necessary. This is particularly true when the software allows for the simulation of stock data, which can be used to anticipate potentially hazardous inventory situations.

8.13 TECHNIQUES USED FOR VARIOUS INVENTORY MANAGEMENT PROBLEMS

The process of inventory management can be perceived as a challenging undertaking. The influence of both the process and results is ubiquitous across all facets of a business. The supply chain may encounter several inventory management challenges that require attention. A comprehensive list of such challenges is available at the provided source.[9]

8.13.1 Challenge #1

Problem—Inconsistent Tracking: Utilizing manual inventory tracking procedures through various software and spreadsheets is a laborious process prone to duplication and is susceptible to inaccuracies. Small enterprises can benefit from a unified inventory monitoring mechanism encompassing accounting functionalities.

Solution—Centralized Tracking: It is recommended to upgrade to tracking software that offers automated functionalities for procurement and reordering. In addition, cloud-based centralized database side by side with inventory management platforms to ensure precise and automatic inventory updates and real-time data backup.

8.13.2 Challenge #2

Problem—Warehouse Efficiency: Managing inventory controls within a warehouse setting is a laborious process encompassing multiple stages, such as the receipt and placement of goods, selection of items, packaging, and dispatch. The objective is to execute these tasks with optimal efficiency.

Solution—Transparent Performance: To address inefficiencies in the warehouse, it is recommended to measure and report performance metrics such as inventory turnover, customer satisfaction, and order processing speed. Then, disseminate this information to both personnel and vendors.

8.13.3 Challenge #3

Problem—Inaccurate Data: It is imperative to maintain accurate knowledge of your inventory at all times. The traditional practice of conducting annual inventory counts with the participation of all staff members is no longer applicable in contemporary times.

Solution—Stock Auditing: Regular stock auditing procedures, such as daily cycle counting, mitigate the risk of human error and yield more precise and current inventory information to facilitate cash flow management. To ensure greater precision in financial data, it is recommended to categorize audits and conduct periodic inventory sampling in a predictable cycle.

8.13.4 Challenge #4

Problem—Changing Demand: The dynamic nature of customer demand is a persistent phenomenon. Maintaining excessive inventory may lead to the accumulation of outdated stock that cannot be sold, whereas maintaining insufficient inventory may result in the inability to meet customer demands. Implementing inventory planning techniques, including prioritization of essential items and utilization of technological tools, can serve as a viable solution to mitigate the impact of fluctuating demand.

Solution—Demand Forecasting: Specific inventory management platforms incorporate tools for forecasting demand. The aforementioned functionality is designed to integrate accounting and sales information to facilitate the anticipation of orders and arrangement of rankings, considering dynamic customer preferences, material availability, and seasonal patterns.

8.13.5 Challenge #5

Problem—Limited Visibility: Inadequate identification or location of inventory within the warehouse may result in incomplete, erroneous, or delayed shipments. Therefore, the acquisition and appropriate inventory allocation are essential in ensuring streamlined warehouse operations and favorable customer satisfaction.

Solution—Add Imagery: Incorporating visual aids, such as images, alongside product descriptions in the inventory database can enhance the efficiency of purchasing and receiving procedures, increase precision, and mitigate the risk of mislaid inventory.

8.13.6 Challenge #6

Problem—Manual Documentation: The utilization of paperwork and manual procedures for inventory management is a laborious and insecure approach. Furthermore, the system's scalability across numerous warehouses with substantial inventory could be improved.

Solution—Go Paperless: It is imperative to provide the appropriate inventory tools to employees for optimal job performance. In addition, the user requires a software solution to automate inventory documentation processes and facilitate digital transactions for invoices and purchase orders.

8.13.7 Challenge #7

Problem—Problem Stock: Specialized plans are required to care for and store perishable and fragile stocks. In addition, high-value inventory necessitates tailored loss-prevention strategies and inventory controls.

Solution—Preventive Control: It is recommended to deploy stock control systems to effectively manage problematic inventory, which may include perishable stock, fragile equipment, or obsolete materials. Per the manufacturer's guidelines, it is also recommended to conduct routine preventative maintenance on machinery and equipment that are kept in storage.

8.13.8 Challenge #8

Problem—Supply Chain Complexity: The dynamic nature of global supply chains necessitates frequent modifications, thereby challenging inventory planning and management operations. The requirement of flexibility and the provision of unpredictable lead times are standard practices by manufacturers and wholesale distributors who exercise control over the shipment of inventory in terms of its timing, destination, and mode of transportation.

Solution—Measure Service Levels: It is imperative to oversee and record supplier information, including but not limited to erroneous shipments, impaired or faulty goods, and unfulfilled delivery schedules. Assessing the performance of suppliers can aid in identifying and resolving supply chain disruptions, simplifying operations, and optimizing logistics.

8.13.9 Challenge #9

Problem—Managing Warehouse Space: Effectively managing space can be a daunting undertaking. The utilization of inventory management platforms in the planning and design of warehouse spaces facilitates improved regulation of the scheduling of new stock deliveries. It can consider significant variables, such as the availability of physical area. Further, explore the distinctions between warehouse management and inventory management.

Solution—Optimize Space: To enhance the efficiency of storage space and inventory flow, utilizing inventory management systems that incorporate warehouse management functionalities is recommended. For example, implement a system of inventory categorization that includes shelf, bin, and compartment levels, and automate the workflows for order picking, packing, and shipping.

8.13.10 Challenge #10

Problem—Insufficient Order Management: Preventing the occurrence of overselling of products and depletion of inventory is a frequently encountered obstacle in the realm of sound inventory management. Using historical and seasonal data patterns can facilitate the precise anticipation of customer orders.

Solution—Automate Reorders: The delay in production caused by inventory being back ordered can result in negative customer experiences. To prevent overselling, it is recommended to utilize inventory management software that can establish automatic reorder points by considering predetermined stock levels and current availability.

8.13.11 Challenge #11

Problem—Increasing Competition: Supply chains that have been globalized are susceptible to unforeseeable economic fluctuations and market dynamics that impact the competition for essential resources. Small enterprises are occasionally confronted with the dilemma of prioritizing between vying for materials that are in high demand or maintaining adequate inventory levels to regulate expenses.

Solution—Safety Stock: It is advisable to uphold safety stock levels to mitigate supply chain disruptions and cope with prolonged lead times that may arise from evolving global competition for raw materials. In addition, effective inventory management facilitates operational flexibility in response to the dynamic nature of global supply chains.

8.13.12 Challenge #12

Solution—Evolving Packaging: Implementing compostable packaging or eliminating packaging to mitigate waste poses novel challenges for warehouse design and storage. This could entail the acquisition of novel equipment or a reduction in the shelf life of certain products.

Solution—Classify Inventory: Developing inventory classifications can be a practical approach to managing dynamic trends, such as implementing packaging strategies to mitigate plastic waste. First, classify stocks based on their packaging type, dimensions, and product. Then, utilize this data to enhance the management of shipping expenses and optimize storage placement.

8.13.13 Challenge #13

Problem—Expanding Product Portfolios: Several e-commerce tactics eliminate the necessity of extensive warehouse distribution centers. The implementation of these tactics facilitates the process of broadening inventory and varying product portfolios. However, it necessitates the utilization of technological advancements and resources for ordering, shipping, and tracking.

Solution—Multi-Location Warehousing: Leverage the capabilities of multi-location warehouse management functionalities to monitor and regulate the growth of inventories effectively. For example, one can optimize inventory management by utilizing automated inventory tracking alerts and scheduling features that facilitate the receipt and storage of schedules while monitoring the location of warehouse inventory and goods in transit.

8.13.14 Challenge #14

Problem—Overstocking: Maintaining excessive inventory levels can pose similar challenges as inadequate inventory levels. In addition, excessive inventory levels can significantly impact a company's cash flow and may result in inventory-related challenges, including storage and loss.

Solution—Leverage Lead Times: It is advisable to consider lead times while placing orders for stocks that are in high demand. Utilize cycle counting data to establish automated reorder points and determine the average lead time to effectively monitor and regulate high-demand inventory, thereby mitigating the risk of stockouts.

8.13.15 Challenge #15

Problem—Inventory Loss: Inventory loss from spoilage, damage, or theft can challenge the supply chain. The process entails the identification, monitoring, and quantification of problematic areas.

Solution—Reduce Human Error: To mitigate the risks of human error, inventory manipu- lation, and shrinkage resulting from theft or negligence, it is recommended to implement inventory control procedures such as blind receiving utilizing barcodes and mobile scanners.

8.13.16 Challenge #16

Problem—Poor Production Planning: Implementing production planning is crucial in mitigating manufacturing delays and cost overruns. Bad execution may hurt sales projections and project timetables.

Solution—Plan Demand: Employing an inventory management system that incorporates sophisticated demand forecasting and reporting capabilities can enable the prioritization of primary inventory. It is advisable to consider the accessibility of the highest performing 20% of stock, which is responsible for 80% of the demand from your clientele.

8.13.17 Challenge #17

Problem—Lack of Expertise: The task of locating proficient inventory managers who possess proficiency in contemporary technology and can enhance inventory strategy can be challenging. More than merely improving the inventory management platform with many functionalities is required. Competent leadership is needed.

Solution—Subcontract Expertise: Consider delegating inventory management tasks to a proficient external specialist. Engage in face-to-face training sessions and offer virtual assistance to assist staff in adhering to optimal procedures for utilizing technical inventory management software functionalities.

8.13.18 Challenge #18

Problem—Poor Communication: Effective communication and collaboration are essential components of success. The need or more need for more information sharing among departments poses a significant challenge in identifying inventory trends and devising strategies for improvement.

Solution—Dashboard Collaboration: Implement dashboards featuring user-friendly interfaces that display up-to-date inventory information in real time. Integrating accounting, sales, and warehouse operations onto a single screen facilitates the elimination of communication barriers.

8.13.19 Challenge #19

Problem—Inefficient Processes: When the inventory is limited, and there is only one warehouse to oversee, implementing low-tech, manual inventory management procedures is not a formidable obstacle. However, as sales volume rises and inventory expands, standard operating systems that could be more efficient, labor-intensive, and low-tech become challenging to scale.

Solution—Productivity Tools: The entirety of the inventory-related data can be conveniently accessed through a portable electronic device. In addition, the utilization of mobile solutions and cloud-based software enables the management of inventory and enhancement of warehouse productivity to be conducted remotely from any location across the globe.

8.13.20 Challenge #20

Problem—Inadequate Software: In order to facilitate intricate logistics, it is imperative that the inventory management software be scaled to incorporate integration with pre-existing business process platforms. The challenging endeavor involves selecting from a plethora of inventory management solutions and acquiring proficiency in a multitude of functionalities that necessitate continuous education and assistance.

Solution—Update Platforms: Adopting a cloud-based inventory management platform offers access to the latest functionalities and confers additional benefits. One can leverage the vendor's proficiency and instruction during the implementation process.

8.14 PHARMACEUTICAL INVENTORY MANAGEMENT SYSTEM

8.14.1 Overview

Irrespective of whether a pharmacy is an independent entity or a constituent of a medical institution, it is tasked with managing a vast and ever-evolving supply chain network. The costs associated with the supply of pharmaceuticals are susceptible to various factors that can lead to an expense escalation and consequent adverse effects on profitability. Therefore, it is imperative to evaluate the inventory forecast of stockings thoroughly. While most pharmaceutical products possess a moderate shelf life, excessive inventory can reduce their worth—poor stocking results in missed sales opportunities. In addition, a pharmacist may be required to address critical factors such as appropriate storage, expiration dates, and other related considerations. Therefore, efficient inventory management is a crucial aspect that pharmacists must prioritize, as per a source.[10] Pharmacy inventory management is a discipline that involves the efficient management of pharmacy inventory. This process or system aids the manager/owner in reducing costs, enhancing operational efficiency, and minimizing overstocking and opportunity loss. To oversee inventory proficiently, an individual must possess comprehensive insight into the current stock, approximate the quantity of stock being held, and forecast future demand.

8.14.2 Importance of inventory management in the pharmaceutical industry

Effective pharmacy inventory management is a crucial prerequisite for ensuring system profitability. Improving inventory management can prevent system disorganization, facilitate better comprehension of market trends, and enable the implementation of cost-saving measures. In addition, identifying significant trends is imperative for pharmacies, as trends tend to be localized and may not align with those of a broader geographic scope.

Moreover, it facilitates the creation of precise forecasts regarding future possibilities. A pharmacy can establish a sales trend by recording and analyzing historical inventory data over several years. It has the potential to provide significant insights into forthcoming sales. An effective inventory management system accurately tracks stocked items and their corresponding prices.

8.14.3 Pharmacy inventory management methods

Inadequate inventory management systems need to notify and restock inventory promptly. Hence, professional inventory management holds significant importance. Several methods exist for managing pharmacy inventory, including but not limited to the following.

8.14.4 The visual method

This inventory management method involves the pharmacist or a designated individual physically inspecting the pharmacy's inventory and comparing the stock on hand to the product list and corresponding quantities. A procurement request is initiated upon inventory depletion below the specified threshold level.

8.14.5 The periodic method

Pharmacists engage in routine stock-taking and quality assessment by cross-referencing the inventory against the product catalog. Then, upon necessity, the pharmacist initiates a procurement request. Multiple variables exert an influence on the inventory management process and necessitate careful contemplation before settling on a particular approach for inventory control. Various factors can influence inventory management, including the nature of the product (e.g., generic or brand), inventory size, policies related to product returns or replacements, inventory shrinkage, unclaimed prescriptions, and other considerations. Generic products are comparatively less expensive than their brand-name counterparts regarding acquisition cost. Typically, smaller pharmacies exhibit a reduced inventory size and may omit certain product lines. The inventory management process should consider manufacturers, vendors, and the pharmacy's product return policies. The reduction in inventory due to various factors results in a noteworthy decline in revenue and therefore warrants consideration. In addition, shrinkage resulting from employee theft or improper handling of medicine stocks necessitates implementing various security measures to mitigate the issue. Hence, major and minor pharmacies must implement a robust inventory management system for their pharmaceutical products.

8.14.6 The perpetual inventory management method

The utilization of this methodology is commonly adopted owing to its notable effectiveness in the management of pharmaceuticals. A computer-based system is employed to observe and record changes in inventory levels. Furthermore, computer systems enable the digital preservation of inventory records that are interconnected with other systems. Any completed purchase order deducts the corresponding quantity from the inventory upon submitting the stock.

8.14.7 The hybrid method

This methodology involves the integration of all three methods. The perpetual inventory system is employed with the yearly physical inventory count to streamline inventory management. Through this methodology, a pharmacist compares the quantity of merchandise on the shelves and the corresponding data recorded in the computerized system. This methodology enables the correction of fluctuations in the levels of supply and demand, as well as the verification of inventory documentation.

8.14.8 Factors affecting pharmacy inventory management

The inventory management process is subject to the influence of various factors, which require thorough consideration before selecting a specific inventory control strategy. Multiple factors can influence inventory management, such as the product's characteristics (e.g., generic or branded item), inventory size, product returns or replacement policies, inventory shrinkage, unclaimed prescriptions, and other relevant considerations. Regarding acquisition cost, generic products are relatively more affordable than their brand-name equivalents. Smaller pharmacies generally demonstrate a diminished inventory magnitude and may exclude specific product categories. The process of managing inventory should consider the policies of product return of the pharmacy and the manufacturers and vendors involved. The decrease in inventory levels, attributed to multiple factors, leads to a significant revenue decline, necessitating careful consideration. Furthermore, the reduction in inventory due to employee pilferage or inadequate management of pharmaceutical stocks mandates the adoption of diverse security protocols to address the concern. Therefore, primary and secondary pharmaceutical establishments must adopt a sturdy inventory management system for their medicinal merchandise.

8.14.9 Shaping pharmacy inventory management

Employing advanced digital technologies enhances inventory management in terms of security, cost effectiveness, and efficiency. Presently, a considerable number of pharmacies utilize a pharmacy inventory management system, a technological solution that aids the pharmacy proprietor in managing inventory and operations in a cost-efficient manner.

Pharmacy inventory management is currently offered in the software-as-a-service model and is powered by artificial intelligence and cloud technologies. This results in a highly adaptable and agile solution with outstanding features and data analysis capabilities. In the contemporary landscape of pharmaceutical medicine, characterized by a proliferation of novel drugs entering the market, it is imperative for pharmacies to implement an inventory management system to mitigate the risks of product leakage, wastage, and missed sales opportunities.

In addition to periodic inventory control methods, there exist other technologies, such as specific scanning devices, which enhance the process by scanning barcode labels located on packages and shelves. The system serves to facilitate electronic order purchases at the individual household level, as well as e-procurement on a larger scale.

8.14.10 IoT in pharmaceutical manufacturing

The pharmaceutical industry's critical responsibilities are maintaining accurate temperatures and tracking and minimizing risk factors during the journey of medicinal products. From a consumer standpoint, it is imperative that the products they procure meet the requisite standards of quality and safety for their intended use. The cold chain system is commonly observed in the pharmaceutical sector, as Jedermann et al. [51] noted in their study on dynamic systems. Rodrigue et al.

[52] define the cold chain as transporting temperature-sensitive commodities along a supply chain by employing thermal and refrigerated packaging methods and strategic logistical planning to ensure the integrity of these shipments. The cold-chain system encompasses procedures, including preparation, storage, transportation, and temperature monitoring of perishable commodities transported from suppliers to consumers [52, 53]. Pharmaceutical products are temperature-sensitive commodities that require lower temperatures to alleviate chemical reactions.

The pharmaceutical industry faces a significant obstacle in logistics, specifically the inadequate and inconsistent transfer of information while transporting pharmaceutical products, resulting in untimely and imprecise delivery. The current state of affairs makes it challenging to achieve instantaneous visibility and traceability of the precise location of commodities throughout the entire supply chain. The IoT, a nascent technology that operates through the Internet, was introduced to address this challenge. The IoT is a worldwide network infrastructure that interconnects tangible and intangible entities with the Internet's cloud-based system. The linkage facilitates the acquisition of real-time data from commodities affixed with sensor apparatuses within the shipping container during transit and subsequently transmits the pertinent updates to the relevant destination. The IoT provides a system for identifying specific objects and determining their current location within the global network. RFID and wireless sensing devices are widely regarded as the principal technologies facilitating the IoT as a system for generating value. In forthcoming times, RFID tags will be furnished with sensors and additional peripherals to facilitate interaction with the network, monitor their ambient surroundings, and acquire geographical data. Pharmaceutical products are affixed with innovative tags with a specific identification number or IP address.

8.14.11 IoT–Pharma architecture

According to Smith [54], sectors such as fresh food, floriculture, pharmaceuticals, and chemicals are classified as temperature sensitive due to their perishable nature. These industries necessitate the implementation of monitoring and tracking systems that operate in real time. Perishable industries face greater complexity in implementing online tracking systems and making real-time decisions. The physical infrastructure necessary to maintain appropriate temperature conditions is a prerequisite for the cold chain, which is contingent upon the characteristics of the products being transported. Specialized physical facilities such as warehouses, loading and unloading facilities, and refrigeration units are necessary to control live temperature. In addition, the logistics procedure must uphold the consignment's consistency and monitor any temperature deviations [52].

To implement the IoT–Pharma system, which is capable of continuously monitoring and tracking the quality and safety of medications throughout the entirety of the PSC, from production to the end consumer, several prerequisites must be taken into account. The effective management of goods necessitates the intelligent regulation of visibility, traceability, accuracy, and controllability at all stages of packaging or transportation—the implementation of RFID technology in containers, as depicted in Figure 8.10, can offer a range of sensing capabilities, energy management features, processing and storage capabilities, and wireless sensor interfaces. RFID technology encompasses both chipless RFID tags and reusable active wireless sensor nodes. The choice between the two depends on the characteristics of the goods being transported and the mode of transportation. The conventional RFID technology is inadequate to handle the comprehensive configurations of the IoT platform. Thus, there is a need for a networked platform that can accommodate various sensor devices and the classification of tags while also supporting heterogeneous radio interfaces and functionalities.

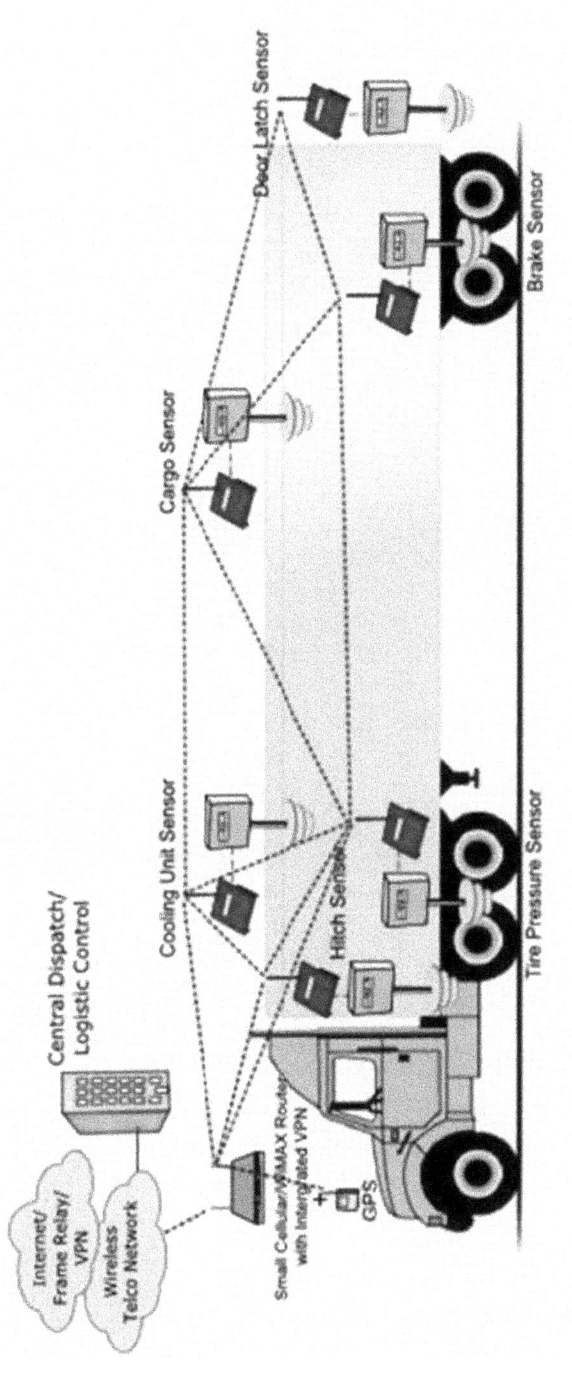

Figure 8.10 IoT-enabled smart container.

The figure denoting an intelligent RFID tag (Figure 8.11) and a two-layer system architecture can be observed in Figure 8.12. This tracking and monitoring system's architecture exhibits distinctiveness in the pharmaceutical goods logistics domain. The RFID layer serves as the initial layer that functions as an asymmetric connection between the tag and reader. The wireless sensor network layer serves as an ad hoc network among master nodes. The

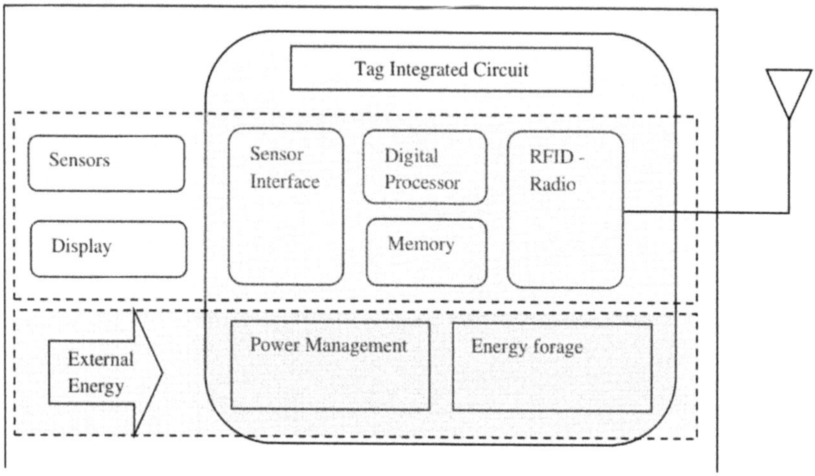

Figure 8.11 Intelligent RFID tag.

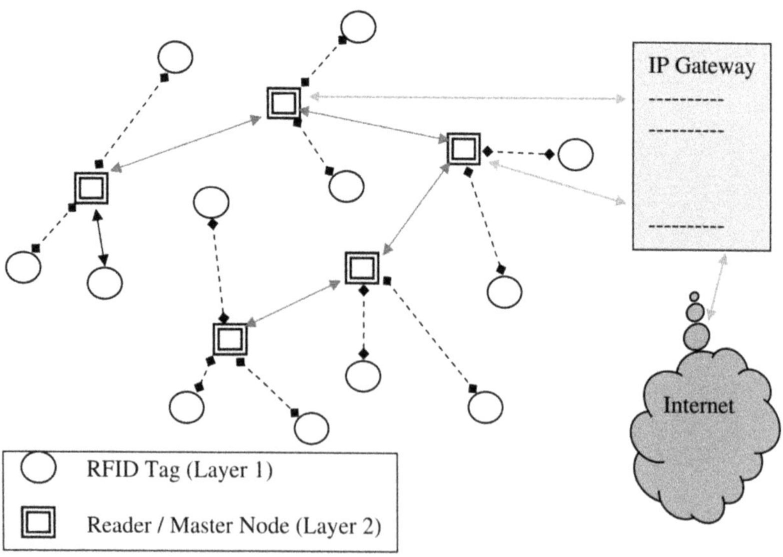

Figure 8.12 RFID tag and reader node with a hierarchical two-layer network.

coordination of RFID tags is carried out by master nodes situated at Layer 2. Layer 2 facilitates the formation of ad hoc networks, which are self-organized among master nodes and function as a wireless sensor network. The RFID layer is a heterogeneous network that features asymmetrical links and offers a wide variety of RFID options. Integrating intelligent RFID tags within pharmaceutical products allows for seamless communication with a designated reader, or master, through the utilization of micro-power wireless connections. The intelligent RFID tag, depicted in Figure 8.3, typically comprises a power management and energy harvesting system, a digital processor, sensor interfaces and memory, and a radio transceiver. The RFID's radio interface includes either near-field coupling or far-field backscattering, which operate within the HF or UHF bands. An example of a technological application is integrating an RFID tag with ultra-wideband (UWB) technologies, which can be used for precise positioning, rapid detection, and time-domain sensing. A chipless title capable of sensing environmental changes, such as temperature and humidity, through RF measurements of impedance changes is a viable option for printed electronics.

Using tags and sensors within this stratum is cost effective and energy efficient when operating under conditions of energy autonomy. The primary node functions as a data collector from the tag and establishes a connection to the Internet cloud via conventional wireless interfaces such as Wi-Fi, WiMAX, GSM/GPRS, and 3G. The primary node ought to function as a high-level wireless sensing site, frequently outfitted with advanced sensors such as imaging and chemical sensors and GPS for geospatial monitoring.

8.15 USE CASES

The implementation of IoT technology within the pharmaceutical industry is a groundbreaking tool. This technology facilitates various aspects of the pharmaceutical industry, including manufacturing, warehousing, packaging, and SCM. As a result, implementing the IoT has enhanced the caliber of pharmaceutical products, diminished inaccuracies, and amplified productivity.

The IoT facilitates various operations within the pharmaceutical industry, from the preliminary to the intermediate stages. The system oversees multiple unit operations, monitors real-time data, enhances product efficiency, and increases visibility to improve operational effectiveness.

Assisting pharmaceutical personnel in devising their daily work schedules and enhancing their comprehension of medications and their constituents, the IoT has had a significant impact on the field of pharmacy.

Contemporary pharmacists depend on information technology (IT) and the IoT to facilitate their professional activities. In addition, the pharmaceutical industry has transitioned to technology-based frameworks due to technological advancements within the research and development sector.

IoT and IT have exerted a significant influence on various sectors of pharmaceutical manufacturing and pharma plants, including the following.

8.15.1 Industrial mechanics

The IoT has been instrumental in enhancing the performance of industrial machinery and mechanics. For example, the utilization of pharmaceutical IoT monitoring sensors facilitates monitoring machinery and consolidating pertinent facility data into a unified dashboard.

Supervisors can detect abnormal conditions and identify the necessary maintenance requirements through observation. In addition, they establish connections with automatic shutoff systems and manage crucial mechanical circumstances.

8.15.2 Material tracking

Material tracking is another benefit of implementing IoT technology in the pharmaceutical industry. The IoT facilitates the connection of devices to the Internet and network, enabling the seamless tracking of material availability in real time by the connected devices. Implementing this system promotes improved inventory management and cost reduction by regulating waste production.

Sensors facilitate the monitoring of the origin of resources, which ensures the preservation of both production efficiency and output quality. Sensors and data-gathering devices are strategically positioned at shipping and receiving stations. The system acquires data from RFID tags and barcodes.

It establishes a correlation between data obtained from various sources. The study centers on scrutinizing production and warehouse facilities to ascertain the consistency of the data.

These tracking mechanisms inform manufacturers about market demand and signal when replenishment is necessary. As a result, this measure effectively maintains optimal supply levels by mitigating the necessity of stocking extra supplies and the potential occurrence of supply depletion during crucial periods.

8.15.3 Logistics

Upon completion of the manufacturing process, the pharmaceutical industry proceeds to convey the finished products from the production site to the marketplace. Sensors are installed to monitor the completed goods and the supply chain.

The embedded computers present on the Internet facilitate the identification of the batch. Furthermore, the IoT facilitates the monitoring of product expiration dates and enables the efficient tracking of recalled batches through mobile devices. Therefore, the elimination of those from the supply chain is facilitated.

8.15.4 Optimize the clinical trials

Before its release into the market, any novel pharmaceutical product must undergo clinical trials. IoT–connected devices are utilized for the management of clinical trials. Using IoT methodologies facilitates monitoring experimental pharmaceuticals impact in real-time control.

The immediate availability of data facilitates the identification of unfavorable responses and associated hazards. Simultaneously, it serves to enhance the caliber of the testing data. Furthermore, IoT devices gather information from the network and transfer it to the integrated software.

This procedure facilitates the identification of the patient's medical conditions and generates alerts. Hence, the complete process of conducting clinical trials and streamlining the research and development domain of the organization is facilitated.

8.15.5 Regulatory compliance

It is imperative that pharmaceutical processes, regardless of their scale, are thoroughly established and documented. In addition, IoT-connected devices transmit data to servers in a continuous manner.

The real-time data are subsequently retained and analyzed to ascertain their compliance with established quality standards. Furthermore, this practice diminishes the quantity of physical documentation and mitigates the possibility of inaccuracies.

8.15.6 Smart equipment

Smart wearable devices and specialized mobile applications are an additional implementation of the IoT within the pharmaceutical industry. These medical devices promote patient adherence and provide incentives.

In previous eras, identifying a patient's medical emergency often occurred belatedly, occasionally during the terminal phases. The IoT employs real-time data monitoring to identify early indicators of a significant medical occurrence.

Through maintaining a systematic record of these medical instruments, medical professionals and pharmaceutical industry representatives can gain insight into the prevailing health conditions of the populace, thereby enabling them to forecast the primary medication needs of the market.

8.15.7 Smart pills and implanted devices

Smart pills are a recent innovation in pharmaceutical science, characterized by their diminutive size. Implanted devices are medical devices surgically inserted into a patient's body.

They aid in identifying even the most minute alteration in the individual's physiological state.

This measure serves to mitigate severe hazards and potential dangers.

8.15.8 Rich insights

The data obtained in real time from the interconnected devices within the IoT can be transmitted to software designed for data analytics. This facilitates a comprehensive analysis of the data. In addition, data analytics software facilitates identifying the least proficient and the most effective areas within a given field.

The IoT facilitates the identification of inefficiencies, thereby enhancing the pharmaceutical process by promoting productivity and profitability.

8.15.9 Better physician engagement

IoT technology enables pharmaceutical industries to establish a more efficient and secure communication channel with physicians by deploying a virtual representative on each desk. The constructive interaction with medical practitioners facilitates efficient time management and enables tailored interventions.

Physician collaboration facilitates the analysis of pharmaceuticals' crucial and efficacious characteristics and properties. As a result, the physician consistently possesses superior insights about patient engagement with drugs.

8.16 CONCLUSION

The IoT facilitates various aspects of the pharmaceutical industry, ranging from preventative maintenance to supply management. Furthermore, excluding pharmaceutical manufacturing, the implementation of IoT technology has resulted in advancements in medical science by providing benefits such as e-prescriptions, online consultations with medical professionals, advanced therapies, and treatments. Therefore, shortly, the IoT will address the conventional limitations and disadvantages of the pharmaceutical industry, ultimately facilitating its peak performance.

NOTES

1 https://news.un.org/en/story/2019/02/1033191
2 http://bccpharmacytech.weely.com/uploads/7/5/0/4/75048
3 https://www.statista.com/study/10708/us?pharmaceutical-industry-statista-dossier/
4 https://www-file.huawei.com/-/media/corporate/minisite/mwc2016/pdf/narrowband-iot-wide-range-of-opportunities-en.pdf?la=en
5 https://www.gzsyjjc.com/documents/referenced/white-papers/the-many-flavors-of-bluetooth-iot-connectivity.pdf

6 www.fda.gov/
7 https://www.investopedia.com/terms/i/inventory-management.asp#toc-what-are-the-four-main-types-of-inventory-management
8 https://www.purchasing-procurement-center.com/principles-of-inventory-management.html
9 https://www.netsuite.com/portal/resource/articles/inventory-management/inventory-management-challenges.shtml
10 https://www.karexpert.com/blogs/pharmacy-inventory-management

REFERENCES

[1] M. Mikulic, "Global spending on medicines in 2010, 2019, and a forecast for 2024," 2020.
[2] W. H. Organization *et al.*, "Access to medicines and health products programme: Annual report 2020.," tech. rep., World Health Organization. Regional Office for Europe, 2021.
[3] E. Osei-Mensah, *The effect of inventory management practices on service delivery at St. Martin's Hospital, Agroyesum, Amansie-West*. PhD thesis, Kwame Nkrumah University of Science and Technology, 2016.
[4] H. Tadeg, E. Ejigu, E. Geremew, and A. Adinew, "Auditable pharmaceutical transactions and services (apts): Findings of the baseline assessment at federal, addis ababa, and teaching hospitals," *Submitted to the US agency for international development by the systems for improved access to pharmaceuticals and services (SIAPS) program. Arlington: Management Sciences for Health*, 2014.
[5] D. Reddy, D. Sai, and D. Prabhu, "A study on the selective controls of inventory management and application of abc xyz control matrix in the cardiology department of a tertiary care hospital," *Journal of Dental and Medical Sciences*, vol. 16, no. 05, pp. 06–9, 2017.
[6] T. G. Gurmu and A. J. Ibrahim, "Inventory management performance of key essential medicines in health facilities of east shewa zone, oromia regional state, ethiopia," *Cukurova Medical Journal*, vol. 42, no. 2, pp. 277–291, 2017.
[7] D. Kritchanchai and W. Meesamut, "Developing inventory management in hospital," *International Journal of Supply Chain Management*, vol. 4, no. 2, pp. 11–19, 2015.
[8] J. R. Stock and D. M. Lambert, *Strategic logistics management*, vol. 4. McGraw-Hill/Irwin Boston, MA, 2001.
[9] Y. Kumar, R. K. Khaparde, K. Dewangan, G. K. Dewangan, J. S. Dhiwar, and D. Sahu, "Fsn analysis for inventory management–Case study of sponge iron plant," *International Journal for Research in Applied Science & Engineering Technology*, vol. 5, no. 2, pp. 53–57, 2017.
[10] S. B. Pund, B. M. Kuril, S. J. Hashmi, M. K. Doibale, and S. Doifode, "Abc-ved matrix analysis of government medical college, aurangabad drug store," *International Journal of Community Medicine and Public Health*, vol. 3, no. 2, pp. 469–472, 2016.
[11] K. Wiedenmayer, R. S. Summers, C. A. Mackie, A. G. Gous, M. Everard, D. Tromp, W. H. Organization, *et al.*, "Developing pharmacy practice: A focus on patient care: Handbook," tech. rep., World Health Organization, 2006.

[12] S. Aheleroff, X. Xu, Y. Lu, M. Aristizabal, J. P. Velásquez, B. Joa, and Y. Valencia, "Iot- enabled smart appliances under industry 4.0: A case study," *Advanced Engineering Informatics*, vol. 43, p. 101043, 2020.
[13] J. Lee, B. Bagheri, and H.-A. Kao, "A cyber-physical systems architecture for industry 4.0- based manufacturing systems," *Manufacturing Letters*, vol. 3, pp. 18–23, 2015.
[14] J. Lee, E. Lapira, B. Bagheri, and H.-A. Kao, "Recent advances and trends in predictive manufacturing systems in big data environment," *Manufacturing Letters*, vol. 1, no. 1, pp. 38– 41, 2013.
[15] Y.-C. Lin, M.-H. Hung, H.-C. Huang, C.-C. Chen, H.-C. Yang, Y.-S. Hsieh, and F.-T. Cheng, "Development of advanced manufacturing cloud of things (amcot)—A smart manufacturing platform," *IEEE Robotics and Automation Letters*, vol. 2, no. 3, pp. 1809–1816, 2017.
[16] T. S. Ing, T. Lee, S. Chan, J. Alipal, and N. A. Hamid, "An overview of the rising challenges in implementing industry 4.0," *International Journal of Supply Chain Management*, vol. 8, no. 6, pp. 1181–1188, 2019.
[17] C. Sun, "Application of RFID technology for logistics on internet of things," *AASRI Procedia*, vol. 1, pp. 106–111, 2012.
[18] L. He, M. Xue, and B. Gu, "Internet-of-things enabled supply chain planning and coordination with big data services: Certain theoretic implications," *Journal of Management Science and Engineering*, vol. 5, no. 1, pp. 1–22, 2020.
[19] K. Ashton *et al.*, "That 'internet of things' thing," *RFID Journal*, vol. 22, no. 7, pp. 97–114, 2009.
[20] L. Atzori, A. Iera, and G. Morabito, "The internet of things: A survey," *Computer Networks*, vol. 54, no. 15, pp. 2787–2805, 2010.
[21] X. Sun, K. Xu, X. Shen, Y. Li, Y. Dai, J. Wu, and J. Lin, "New hierarchical architecture for ubiquitous wireless sensing and access with improved coverage using CWDM-ROF links," *Journal of Optical Communications and Networking*, vol. 3, no. 10, pp. 790–796, 2011.
[22] J. H. Watanabe, "Examining the pharmacist labor supply in the united states: Increasing medication use, aging society, and evolution of pharmacy practice," *Pharmacy*, vol. 7, no. 3, p. 137, 2019.
[23] M. P. Singh, *The pharmaceutical supply chain: A diagnosis of the state-of-the-art*. PhD thesis, Massachusetts Institute of Technology, 2005.
[24] L. Hunt III, I. B. Murimi, J. B. Segal, M. J. Seamans, D. O. Scharfstein, and R. Varad-Han, "Brand vs. generic: Addressing non-adherence, secular trends, and non-overlap," arXiv preprint arXiv: 1907.05385, 2019.
[25] P. Morris and E. Sweeney, "Responding to disruptions in the pharmaceutical supply chain," *Clinical Pharmacist*, vol. 11, no. 2, 2019.
[26] N. Shah, "Pharmaceutical supply chains: Key issues and strategies for optimisation," *Computers & Chemical Engineering*, vol. 28, no. 6–7, pp. 929–941, 2004.
[27] P. Krishnamurthy, A. Prasad, *et al.*, *Inventory strategies for patented and generic products for a pharmaceutical supply chain*. PhD thesis, Massachusetts Institute of Technology, 2012.
[28] K. M. Iacocca and S. Mahar, "Cooperative partnerships and pricing in the pharmaceutical supply chain," *International Journal of Production Research*, vol. 57, no. 6, pp. 1724–1740, 2019.

[29] F. J. Beier, "The management of the supply chain for hospital pharmacies: A focus on inventory management practices," *Journal of Business Logistics*, vol. 16, no. 2, p. 153, 1995.
[30] O. Aigbogun, Z. Ghazali, and R. Razali, "Resilience attributes of halal logistics on the pharmaceutical supply chain.," *Global Business and Management Research: An International Journal*, vol. 7, no. 3, pp. 268–284, 2015.
[31] E. A. R. D. Campos, I. C. D. Paula, R. N. Pagani, and P. Guarnieri, "Reverse logistics for the end-of-life and end-of-use products in the pharmaceutical industry: A systematic literature review," *Supply Chain Management: An International Journal*, vol. 22, no. 4, pp. 375–392, 2017.
[32] J. De Vries, "The shaping of inventory systems in health services: A stakeholder analysis," *International Journal of Production Economics*, vol. 133, no. 1, pp. 60–69, 2011.
[33] E. S. Schneller and L. R. Smeltzer, *Strategic management of the health care supply chain*. Jossey-bass, 2006.
[34] H. Rivard-Royer, S. Landry, and M. Beaulieu, "Hybrid stockless: A case study: Lessons for health-care supply chain integration," *International Journal of Operations&Production Management*, vol. 22, no. 4, pp. 412–424, 2002.
[35] S. D. Lapierre and A. B. Ruiz, "Scheduling logistic activities to improve hospital supply systems," *Computers& Operations Research*, vol. 34, no. 3, pp. 624–641, 2007.
[36] C. I. Papanagnou and O. Matthews-Amune, "Coping with demand volatility in retail pharmacies with the aid of big data exploration," *Computers& Operations Research*, vol. 98, pp. 343–354, 2018.
[37] J. Holdowsky, M. Mahto, M. E. Raynor, and M. Cotteleer, "Inside the internet of things (IoT)," *Deloitte Insights*, vol. 21, pp. 1065–1079, 2015.
[38] J. Holdowsky, M. Mahto, M. Raynor, and M. Cotteleer, "A primer on the technologies building the iot," *Inside the Internet of Things (IoT)*, vol. 10, pp. 15–32, 2015.
[39] Y. Liu, K.-F. Tong, X. Qiu, Y. Liu, and X. Ding, "Wireless mesh networks in IoT networks," in *2017 International workshop on electromagnetics Applications and student innovation competition*, pp. 183–185, IEEE Access, 2017.
[40] V. D. Hunt, A. Puglia, and M. Puglia, *RFID: A guide to radio frequency identification*. John Wiley & Sons, 2007.
[41] A. Kerr and A. Orr, "Iot-based inventory tracking in the pharmaceutical industry," 2020, https://hdl.handle.net/1721.1/126490
[42] H. Lawrence, *Introduction to bluetooth technology market operation profiles and services*. Lightning Source Inc, 2004.
[43] N. Carroll, "Accounting for inventory and cost of goods sold," *Financial management for pharmacists*, 2nd ed. Williams & Wilkins, 1998.
[44] S. P. Desselle, D. P. Zgarrick, and G. L. Alston, *Pharmacy management: Essentials for all practice settings*. McGraw-Hill, 2005.
[45] A. Bouldin, E. Holmes, D. Garner, and A. Devoe, "Purchasing and managing inventory," Chrisholm-Burns MA, Vaillancourt AM, Shepherd M. *Pharmacy management, leadership, marketing and finance*. Jones & Bartlett Publishers, LLC, 2011.
[46] D. Huffman, "Purchasing and inventory control," *Effective pharmacy management*, 8th ed. National Association of Retail Druggists, 1996.
[47] S. Silbiger, *The ten-day mba*, rev. ed. Quill William Morrow, 1999.

[48] D. McCaffrey, M. Smith, B. Banahan, D. Frate, and F. Gilbert, "A continued look into the financial implications of initial noncompliance in community pharmacies: An unclaimed prescription audit pilot," *Journal of Research in Pharmaceutical Economics*, vol. 9, pp. 33–58, 1998.
[49] T. Awaya, K.-I. Ohtaki, T. Yamada, K. Yamamoto, T. Miyoshi, Y.-I. Itagaki, Y. Tasaki, N. Hayase, and K. Matsubara, "Automation in drug inventory management saves personnel time and budget," *Yakugaku Zasshi*, vol. 125, no. 5, pp. 427–432, 2005.
[50] A. S. of Health-System Pharmacists *et al.*, "ASHP statement on bar-code verification during inventory, preparation, and dispensing of medications," *American Journal of Health-System Pharmacy: AJHP: Official Journal of the American Society of Health-System Pharmacists*, vol. 68, no. 5, pp. 442–445, 2011.
[51] R. Jedermann, L. J. A. Congil, M. Lorenz, J. D. Gehrke, W. Lang, and O. Herzog, "Dynamic decision making on embedded platforms in transport logistics–A case study," in *Dynamics in Logistics First International Conference, LDIC 2007*, Bremen, Germany, August 2007, *Proceedings*, pp. 191–198, Springer, 2008.
[52] J.-P. Rodrigue and T. Notteboom, "3.1–transportation and economic development," *The geography of transport systems*, 2013.
[53] L. Li, *Managing supply chain and logistics: Competitive strategy for a sustainable future*. World Scientific Publishing Company, 2014.
[54] J. N. Smith, "Specialized logistics for a longer perishable supply chain," *World Trade*, vol. 18, no. 11, p. 46.

Chapter 9

Decentralized file sharing system based on IPFS and blockchain

Mahesh Vanam, Margani Rohith, N. Sai Chandu, Mudavath Sai Teja, and Bhanu Chander
Indian Institute of Information Technology Kottayam, Kottayam, India

9.1 INTRODUCTION

File sharing has become an indispensable aspect of our lives in this modern digital era. We share files for various reasons, such as collaborating with colleagues, exchanging information, or simply sharing media files with friends and family. However, the current file sharing systems rely on centralized services that manage, store, and access our files. This centralized approach poses several limitations and risks; to overcome these limitations, decentralized file sharing systems have emerged as a viable alternative. A decentralized system enables users to share files without relying on centralized services. This approach ensures that the file sharing system is secure, censorship-resistant, and accessible to all users, regardless of their geographic location. Decentralized systems also provide greater control over the data, enabling users to maintain their privacy.

Decentralized file sharing has gained popularity in recent years due to its benefits in terms of security, privacy, and efficiency. Two technologies that have shown a lot of promise in this field are blockchain and InterPlanetary File System (IPFS), while elliptic curve cryptography (ECC) encryption can enhance security. Blockchain technology provides a secure, transparent, and tamper-proof method for recording transactions and ensuring data integrity. In a decentralized file sharing system, blockchain can be used to manage the file sharing process, enforce permissions and conditions, and ensure authenticity and integrity of the files. IPFS is a distributed file system that provides efficient, decentralized storage and retrieval of files based on content addressing. By using IPFS in a decentralized file sharing system, files can be stored and retrieved efficiently and reliably, without the need for a central server or intermediary. ECC encryption is a public key encryption method that uses the mathematics of elliptic curves to create keys that are more secure and efficient than traditional encryption methods.

By integrating ECC encryption into a decentralized file sharing system, the system can ensure that files are protected from unauthorized access or interception during transfer, providing an additional layer of security. Combining blockchain, IPFS, and ECC encryption provides a highly secure, efficient, and distributed method for sharing files without relying on a

DOI: 10.1201/9781003474524-9

central server or intermediary. The system can be designed to provide specific permissions and conditions for file sharing while ensuring data integrity, authenticity, efficient storage and retrieval, and resistance to censorship and data loss. This technology has the potential to transform the way data is stored and exchanged in various industries.

9.1.1 Blockchain

Blockchain is a distributed digital ledger that enables secure, transparent, and tamper-proof transactions between parties without the need for intermediaries. It was first introduced in 2008 as the underlying technology behind Bitcoin, the world's first decentralized digital currency. However, the potential applications of blockchain technology extend far beyond cryptocurrency, and it is now being adopted across various industries, from finance and healthcare to supply chain management and voting systems. The main concept of blockchain is to create a network of nodes that work together to verify and store a record of transactions.

Once a transaction is added to the blockchain, it cannot be altered or deleted, making the ledger tamper-proof and transparent. This is achieved through a consensus mechanism, where a network of nodes must agree on the validity of a transaction before it can be added to the blockchain. The blocks in a blockchain are cryptographically linked together, forming a secure and tamper-proof chain of transactions, documents, and other information. This makes it nearly impossible for a bad actor to manipulate the data or forge new transactions. Moreover, the decentralized nature of the blockchain means that there is no central point of control, making it resilient to attacks and failures. Blockchains have found numerous applications beyond their original use case in cryptocurrencies. They are being explored for use in supply chain management, voting systems, identity verification, and many other fields where transparency, security, and immutability are crucial. The potential of blockchain technology is immense and its impact on various industries is still unfolding.

9.1.2 IPFS

IPFS, or the Interplanetary File System, is a distributed storage system that enables the storage of files and versions of data over time. It works similarly to Git and BitTorrent by keeping track of them on a decentralized network. With IPFS, users can directly interact with a secure and global peer-to-peer (P2P) network, allowing for a permanent new internet and enhancing the way we use existing web protocols like HTTP. Since its introduction in 2016, IPFS has seen significant improvements and adoption by both individuals and businesses, allowing users to share files and data without limitations. It is particularly useful for large files that may require high bandwidth to upload and/or download over the internet. IPFS uses a distributed hash table

(DHT) to store data. Once we have a hash, we ask the peer network who has the content located at that hash, and we download the content directly from the node that has the information we need. However, the popularity and effectiveness of IPFS as a distributed file system also create security and access control concerns. In a distributed network like IPFS, when an item is loaded onto the network, anyone who has access to the file's hash address can access its content.

This is where blockchain comes in as a decentralized data management platform that provides immutability, making it an excellent choice to assist with file traceability metadata on a distributed file system like IPFS. IPFS is a protocol and peer-to-peer network for storing and sharing data in a distributed file system. It uses content-addressing to uniquely identify each file in a global namespace connecting all computing devices, and uses DHT to spread the data across a network of computers, effectively coordinated to enable efficient access and analysis among nodes. Unlike location-based addressing, content-based addressing makes it difficult to remove data from the network once it has been uploaded because it can be pinned on multiple servers. The backbone that holds everything together is IPFS, a hypermedia protocol designed to make the web more resilient by addressing data by its content instead of its location. To achieve this, IPFS uses CIDs instead of URLs, which point to the server where the data is hosted.

9.1.3 ECC

Elliptic-curve cryptography (ECC) is a type of asymmetric/public-key cryptography that is used to secure digital communications and transactions. The core concept of ECC is to use mathematical functions to generate two related keys, one public and one private. The public key is used to encrypt data, while the private key is used to decrypt it. Unlike traditional cryptography methods, ECC provides the same level of security with shorter key lengths, making it a more efficient and effective method of encryption. One of the main benefits of ECC is its ability to provide strong encryption while using fewer resources.

Elliptic Curve equation, $y^2 = x^3 + ax + b$, For Operation values will be given as $E_p(a,b)$. Where, p will be the maximum permitted value toward +ve X-Axis direction, and a,b are the coefficients in the elliptic curve equation.

When data or information is shared over the Internet, it goes through numerous network devices, which are part of the public Internet. Unfortunately, this presents an opportunity for hackers to intercept and steal data. To prevent this, users can use software or hardware that ensures secure data transmission. This process is known as encryption in network security. Encryption involves converting human-readable text, also known as plaintext, into unreadable text called ciphertext. Essentially, this means transforming human-readable data into something that looks random. To achieve this, encryption uses cryptographic keys, which are sets of mathematical values agreed upon by the sender and receiver. The recipient uses the same key to decrypt the data and convert it back into readable plaintext. Encryption keys can be made more complex to enhance security. This approach makes it less likely for a third party to crack the encryption by using a brute force attack. A brute force attack involves guessing random numbers until the correct combination is discovered. Additionally, encryption is also used to protect passwords by scrambling them to make them unreadable to hackers.

Problem Statement The decentralized file sharing system will enable users to share files without relying on centralized services. The system will be designed to be censorship resistant, ensuring that users can share and access files without any restrictions. The system will also provide greater control over the data, enabling users to maintain their privacy and ensure that their data is secure. The goal of this project is to design and implement a decentralized file sharing system that is based on IPFS and blockchain technology. The decentralized file sharing system aims to address the limitations and risks associated with centralized file sharing systems, such as privacy concerns, security issues, and the need to trust third-party services. The system will leverage the benefits of IPFS and blockchain to provide users with a reliable, efficient, and secure file sharing solution. The project will involve designing and implementing a distributed file storage system using IPFS. The system will use content addressing to ensure that data is always available and can be accessed from anywhere in the world. The project will also involve designing and implementing a decentralized network using blockchain technology to ensure that the system is secure and resistant to attacks. The system will be developed to provide users with a seamless and intuitive file sharing experience, while also ensuring that their data remains safe and secure.

9.2 LITERATURE REVIEW

This chapter aims to provide an in-depth analysis of the existing research and developments in the field by exploring various aspects of decentralized file sharing, including the underlying technologies such as IPFS and blockchain, the challenges faced in developing such systems, and the potential benefits they offer. It will also examine the current state of the art in the field, identify gaps in the research, and suggest potential areas for future investigation. By providing a comprehensive overview of the literature, this chapter will serve as a foundation for the development of a novel and effective decentralized file sharing system based on IPFS and blockchain.

Rui Guo et al. proposed a secure storage system for electronic health records (EHRs) using the Hyperledger blockchain in edge nodes. The proposed system aims to address the security and privacy concerns related to EHRs by ensuring that only authorized individuals have access to the data.

The authors use a private blockchain based on Hyperledger Fabric and design a smart contract to enforce access control policies. The system is deployed on edge nodes to reduce latency and improve scalability. The authors evaluated the proposed system's performance and demonstrated that it can efficiently handle EHR data while ensuring secure and authorized access. Overall, the chapter presents a promising solution for secure storage and management of EHRs using blockchain technology. **Masayuki Fukumitsu et al.** proposed a secure P2P storage scheme that uses secret sharing and blockchain technology to protect data confidentiality and integrity. The authors believe that traditional centralized storage systems have vulnerabilities that make them vulnerable to data breaches and attacks. They proposed a distributed approach that uses secret sharing to divide data into fragments and distribute them across multiple nodes. The nodes are connected via blockchain network to ensure data integrity and prevent tampering. The authors present a prototype implementation of the proposed scheme and evaluate its security and performance. They demonstrate that the proposed scheme can provide secure and efficient storage for sensitive data. Overall, the chapter presents a promising solution for secure P2P storage using a combination of secret sharing and blockchain technology. **N. Nizamuddin et al.** proposed a decentralized document version control system that uses Ethereum blockchain and IPFS. The authors believe that traditional centralized document version control systems have limitations, such as single point of failure and lack of transparency. They proposed a decentralized approach that uses Ethereum smart contracts to manage document versions and IPFS to store the documents. The authors describe the system architecture and present a prototype implementation. They evaluated the proposed system's performance and demonstrated that it can efficiently manage document versions while ensuring transparency and security. The chapter highlights the advantages of using a decentralized approach and emphasizes the potential of blockchain and IPFS for document version control. Overall, the chapter presents a promising solution for decentralized document version control using ethereum blockchain and IPFS. **Shangping Wang et al.** proposed a blockchain-based framework for data sharing with fine-grained access control in decentralized storage systems. The authors argue that traditional decentralized storage systems have limitations in managing data access control, which may lead to data breaches and unauthorized access. They propose a framework that uses blockchain technology to manage access control policies and ensure data confidentiality and integrity. The authors describe the system architecture and present a prototype implementation. They evaluate the proposed system's performance and demonstrate that it can efficiently manage data access control while ensuring security and privacy. The chapter highlights the advantages of using a blockchain-based approach for data sharing and emphasizes the potential of blockchain for managing access control policies in decentralized storage systems. Overall, the chapter presents a promising solution for data sharing with fine-grained access control in decentralized storage systems using blockchain technology.

Onur Demir et al. proposed a decentralized file sharing framework that ensures secure and private sharing of sensitive data. The proposed system is built on top of blockchain technology, which provides immutability, transparency, and security, and utilizes IPFS for distributed and efficient storage of files. The framework includes a smart contract-based system that manages the sharing process and allows users to share files with specific permissions and conditions, such as time-limited access and payment requirements. The system uses ECC encryption to secure the files, making it resistant to malicious attacks and data breaches. The authors demonstrate the feasibility and effectiveness of the proposed system through experiments and evaluations, showing that it provides low latency, high throughput, and is highly resistant to network outages and malicious attempts. Overall, the proposed decentralized file sharing framework offers a secure and efficient solution for sharing sensitive data without the need for a central authority or intermediary. Increased security, privacy, and efficiency are just a few of the potential advantages of decentralized file sharing systems built on IPFS and blockchain that have been identified by the poll. But the literature also suggests that there are a number of difficulties in creating such systems, including problems with scalability and user acceptance. This literature analysis has provided a solid theoretical framework for the creation of an innovative and successful decentralized file sharing system based on IPFS and blockchain by assessing the state of the art and highlighting research needs. Overall, this research represents a significant step toward creating a decentralized and more secure internet infrastructure.

9.3 PROPOSED WORK

9.3.1 Architecture

In this chapter, we propose an architecture for decentralized file sharing systems using blockchain and IPFS that can provide secure and efficient storage for sensitive information without relying on a centralized server or risking data tampering. By using encryption and access control mechanisms, users can ensure that their data is protected and only accessible by authorized parties see Figures 9.1 and 9.2. This eliminates the need for centralized systems and also overcomes the disadvantages of centralized systems like single point of failure and data tampering. Here we are using IPFS for storing large amounts of data as blockchain increases the storage costs and leads to slower transaction times.

9.3.1.1 Uploading process

Steps in the uploading process:

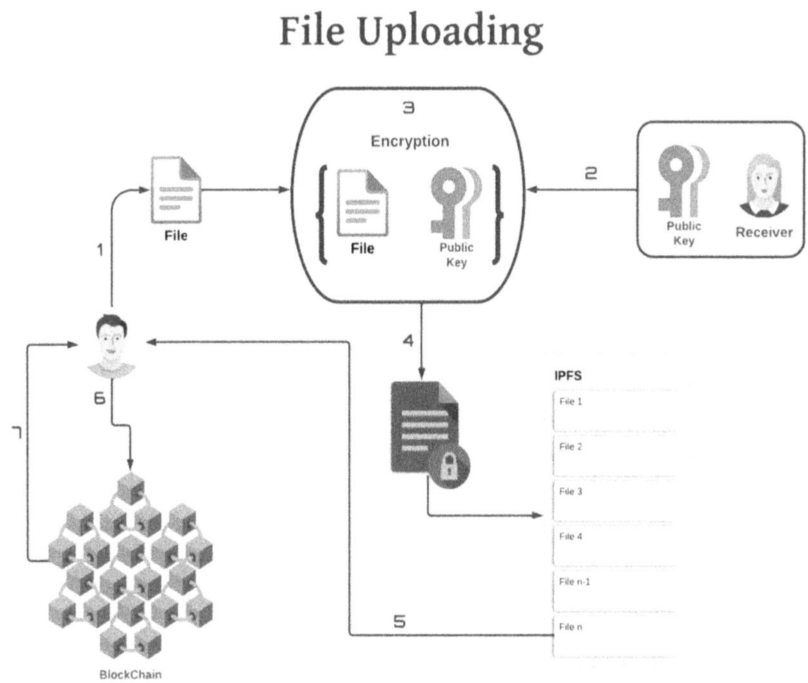

Figure 9.1 Pictorial representation of file uploading.

- Step 1: Sender chooses the file to send to the receiver.
- Step 2: Receiver generates a pair of keys called public and private and sends the public key to the sender. Public key is used for encrypting the file and private key is used for decrypting the file.
- Step 3: File is encrypted with public key using ECC encryption.
- Step 4: The encrypted file is uploaded to IPFS.
- Step 5: Sender gets the hash of the file uploaded into the IPFS
- Step 6: A smart contract is created and the IPFS hash stored in the blockchain.
- Step 7: Sender gets the transaction details of the hash uploaded.

9.3.1.2 Downloading process

Steps for downloading a file:

- Step 1: Sender sends the transaction details to receiver.
- Step 2: Receiver searches the blockchain network for IPFS hash using the transaction details.
- Step 3: Receiver gets IPFS hash (hash of the file uploaded).
- Step 4: Receiver requests the file using IPFS hash from IPFS.

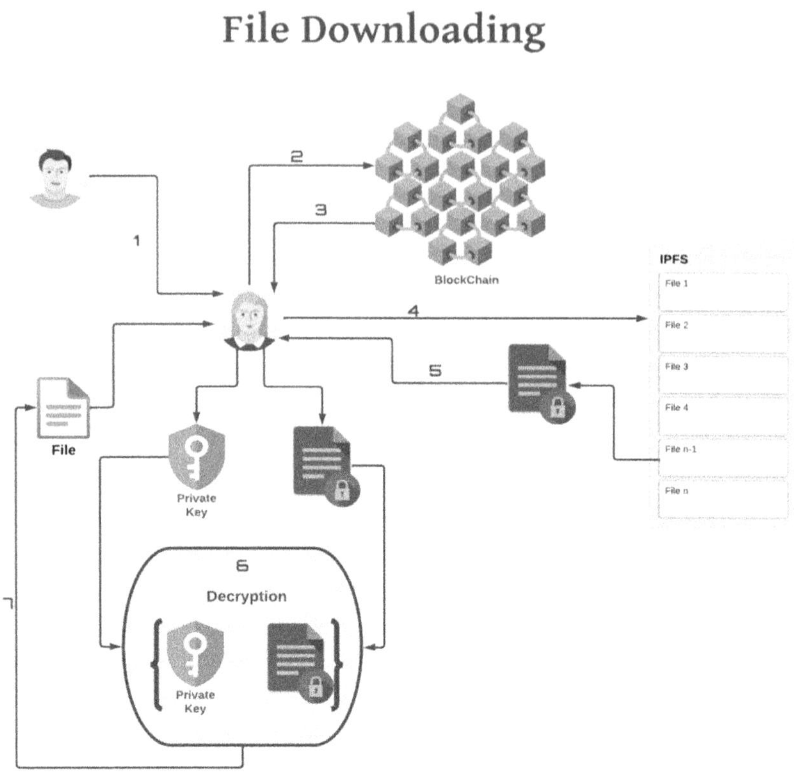

Figure 9.2 Pictorial representation of file downloading.

- Step 5: Receiver gets the encrypted file linked to IPFS hash.
- Step 6: Receiver decrypts the encrypted file using their private key.
- Step 7: Receiver gets the decrypted original file.

9.4 RESULTS AND DISCUSSION

In this work, we conducted an experiment to compare the performance of ECC and Rivest–Shamir–Adleman (RSA) encryption algorithms in a decentralized file sharing system based on IPFS and blockchain. We evaluated the two algorithms based on their speed and security, using a sample set of files to measure the time taken to encrypt and decrypt the files using each algorithm. Our results show that the ECC encryption algorithm outperforms RSA encryption algorithm in terms of speed, taking significantly less time to encrypt and decrypt files compared to RSA. However, the RSA encryption algorithm offers stronger security compared to ECC. This is because RSA uses a larger key size compared to ECC, making it more difficult to break the encryption. Therefore, the choice of encryption algorithm to be used in a decentralized file sharing system should depend on the priority given to speed and security.

- Receiver generates a key pair as shown in Figure 9.3 and sends it to sender through any medium.
- Sender encrypts the file with the receiver's public key and uploads it to IPFS to get IPFS hash as shown in Figure 9.4.
- Sender gets the IPFS hash after uploading the encrypted file to IPFS and initiates a transaction along with IPFS hash to store it in the blockchain as shown in Figures 9.5 and 9.6 and sends the transaction hash to the receiver for file retrieval.
- Receiver retrieves the file from IPFS with IPFS hash and decrypts it with his/her private key to get the contents of the file as shown in Figure 9.7.

File size	*ECC encryption*	*ECC decryption*	*RSA encryption*	*RSA decryption*
1 MB	0.006149	0.004147	4.408545	8.6078
5 MB	0.029446	0.01707	97.24522	86.67526
10 MB	0.0437	0.03358	142.6064	112.0478
20 MB	0.093426	0.07398	~300	~280
50 MB	0.311639	0.27005	~800	~720
100 MB	0.61875	0.5595	~1800	~1500
500 MB	3.0752	2.83605	~9700	~8200
1 GB	6.53764	5.71382	~24,500	~19,300

Symmetric algorithm security (in bits)	*RSA with key size (in bits)*	*ECC with key size (in bits)*
80	1024	160–223
112	2048	224–255
128	3072	256–383
192	7680	384–511
256	15,360	512+

CITATION: https://cheapsslsecurity.com/p/ecc-vs-rsa-comparing-ssl-tls-algorithms

9.5 CHALLENGES

1. Implementing strong encryption mechanisms to ensure that user data and files are protected against unauthorized access and cyberattacks.

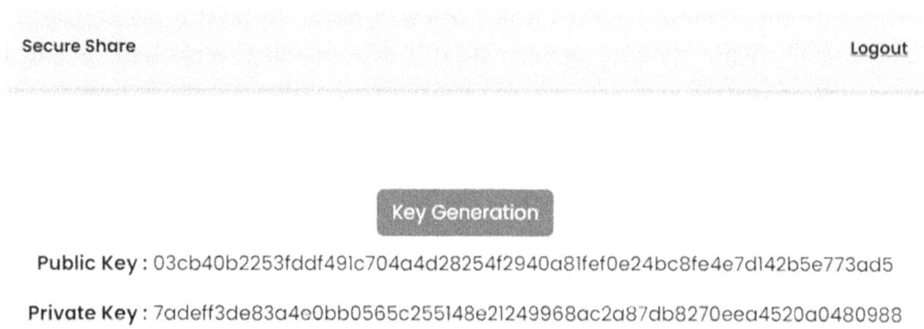

Figure 9.3 Key generation in the process of file sharing.

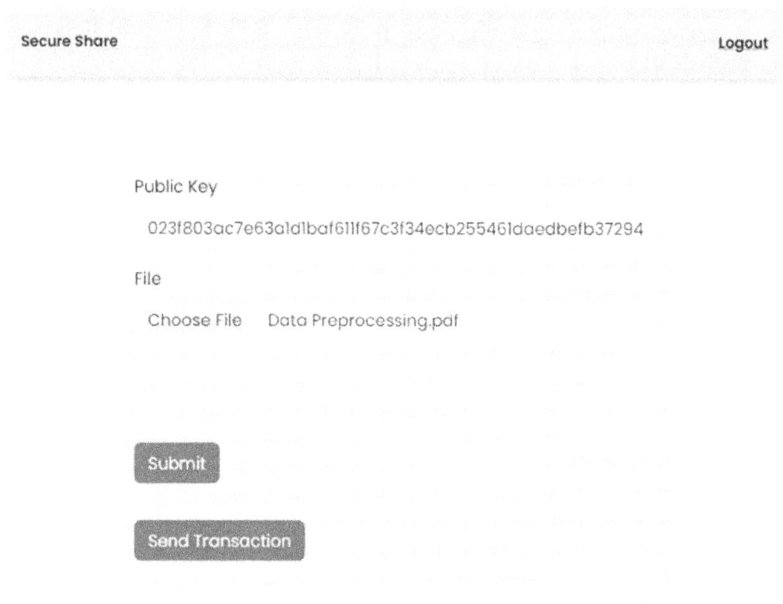

Figure 9.4 Attaching public key with file uploading.

2. Ensuring that the system complies with relevant legal and regulatory frameworks, particularly those related to data protection and privacy.
3. Disclosure of sensitive information, such as controlled unclassified information and personally identifiable information, which can lead to financial loss, violation of sharing agreements, legal action, and loss of reputation.
4. Designing a user-friendly interface and experience that is easy to use and navigate, even for users who are not familiar with blockchain or IPFS.
5. Incorrect implementation can lead to leakage of the ECC private key in some cases. Such leakage can occur when the results are calculated incorrectly and when the input is not on the selected curve.

Figure 9.5 IPFS hash after uploading the encrypted file to IPFS.

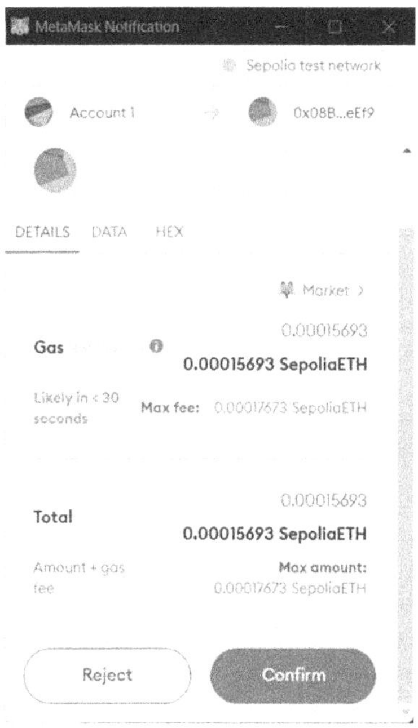

Figure 9.6 A transaction along with IPFS hash to store it in the blockchain.

Limitations

1. Because the public key is not authenticated, no one can guarantee that the public key belongs to the designated person. Therefore, users must verify that their public key belongs to them.
2. If an individual loses their private key, they cannot decrypt the messages they receive.

Figure 9.7 Blockchain sends the transaction hash to the receiver for file retrieval.

3. If a malicious actor determines a person's private key, an attacker can read that person's messages.
4. While blockchain technology is known for its security features, it is not immune to attacks. The system must be designed to address potential security risks such as 51% attacks, Sybil attacks, and others.

9.6 CONCLUSION

In this chapter, we presented the idea of a secure file sharing system based on permissionless blockchain and IPFS. A simple, affordable, easy-to-use, and highly secure system is proposed to solve data security issues such as data integrity, authenticity, and unavailability. The proposed project is particularly relevant in today's digital world where sensitive information is shared and stored on centralized platforms that are vulnerable to attacks. By creating a decentralized system, the project provides a much-needed alternative that prioritizes security and privacy. The project has the potential to revolutionize the way we store and share sensitive information, and could have a significant impact on various industries such as healthcare, finance, and government, where security and privacy are very important in this current world. The proposed project helps in sharing information with only users required; using cryptography to increase the security of the file where the receiver only has the key to decrypt the encrypted file which is encrypted with the public key of the receiver by the sender of the file. These features of our project provide a more secure way of sharing sensitive data in a simple, affordable, easy-to-use, and highly secure system. This work provides a better way of sharing sensitive data with users without anyone having sole control over the data.

The presented work on decentralized file sharing systems offers a promising solution for secure and efficient file sharing with the potential to transform the way data is stored and exchanged in various industries. It addresses the limitations of traditional file sharing methods and provides users with a more secure and reliable method for sharing files. We believe that our system has the potential to revolutionize the way data is shared and stored, and we hope that this chapter will inspire further research and development in this area.

BIBLIOGRAPHY

Demir, O., & Kocak, B. (2019). A decentralized file sharing framework for sensitive data. In Younas, M., Awan, I., & Benbernou, S. (Eds.), *Big Data Innovations and Applications* (pp. 142–149). Springer International Publishing, Cham.

Fukumitsu, M., Hasegawa, S., Iwazaki, J. Y., Sakai, M., & Takahashi, D. (2017). A proposal of a secure P2P-type storage scheme by using the secret sharing and the blockchain. In *2017 IEEE 31st International Conference on Advanced Information Networking and Applications (AINA)* (pp. 803–810).

Guo, R., Shi, H., Zhao, Q., & Zheng, D. (2018). Secure attribute-based signature scheme with multiple authorities for blockchain in electronic health records systems. *IEEE Access*, 6, 11676–11686.

Huang, H. S., Chang, T. S., & Wu, J. Y. (2020). A secure file sharing system based on IPFS and blockchain. In *Proceedings of the 26th ACM Symposium on Access Control Models and Technologies* (pp. 96–100). Association for Computing Machinery, New York, NY, USA.

Nizamuddin, N., Salah, K., Azad, M. A., Arshad, J., & Rehman, M. H. (2019). Decentralized document version control using Ethereum blockchain and IPFS. *Computers & Electrical Engineering*, 76, 183–197.

Steichen, M., Fiz, B., Norvill, R., Shbair, W., & State, R. (2018). Blockchain-based, decentralized access control for IPFS. In *2018 IEEE International Conference on Internet of Things (iThings) and IEEE Green Computing and Communications (GreenCom) and IEEE Cyber, Physical and Social Computing (CPSCom) and IEEE Smart Data (SmartData)* (pp. 1499–1506).

Wang, S., Zhang, Y., & Zhang, Y. (2018). A blockchain-based framework for data sharing with fine-grained access control in decentralized storage systems. *IEEE Access*, 6, 38437–38450.

Chapter 10

FAWT-based advanced multiboost learning algorithm for driver fatigue detection using brain EEG signals

P. Bora
Mayang Anchalik College, Marigaon, India

A. Saikia and J. Deka
Cotton University, Guwahati, India

A. Hazarika
Mayang Anchalik College, Marigaon, India

10.1 INTRODUCTION

As per global reports, 14–20% of total road accidents are caused by driver fatigue [1]. The fatigue refers to extreme tiredness, weariness, or exhaustion which are mainly due to mental or physical exertion or illness. Fatigue reduces the focus and concentration of a driver, increasing the risk to the driver and other passengers, vehicle drivers, cyclists, and pedestrians in the context of injuries and fatalities. Therefore, it is essential to develop an advanced machine learning method that could be embedded in sensor-based in-built car detection systems in an effective manner for fatigue detection and alerting the driver to avoid unwanted road accidents [2, 3]. Such an advanced algorithm would have great impact on an Internet of things (IoT)–based decision-making platform. With the development of various artificial intelligent methods, so far various algorithms have been reported with special emphasis on model accuracy, ease of implementation, and better utility in a real-world setup. Although various methods reported significant performances, they often suffer various limitations including model complexity, high dimensionality of features space, and involvement of multiple processing stages that obscure them in implementing real-world applications. Therefore, recent research focuses on an initial learning framework that mainly deals with proper utilization of resources and extraction of features that would carry inherent physiological information of signals from the source thereof.

 DOI: 10.1201/9781003474524-10

10.2 STATE-OF-ART-METHODS

Various fatigue detection methods reported over the decades employed various physiological features (amplitude, frequency, etc.) [4] of electroencephalogram (EEG) [3], electrooculography (EOG) [5], electrocardiogram (ECG) [6], electromyography (EMG) and driver's facial expressions during yawning and blinking [2]. The method in Lai et al. [7] used subjects' self-reported fatigue and method in Chai et al. [3] used video measurements of facial expressions, reaction time, steering error, and lane departure. But self-report–based approaches are time consuming to validate symptoms [3]. In addition, subjective feedback may be biased and therefore unreliable. Indirect video-based approaches require recording the driver's face during road conditions, which can compromise the driver's privacy. Recently, many effective methods are reported for classification of drowsiness of drivers. Sheykhivand et al. [8] suggested an automated algorithm that can be used in a real-time driver drowsiness detection system based on compressed sensing (CS) theory and deep neural network (DNN). Samplewise analysis of important features for classification to detect driver fatigue is explained in [9], which used convolutional neural networks (CNN) to identify drowsiness patterns with a compact structure using convolution to process the EEG signals. In an another study, a single EEG electrode placed behind the ear to detect diver fatigue achieved 78.3% accuracy with cross subject validation [10]. In this consequence a hand-crafted framework has been adopted by Sengul et al. [11]. To detect human drowsiness during driving a scheme named passive computer interface (pBCI) has been modelled using multichannel EEG [12]. Furthermore, Li et al. [13] developed a hybrid ensemble CNN framework to decrypt EEG signals to detect driver fatigue apart from the commonly used CNN model. Some additional methods include yawning state detection [14], drowsiness detection using independent multimodal blood volume pulses (BVP), and blink and yawn signals [15]. Local binary pattern (LBP) includes a combination of facial expressions [16], eye state detection [17], and deep learning based methods [18, 19]. Among various methodologies, those based on EEG are seen to be more effective. A basic fact of the EEG–based method is that EEG contains the inherent information content of neurophysiological brain activity. It can be considered as a good indicator of fatigue [20]. Typically EEG is recorded from planar electrodes placed on the scalp. These are classified as gamma (30–42 Hz), beta (13–30 Hz), alpha (8–13 Hz), theta (4–8 Hz), and delta (0.5–4 Hz) waves. It is worth mentioning that beta waves' presence may indicate a person's alertness or may also be present in the early stages of sleep. Theta and delta are associated with early and deep sleep, and alpha indicates a relaxed state, showing early signs of fatigue. It is worth noting that changes in heart rate variability [21] and brain activity [22] are related to fatigue. Due to inherent limitations of various aforesaid methods, recent research focuses to develop a better version or new method with proper utilization of EEG signals.

Some approaches used time, time-frequency, wavelets, statistical functions, and learning strategies [23]. For example, Vuckovic et al. [24] evaluated EEG signal time series of interhemispheric and intrahemispheric transverse spectral densities and used them in an artificial neural network (ANN) to categorize driver drowsiness or attentiveness. Hu et al.'s method [25] transformed EEG signals in the time domain into alpha, theta, beta, and delta bands to assess frequency domain features for the classification task. It evaluated functional spectrum-based frequency-domain features embedded into support vector machines (SVMs). In Zhang et al. [26], power spectral density (PSD) and sparse representation-based method is introduced to estimate the driver's arousal level. Wang et al. [27] classified fatigue using wavelet entropy and spectral entropy features. Besides, some other methods employed sample entropy [28], permutation entropy [29], fuzzy entropy [30], and approximate entropy [21].

Usually various earlier methods employed combined statistical approaches in order to enhance the overall performance. Luo et al. [2] estimate multiscale entropy features using an adaptive factor algorithm. Chai et al. [3] employed blind source separation techniques with autoregressive models for extracting multivariate features. The method used by A. Chaudhuri and A. Routray [31] adopted current dipoles in the EEG wavelet search space for finding chaotic entropy features to detect the fatigue states. Dynamic binary and ternary patterns based on discrete wavelets (DWT) are employed in Tuncer et al. [32] and evaluated features are embedded ANN, rotational forest (RoF), SVM, and various flat classifiers, including k-nearest neighbors (k-NN). In Mu et al. and Chin et al. [33, 34], fuzzy entropy-based support systems are developed. Kaur and Singh's method [22] employed empirical mode decomposition (EMD)–based statistical features for classification of fatigues. Some methods such as Chai et al. and Fu et al. [35, 36] used deep neural networks, that is, sparse-deep belief networks (DBN). Such a model often combines supervised and unsupervised learning, and hidden Markov models (HMM) for EEG–based fatigue detection. Despite remarkable achievement in various stages of their algorithms, they have some theoretical bottlenecks and several limitations. For instance, EMD–based methods [37] often suffer from a mode-mixing issue. It is seen that previous methods employed single or multi-modality patterns viz EEG and EOG [38], EEG, EOG, and EMG [15], EEG, ECG, EOG, and functional near-infrared spectroscopy (fNIRS) [6], and EEG and EMG. Such multiple modalities–based methods are based on extraction of features from multiple input feature searched spaces and their combination into single form of vectors. As a result, they often requires multiple processing stages. Instead of considering such high-dimensional input space, it would be better to consider a single modality profile that truly reflects the underlying phenomena through analyzing the significant attributes employing a well-defined mathematical framework.

It is worth mentioning that earlier methods also reported better recognition performances by deploying various decision models such as as SVM [2], ANN, RF, k-NN [32], RoF, random forest (RF), decision tree (DT), classification and regression tree (CART) [39], deep learning, C4.5, LAD–tree, and others. Nonetheless, input features searched space techniques have more impact on algorithm performances as compared to decision models. It also to be mentioned that various cited methods have the limitations of overfitting, dimensionality, and instability issues. For example, deep learning methods that work as black box provide high degree of freedom due to its inherent layers [37]. Despite wide popularity, ANN–based methods are based on trail and error for proper setting of optimal parameters. They may also suffer computational complexity issues in case of high-order training space. SVM requires kernel parameter setting for good learning ability, which needs multiple processing steps.

The previous study shows the urgent needs of some advanced learning-based algorithm either by adopting a new feature extraction technique or using a generalized version of performance-boosting models or both for a better method of fatigue detection. In such cases, conventional models needed to be embedded with performance-boosting strategies like boosting and multiboost [40]. Then it could overcome many limitations associated with earlier approaches.

In this chapter, an advanced ensemble learning method is addressed by employing an advanced version of wavelet transformation, namely, flexible analytic wavelet transform (FAWT) [41] for classification of various states of fatigue for viable implementation in sensor-based setups. The major contributions of the proposed method are as follows: 1. It first developed FAWT–based feature extraction framework which first decomposes signals into low- and high-pass channels with proper values of model parameters. It is noted that FAWT is more robust in extracting inherent information in the context of features from non-stationary signals [42] that could lead to develop a more effective method. 2. In this section, publically available tested EEG data are subjected into the FAWT method. After analysis of signals using the FAWT, the low-dimensional statistical feature set is evaluated. The features set is subjected to statistical significant test to find the comprehensive measures that could replicate the physiological information of signals. 3. The chapter then proposed a FAWT–based Adaboost learning algorithm for classification of fatigue states. The adopted algorithm is validated in offline mode and demonstrated the effectiveness for real-time implementations. 4. In the last section, the performances of the proposed methods are also compared to various earlier methods in the context of modalities profiles, type of features, complexity, limitations, and performances.

10.3 METHOD

10.3.1 EEG dataset

In order to formulate as well as validate the proposed method, an open source EEG dataset [2] is employed. The data in the database were collected using a platform environment INCLUDING a static ZY-31D vehicle driving simulator (Beijing-China Joint Teaching Equipment Co. Ltd.). The acquisition setup includes three 24-inch monitors and software teaching system ZM-601 V9.2 for driving simulations. It includes a 32 electrodes EEG–collecting cap, computer system with 7 × 64 window size, EEG recording and processing software Neuroscan 3.2, and a signal analysis platform based on MATLAB. This study included signals from 16 valid individuals of age range 17–25 years. Before the experiment, it was ensured that no one got sick during the week. They also made sure that the individuals concerned had enough sleep the night before the experiment without ingesting energy drinks, alcohol, tea, and so on. In addition, for smooth coordination during the experiment, the persons concerned were well informed about the observational procedure, setup, and setting of electrodes. After a successful recovery, the laboratory assistant started collecting 5 minutes of EEG data using the software. These signals were assigned as normal EEG data. Subjects were then placed in a simulated driving state and asked to drive for a while. The different states of fatigue assigned to individuals at different times refer to L is subjective fatigue scale and the Borg CR-10 scale. Although the experiment was considered effective, the results indicated that the candidate was in fatigue, and an EEG was recorded for the same period, called the fatigue state, and then data acquisition was supposed to be completed.

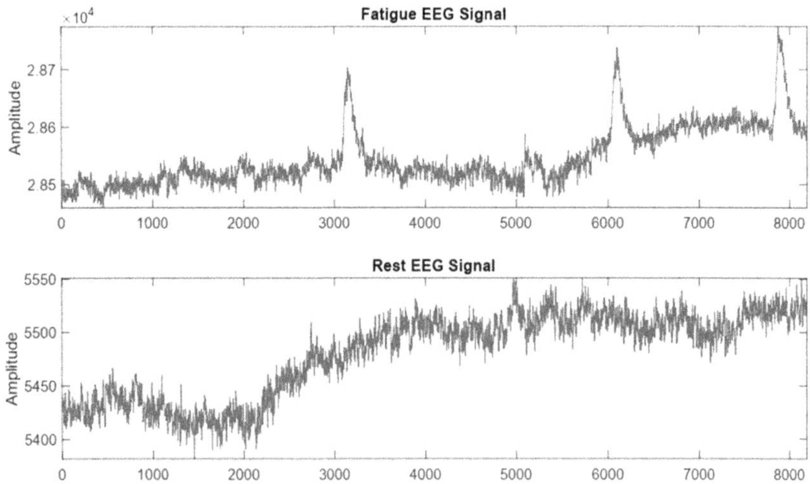

Figure 10.1 EEG signals of fatigue and rest state of driver.

The signal (Figure 10.1) was preprocessed by the Neuroscan 3.2 software platform with 1 kHz sampling frequency, 0.1545 Hz bandwidth, and 50 Hz cutoff frequency. The preprocessing stages remove various noises such drift, noise, electrooculogram and also do baseline correction. Each candidate's results were then split into two groups, normal and fatigue.

10.3.2 Feature extraction using FAWT

In order to extract relevant features from signals by removing redundant information, an advanced version of DWT commonly known as FAWT [41, 42] is used. It provides wide coverage of time-frequency scale of signal. It contains Hilbert-transformed pairs of atoms that makes it suitable for signal analysis involving oscillations. This analysis requires to maintain input control parameters, namely, quality factor (Q), number of decompositions (J), and redundancy (r), where Q limits the frequencies of the parent wavelet defined by the frequency ratio and constant parameter β [42]. The control parameters are defined as follows:

$$Q = \frac{\omega_0}{\nabla\omega}, \beta = \frac{2}{Q+1} \tag{10.1}$$

ω_0 and $\nabla\omega$ are the central frequency and bandwidth of the signal, and r controls the emplacement of wavelet time. The redundant parameter r controls the temporal position of the wavelet. FAWT decomposes the signal using an iterative filter bank containing high-pass and low-pass channels. FAWT allows us to specify the correct choice of expansion coefficients Q and r by versatile tuning of the parameters viz. positive constants β and e, f, g, h. Parameters e and f are set for high-pass filter upsampling and downsampling, and g and h are set for low-pass channel upsampling and downsampling. It provides iterative J-decomposition stages, each stage consisting of a low-pass (LP) and high-pass (HP) channel that isolates negative and positive frequencies, respectively. The frequency responses $H(\omega)$ and $G(\omega)$ for HP and LP are as follows:

$$H(\omega) = \begin{cases} (ef)^{1/2} |\omega| < \omega_p \\ (ef)^{1/2} \theta\left(\dfrac{\omega - \omega_p}{\omega_s - \omega_p}\right), \omega_p \le \omega \le \omega_s \\ (ef)^{1/2} \theta\left(\dfrac{\pi - (\omega - \omega_{p)}}{\omega_s - \omega_p}\right) - \omega_s \le \omega \le -\omega_p \\ 0 |\omega| \ge \omega_s \end{cases} \tag{10.2}$$

$$G(\omega)=\begin{cases}(gh)^{1/2}\,\theta\left(\dfrac{\pi-\omega-\omega_0}{\omega_1-\omega_0}\right),\omega_0\le\omega\le\omega_1\\ (gh)^{1/2}\;\omega_1<\omega<\omega_2\\ (gh)^{1/2}\,\theta\left(\dfrac{\omega-\omega_2}{\omega_3-\omega_2}\right)\omega_2\le\omega\le\omega_3\\ 0\;\omega\epsilon\left[(0,\omega_0)\cap(\omega_3,2\pi)\right]\end{cases} \tag{10.3}$$

Various parameters associated with above filter banks are $\omega_p=\dfrac{(1-\beta)\pi+e}{e}$, $\omega_s=\dfrac{\pi}{f}$

$\omega_0=\dfrac{(1-\beta)\pi+e}{g}, \omega_1=\dfrac{e\pi}{fg}, \omega_2=\dfrac{\pi-e}{g}, \omega_3=\dfrac{\pi+e}{g}, \epsilon\le\dfrac{e-f+\beta f}{e+f}\pi$. It is to be mentioned that in order to avoid losing information from the features the value or r needs to be greater than one as per the following rule:

$$r=\left(\frac{g}{h}\right)\frac{1}{1-\frac{e}{f}} \tag{10.4}$$

The limiting value of β for perfect reconstruction needs is as follows, from where the Q is calculated.

$$1-\frac{e}{f}\le\beta\le\frac{g}{h} \tag{10.5}$$

With the proper choice of parameters as defined earlier, seven levels of decomposition are performed and subsequently sub-bands signals are reconstructed in descending frequency order of the two-state EEG signal. As defined in Gupta et al. [43], the value of $e = f$ is set to ¾ (dilation factor) with fixed value of r and Q (see, (10.1) (10.4)) and the value of $g = h$ is fixed as ½. FAWT provides higher order feature searched space in terms of sub-band components for which it is further subjected to statistical analysis in order to extract low-order features, specifically, mean absolute, mean power, standard deviation, ratio of the absolute mean, skewness and kurtosis of coefficients in each sub-band. Furthermore, in total were extracted 95 measures imported to linear discriminant analysis (LDA) for deriving best decision surface. LDA minimizes the intra-class variance and maximizes between-class variance to achieve optimal discriminant properties [44, 45]. Finally, features are tested for $0{:}05 < p$ with confidence level 95% and final features matrix is used for classification purposes.

10.4 RESULTS AND ANALYSIS

Conventional machine learning models work as weak learners that also suffer many theoretical limitations despite having good feature extraction frameworks. In such cases, ensemble meta-algorithms that adopt different boosting strategies, namely, adaptive boosting (AdaBoost) and multiboosting (multiBoost) provide better performances. The objective of such learners is to combine multiple weak learners that carry inherent information in the same or different formats and to provide better inferential models that reduce the variance in results by controlling the over-fitting issue of individual weak learner [46], for example, CART. The multiple model–based methods summarize the outputs of all sub weak classifiers for overall conclusion. Such approach uses voting (for classification) or averaging (for numeric prediction) to syndicate the output of the individual model. It is to be mentioned that the boosting strategy emphasizes weight assignment to the models based on its confidence rather than giving equal weight to all models. AdaBoost [40] is commonly used as boosting algorithm in regression (e.g., C4.5 tree) and classification problems. It is simple, flexible and easy to implement.

The algorithm for a given data pattern is defined as follows: for a given training data $\left\{\mathrm{f}\left(\mathrm{x}_{\mathrm{i}};\mathrm{y}_{\mathrm{i}}\right)\right\}_{i=1}^{N}$, where, N is the number of iterations, $x_i \in \mathbb{R}^K$ and $y_i \in \{1, 1\}$, there are large number of weak classifiers, denoted $f_m(x) \in \{1, 1\}$, and a 0–1 loss function I, defined as

$$I\left(f_m\left(x\right), y\right) = 0, \text{if } f_m\left(x_i\right) = y_i 1, \text{if } f_m\left(x_i\right) \neq y_i$$

ALGORITHM 1 ADABOOST ALGORITHM

for i = 1 to N, $\omega_i^{(1)}$

for m = 1 to M, **do**

Fit weak classifier m to minimize the objective function:

$$\epsilon_m = \frac{\sum_i 1^N \omega_i^{(m)} I\left(f_m\left(x_i\right) \neq y_i\right)}{\sum_i \omega_i^{(m)}}$$

Where $(f_m(x_i) \neq y_i) = 1$ if $f_m(x_i) \neq y_i$ and 0 otherwise

$$\alpha_m = \ln \frac{1 - \epsilon_m}{\epsilon_m}$$

for all *i* **do**

$$\omega_i^{m+1} = \omega_i^{(m)} e^{\alpha_m I\left(f_m\left(x_i\right) \neq y_i\right)}$$

end forend for

The final model output is the linear combination of the weak classifiers as defined below.

$$g(x) = \text{sign}\left(\sum_{i=1}^{N} \alpha_m f_m(x)\right) \tag{10.7}$$

This greedy algorithm builds up incrementally a strong classifier $g(x)$ by optimizing the weights for, and adding, one weak classifier at a time. It is also known as generalized version, in other words, M-array classifiers. It is known as multiBoost, which learns parallel learning models in order to stabilize the over-fitting issue. This chapter explores the use of multiBoost-based meta algorithm with various conventional weak learners in classifying fatigue and rest data for establishing effective classification algorithm. In order to investigate the performance of the proposed method, various bookmarkers including accuracy (Ac), misclassification error (ME), true positive rate (TPR), false positive rate (FPR), precision (PRe), and statistical parameters F-score (Fsc) and kappa (κ) are used, which are as follows:

$$ME = \frac{FP + FN}{TP + TN + FP + FN} \tag{10.8}$$

$$Ac = \frac{TP + TN}{TP + TN + FP + FN} \tag{10.9}$$

$$TPR = \frac{TP}{TP + FN}; FPR = \frac{FP}{TN + FP} \tag{10.10}$$

$$P\text{Re} = \frac{TP}{TP + FP}; FSc = \frac{(\beta^2 - 1).P\text{Re}.\text{Re}}{\beta^2.P\text{Re} + \text{Re}} \tag{10.11}$$

$$\kappa = \frac{P_0 - P_e}{1 - P_e} \tag{10.12}$$

Various parameters used in the preceding equations are defined as follows: the term ME indicates percentage of misclassified instances out of all instances, Ac indicates total cases correctly identified by model, and TPR and FPR indicate total correct positive cases (i.e., fatigue) and negative case (i.e., rest) classified to total positive and negative. The term PRe referred to the ratio of instances correctly classified as positive to all instances classified as positive whereas F-score (FSc) measures the balance while $\beta = 1$ and favours PRe if $\beta > 1$ and recall (Re) otherwise. P0 and Pe represent observed agreement and agreement expected by chance. The remaining terms used are abbreviated as false positive (FP) and negative (FN), and true negative (TN) and positive (TP). All the parameters are evaluated from formulated confusion matrix. Besides all such parameters, two other parameters, namely, area under curve (AUC) and Kappa coefficient (κ) are also estimated. The AUC indicates the degree or measure of separability whereas the kappa κ indicates the agreement or disagreement of measurements (κ = 1 or 0).

For analyzing the performance of the model, K-fold cross-validation is used and data set is partitioned into 10 subsets. One set is used for training and the remaining data is used for feature extraction. It is to be mentioned that each folded dataset contains an equal portion of Fatigue (FAT) and Rest (REST).

The mean performances of various models are presented in graphical forms in Figure 10.2. As seen from the Figure 10.2 the boosting inspired models provide the most prominent results in terms of various parameters including FSc, AUC, and κ. It is to be mentioned that the higher value of these parameters in addition to the accuracy indicates the effectiveness of these inspired models. Additionally, all the models provide uniform parameter values in categorizing the FAT and REST respectively. Average recognition rates of SVM, ANN, and rotation forest with multiBoost ensemble are 97:20%, 96:90%, and 96:70% and 97:90%, 96:50%, and 96:30% respectively.

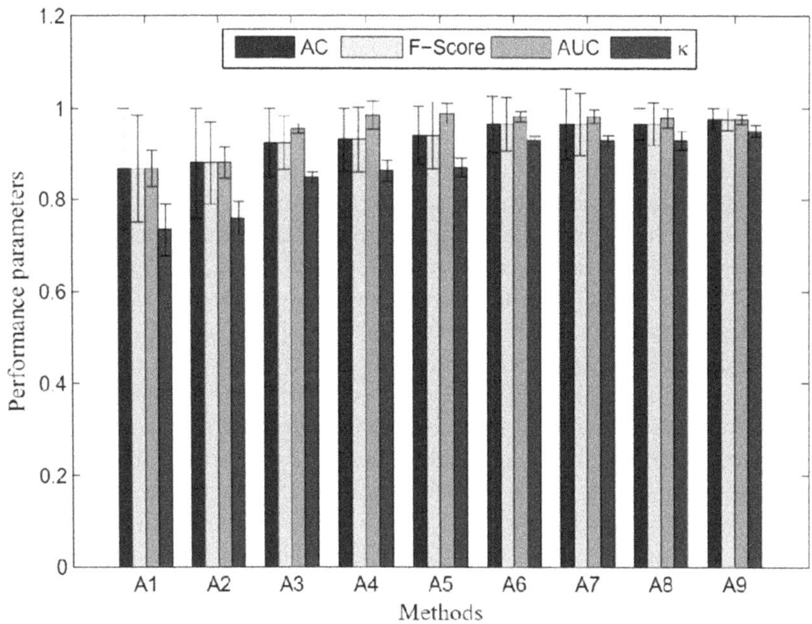

Figure 10.2 Performance parameter measures of various methods viz. A1-extra tree A2-random tree A3-KNN A4-REP tree A5-CART A6-C45 A7-random forest A8-ANN A9-SVM.

Extra tree provides the lowest recognition rates of 80:60% and 86:30%. It is also seen that uniform results are observed in classifying REST groups, presumably because the REST group has a higher identification set compared to the set of characteristics that belong to the FAT group. Importantly, inspired SVM with FSc carry out the optimal level of accuracy of 97:50%, error, AUC, and κ of 2:50%, 0:975, and 0:950 respectively. It is worth noting the facts that the model performance depends on relevant feature extraction framework that effectively evaluates comprehensive low-order features that replicate the behaviour of underlying phenomena and choice of proper decision models.

This study therefore includes FAWT and statistical data-driven approaches, followed by inspired decision models. Deep learning models such as DBN, recurrent neural networks, convolutional neural networks, deep neural networks, deep Boltzmann machines, and others often provide better performances. However, such models suffer difficulties of the dimensionality issue due to multiple layer structures which limits their applications in a real-world setup [37]. The robustness and stability of the proposed method is indicated by the promising results. The algorithm is implemented in MATLAB and mean processing time is 10–15.67 s. As seen from the evidence, the proposed method outweighs the challenges of various state-of-the art methods. In the following section, the comparative analysis is also discussed in a detailed manner so as to explore the efficacy of the proposed method thereof.

10.5 COMPARATIVE STUDY

Table 10.1 highlights many earlier methods along with various attributes including feature types, patterns, and performance for better comparison. This study included two-class and three-class classification methods, two-class methods using single modality profile EEG [2, 3, 21, 22, 32, 34], eye vision [18] and video [16] and multiple modality profiles [6, 38], and three-class classification used multiple modalities [15, 31, 47]. Unlike various methods, specifically, three-class methods, this study focuses on utilization of single modality profile EEG. However, comparison of the outcomes of the proposed methods with all cited methods in respect to various inherent model attributes help to understand of the effectiveness of proposed feature extraction framework and the decision module adopted herein. Many methods, as mentioned in Section 10.2 and in Table 10.1, adopted multi-scale entropy [2]; autoregressive coefficient [3]; binary pattern–based features [32]; entropies—that is, spectrum, approximate, sample and fuzzy entropy (SpEn, ApEn, SamEn, FuzEn) [21]; EMD–based mode function (EMD-EMF) [37]; fast Fourier transformation (FFT) using EEG signals for detection of fatigue, namely FAT, REST, mild fatigue (M-FAT), excessive fatigue (EX-FAT), sleep-deprive (SLEEP-DEP), and so on; including normal (NOR) and

Table 10.1 Summaries of various prior methods and their comparison with the proposed method

Method	*Signal*	*Feature*	*Classifier*	*State*	*Ac[%]/AUC*
Chai et al. [3]	EEG	ICA-AR	BNN	FAT, ALERT	88.20/0.930
Mu et al. [21]	EEG	(Sp,Ap,Sam,fuz)En.	SVM	FAT,REST	87.69/-
Yin et al. [34]	EEG	Fuzzy-entropy	SVM	FAT,NOR	95.00/-
Khushaba et al. [38]	EEG,,EOG	Fuzzy-mutual-DWT KNN, k-SVM	LDA,LSVM,	DROWSINESS, FAT	95–97/-
Zhang et al. [15]	EEG,EOG, EMG	Approx. entropy	ANN	NOR, MILD,EX-FATNOR, MILD,EX-FAT	~96.50/0.990
Chaudhuri and Routray [31]	EEG	ApEn, SamEn,	PCA-SVM	11 FAT STATES	86.00/-
Ahn et al. [6]	EEG,ECG etc.	EEG,ECG etc.	LDA	REST, SLEEP-DEP	84.50/-
Luo et al. [2]	EEG	multi-scale En	SVM	FAT, NOR	95.37/ -
Zhang and Hua [16]	Video	LBP	SVM, LBP-SVM Boost-LBP-SVM	FAT, NOR	85.85/-
Mandal et al. [18]	Eye vison	Fused-feature	PCA-SVM	2EYE STATES	95.18/-
Zhao et al. [19]	Video (eye, mout)	texture	DBN	DROWSINESS	96.70/-
Tuncer et al. [32]	EEG	DWT-(DCBP,MTTP)	KNN,RF,ANN, SVM	FAT,REST	97.29/-
Kaur and Singh [22]	EEG	EMD-IMF	ANN	DROWSY,AWAKE	88.22/-
Fu et al. [47]	EEG,EMG, etc.	Contextual	HMM	ALERT, M-FAT,FAT	89.20/-
Our	**EEG**	**FAWT-Features**	**multiboost-SVM**	**FAT,REST**	**97.90,97.10/0.975**

reported mean recognition rates. A major focus of most of the methods is performance enhancement with less focus on other aspects of the algorithm, especially the input feature searched space, are less explored. Also DWT provides approximate and detailed coefficients that are often high dimensional [48]. Additionally, choosing of the proper mother wavelet plays an important role since it requires the subjective knowledge and morphology of signal. In such circumstances, selection of specified components and its derivatives as features may not carry inherent signal information. The EMD–based methods often suffer a mode-mixing problem [37] that also needs to be considered. In the case of ANN–based approaches, proper parameter setting with proper selection models is very essential to be evaluated from the trial-and-error strategy [49]. In such circumstances, feature extraction and processing framework–based FAWT is seen as more suitable due to its theoretical advantages. The effectiveness and efficacy of such method is also conferred from the subsequent analysis and inferences. Unlike the advocated methods, some prior methods used different modalities data such as driver facial expression [2] and video recording [7] for feature extraction with special emphasis on performance enhancement, which make them unrealizable in practical domains for common users. On the other hand, statistical measures, namely entropies [21], band power, and fractal density, and combined them randomly into single form of pattern identification matrix which was subsequently embedded into the decision module to detect fatigue. Such a unified model can provide good inferences on a limited dataset but can not ensure the same or better performance on wide varieties of datasets. Although some methods provided promising outcomes with adoption of multiple stages in developing the algorithm, that also increases the computational bottlenecks. Combined methods such as EEG + EOG [38], EEG + EOG + EMG [15], EEG + ECG + EOG + fNIRS [6], and EEG + EMG + respiration [47] suffer major issues in handling of large input feature searched spaces, extraction, and in concatenating them into a single form of vector. Some methods use duration or percentage of closed eyes, frequency of mouth features yawning, and multi-posture visual acuity [18]. It is worth noting that EEG comprehensively reflects brain activity directly related to fatigue [22, 32], which is also inferred indirectly from the outcomes of this study. The proposed method provides single model–based feature searched space framework for consistent and better performance.

Conventional models including SVM, SVM-RBF, ANN, k-NN, EMD, ICA-autoregressive, kernel SVM (k-SVM), LBP-SVM, principal component analysis-SVM (PCA-SVM), and RF models often perform as weak learners due to their inherent characteristics. For instance, in case of SVM-RBF, it is essential to set proper value of kernel parameter (γ) while doing training of data. In case of SVM, instability and over fitting are two major concerns. Major issues in deep learning models such as DBN [50], CNN, and BNN are complex layer structures, higher degree of freedom, and lake model interpretation. In such contrast, the proposed methods, called as FAWT–multiBoost ensemble method is well defined with effective feature framework that handle single modality data profile for effective classification of fatigue which is also seen as superior to many state-of-the art methodologies. The advocated method not only provides better accuracy than the methods mentioned in some sources [2, 3, 18, 21, 22, 31, 32, 34], but also even higher than the methods described in [6, 15, 16, 19, 38, 47, others]. Although our methods show very close performance with that of the Tuncer et al. method [32], our algorithm adopted well-defined FAWT–based feature extraction framework and strong learning model. In order to explore the effectiveness of the proposed method, FAWT–based features are embedded into multiple inspired decision modules and subsequent analysis as depicted in Figure 10.2 and in Table 10.1, which reveals that fact. Our method first develops FAWT feature extraction framework and then extracts comprehensive statistical measures, followed by significant test at $p < 0{:}05$, which reflects underlying phenomena associated with the signal profiles. Then the features are tested with various learning methods. Importantly, low diversity in the outcomes reveals the effectiveness of features as well as of the algorithm. The promising results with minimum diversity investigated over multiple folded dataset ensures getting good performance on large databases, which will be investigated later.

10.6 CONCLUSION

This chapter comprehensively explores the various stages of the proposed method integrating an advanced flexible analytic wavelet transform feature extraction framework and boosted learning model for better detection of various states of fatigue using single modality EEG signals of brains. In deriving the single decision-making platform, initially FAWT–based feature extraction framework is developed with proper setting parameters in FAWT. Afterward, filtered EEG signals are subjected to the FAWT and low-order statistical measures were evaluated with $p < 0{:}05$. Then low-dimensional significant statistical features were subjected to various strong learners, for instance, boosting inspired-SVM, CART, k-NN, ANN, RF, RoF, REP–tree, LAD–tree, and C4.5. The classification performances were investigated in terms of various markers such as Ac, ME, TPR, FPR, FSc, AUC, and κ. The results were further compared with the various methods akin to this method. Quantitatively, this method achieves an optimum accuracy in case of multiBoost-SVM, which is 97:90% in categorizing REST with an average accuracy of 97:50% over both states. Promising results and subsequent thorough analysis indicate the effectiveness of the advocated method. Future work will focus on consideration of wide varieties of datasets and then on implementing a wearable prototype device for real-world application.

ACKNOWLEDGEMENT

The authors would like to thank anonymous reviewers for their valuable suggestions that helps to improve the chapter. The authors would also like to thank Cotton University fraternity for various supports. This work was supported by Inhouse research project, Cotton University with the Ref:CU/Dean/R & D/2019/05/1995, 4/3/222 Assam, India.

BIBLIOGRAPHY

[1] R. F. S. Job, A. Graham, C. Sakashita, and J. Hatfield, "Fatigue and Road Safety: Identifying Crash Involvement and Addressing the Problem within a Safe Systems Approach", *The Handbook of Operator Fatigue*, vol. 1: Ashgate Publishing Limited, 2012, pp. 349–363.

[2] Haowen Luo, Taorong Qiu, Chao Liu, and Peifan Huang. "Research on fatigue driving detection using forehead EEG based on adaptive multiscale entopy", *Biomed. Signal Process. Control*, vol. 51, pp. 50–58, 2019.

[3] R. Chai, R. Ganesh, T. N. Naik S. H. Nguyen Y. Ling A. Craig Tran, and H. T. Nguyen, "Driver fatigue classification with independent component by entropy rate bound minimization analysis in an EEG-based system", *IEEE J. Biomed. Health Inform.*, vol. 21, no. 3, pp. 715–724, 2016.

[4] S. Ye and Y. Xiong Bill, "An innovative nonintrusive driver assistance system for vital signal monitoring", *IEEE J. Biomed. Health Inform.*, vol. 18, pp. 1932–1939, 2014.

[5] S. Hu and G. Zheng, "Driver drowsiness detection with eyelid related parameters by support vector machine", *Expert Syst. Appl.*, vol. 36, pp. 7651–7658, 2009.

[6] S. Ahn, T. Nguyen, H. Jang, J.G. Kim, and S.C. Jun, "Exploring neurophysiologicalcorrelates of drivers mental fatigue caused by sleep deprivation usingsimultaneous EEG, ECG, and fNIRS ata", *Front. Hum. Neurosci.*, vol. 10, pp. 128–141, 2016.

[7] J.-S. Lai, D. Cella, S. Choi, D. U. Junghaenel, C. Christodoulou, R. Gershon, and A. Stone, "How item banks and their application can influence measurement practice in rehabilitation medicine: A PROMIS fatigue item bank example", *Arch. Phys. Med. Rehabil.*, vol. 92, pp. S20–S27, 2011.

[8] Sobhan Sheykhivand, Tohid Yousefi Rezaii, Saeed Meshgini, Somaye Makoui, and Ali Farzamnia, "Developing a deep neural network for driver fatigue detection using EEG signals based on compressed sensing ", *Sustainability*, vol. 14, no. 5 p. 2941, 2022.

[9] Jian Cui, Zirui Lan, Olga Sourina, and Wolfgang Mller-Wittig, "EEGbased cross-subject driver drowsiness recognition with an interpretable convolutional neural network ", *IEEE Trans. Neural. Netw. Learn. Syst*, vol. 4, pp. 162–186, 2022.

[10] Sagila Gangadharan and A. P. Vinod, "Drowsiness detection using portable wireless EEG", *Comput. Methods Programs Biomed.*, vol. 214, p. 106535, 2022.

[11] Sengul Dogan, Ilknur Tuncer, Mehmet Baygin, and Turker Tuncer, "A new hand-modeled learning framework for driving fatigue detection using EEG signals", *Neural. Comput. Appl.*, vol. 16, pp. 1–18, 2023.
[12] Saad Arif, Saba Munawar, and Hashim Ali, "Driving drowsiness detection using spectral signatures of EEG-based neurophysiolgy", *Front. Physiol.*, vol. 14, p. 463, 2023.
[13] Ruilin Li, Ruobin Gao, and Ponnuthurai Nagaratnam Suganthan, "A decomposition-based hybrid ensemble CNN framework for driver fatigue recogniton", *Inf. Sci.*, vol. 624, pp. 833–848, 2023.
[14] M. Omidyeganeh et al., "Yawning detection using embedded smart camras", *IEEE Trans. Instrum. Meas.*, vol. 65, no. 3, pp. 570–582, 2016.
[15] C. Zhang, X. Wu, X. Zheng, and S. Yu, "Driver drowsiness detection using multi-channel second order blind identificatons", *IEEE Access*, vol. 7, pp. 11829–11843, 2019.
[16] Y. Zhang and C. Hua, "Driver fatigue recognition based on facial expression analysis using local binary pattrns", *Optik*, vol. 126, no. 23, pp. 4501–4505, 2015.
[17] H. Kalbkhani, M. G. Shayesteh, and S. M. Mousavi, "Efficient algorithms for detection of face, eye and eye state", *IET Comput. Vis.*, vol. 7, no. 3, pp. 184–200, Jun. 2013.
[18] B. Mandal, L. Li, G. S. Wang, and J. Lin, "Towards detection of bus driver fatigue based on robust visual analysis of eye state", *IEEE Trans. Intell. Transp. Syst.*, vol. 18, no. 3, pp. 545–557, 2017.
[19] L. Zhao, Z. Wang, X. Wang, and Q. Liu, "Driver drowsiness detection using facial dynamic fusion information and a DBN", *IET Intell. Transp. Syst.*, vol. 12, no. 2, pp. 127–133, 2018.
[20] G. Sikander and S. Anwar, "Driver fatigue detection systems: A review", *IEEE Trans. Intell. Transp.*, vol. 20, no. 6, pp. 2339–2352, 2019.
[21] Z. Mu, and J. Hu, and J. Min, "Driver fatigue detection system using electroencephalography signals based on combined entropy features", *Appl. Sci.*, vol. 7, no. 150, pp. 1–17, 2017.
[22] R. Kaur, and K. Singh, "Drowsiness detection based on EEG Signal analysis using EMD and trained neural network", *Int. J. Sci. Res.*, vol. 10, pp. 157–161, 2013.
[23] T. Tuncer, S. Dogan, and A. Subasi, "EEG-based driving fatigue detection using multilevel feature extraction and iterative hybrid feature selection", *Biomed. Signal Process. Control*, vol. 68, no. 10, pp. 102591, 2021.
[24] A. Vuckovic, V. Radivojevic, A. C. Chen, and D. Popovic, "Automatic recognition of alertness and drowsiness from EEG by an artificial neural network", *Med. Eng. Phys.*, vol. 24, no. 5, pp. 349–360, 2002.
[25] S. Hu, G. Zheng, and B. Peters, "Driver fatigue detection from electroencephalogramspectrum after electrooculography artefact removal", *IET Intell. Transp.*, vol. 7, no. 1, pp. 105–113, 2013.
[26] Z. Zhang, D. Luo, Y. Rasim, Y. Li, G. Meng, J. Xu, and C. Wang, "A vehicle active safetymodel: Vehicle speed control based on driver vigilance detection usingwearable EEG and sparse representation", *sensor*, vol. 16, no. 2, pp. 242, 2016.
[27] Q. Wang, Y. Li, and X. Liu, "Analysis of feature fatigue EEG signals based on waveletentropy", *Int. J. Pattern Recognit. Artif. Intell.*, vol. 26, p. 1854023, 2018.

[28] J.S. Richman, D.E. Lake, and J.R. Moorman, "Sample entropy", *Methods Enzymol.*, vol. 384, pp. 172–184, 2004.
[29] C. Bandt, and B. Pompe, "Permutation entropy: A natural complexity measure fortime series", *Phys. Rev. Lett.*, vol. 88, no. 17, pp. 174102, 2002.
[30] W. Chen, Z. Wang, H. Xie, and W. Yu, "Characterization of surface EMG signal basedon fuzzy entropy", *IEEE Trans. Neural Syst. Rehabil. Eng.*, vol. 15, no. 2, pp. 266–272, 2007.
[31] A. Chaudhuri and A. Routray, "Driver fatigue detection through chaotic entropy analysis of cortical sources obtained from scalp EEG signals", *IEEE Trans. Intell. Transp. Syst.*, vol. 21, no. 1, 2019.
[32] T. Tuncer, S. Dogan, F. Ertam, and A. Subasi, "A dynamic center and multi threshold point based stable feature extraction network for driver fatigue detection utilizing EEG signals", *Cog. Neurodynm.*, vol. 15, pp. 223–237, 2021.
[33] Z. Mu, J. Hu, and J. Min, \Driving fatigue detection based on EEG signals of forehead area", *Int. J. Pattern Recognit. Artif. Intell.* vol. 31, no. 05, p. 1750011, 2017.
[34] J. Yin, J. Hu, and Z. Mu, "Developing and evaluating a mobile driver fatigue detection network based on electroenchephalograph signals", *Healthcare Tech. Letts.*, vol. 4, no. 01, pp. 34–38, 2017.
[35] R. Chai et al., "Improving EEG-based driver fatigue classification using sparse-deep belief networks", *Frontiers Neurosci.*, vol. 11, no. 103, pp. 1–14, 2017.
[36] R. Fu, H. Wang, and W. Zhao, "Dynamic driver fatigue detection using hidden Markov model in real driving condition", *Expert Syst. Appl.*, vol. 63, pp. 397–411, 2016.
[37] A. Hazarika, L. Dutta, M. Barthakur, and M. Bhuyan, A multiview discriminant feature fusion-based nonlinear process assessment and diagnosis: application to medical diagnosis", *IEEE Trans. Instrument. Meas.*, vol. 68, no. 7, pp. 2498–2506, 2018.
[38] R. N. Khushaba, S. Kodagoda, and S. Lal, G. Dissanayake, "Driver drowsiness classification using fuzzy wavelet-packet-based feature-extraction algorithm", *IEEE Trans. Biomed. Eng.*, vol. 58, pp. 121131, 2011.
[39] J. Xia, M. Dalla Mura, J. Chanussot, P. Du, and X. He, \Random subspace ensembles for hyperspectral image classification with extended morphological attribute profiles", *IEEE Trans. Geosci. Remote Sens.*, vol. 53, no. 9, pp. 4768–4786, 2015.
[40] A. R. Hassan, " Computer-aided obstructive sleep apnea detection using normal inverse Gaussian parameters and adaptive boosting", *Biomed. Signal Process. Control*, vol. 29, pp. 22–30, 2016.
[41] I. Bayram, "An analytic wavelet transform with a flexible time-frequency covering", *IEEE Trans. Signal Proces.*, vol. 61, no. 5, p. 11311142, 2013.
[42] V. Gupta, M. D. Chopda, and R. B. Pachori, "Cross-subject emotion recognition using flexible analytic wavelet transform from EEG signals", *IEEE Sensor J.*, vol. 19, no. 6, pp. 2266–2274.
[43] V. Gupta, T. Priya, A. K. Yadav, R. B. Pachori, and U. R. Acharya,"Automated detection of focal EEG signals using features extracted from flexible analytic wavelet transform", *Pattern Recogn. Let.*, vol. 94, p. 180188, 2017.
[44] A. Hazarika, M. Barthakur, L. Dutta, and M. Bhuyan, "F-SVD based algorithm for variability and stability measurement of Bio-Signals, feature extraction and

fusion for pattern recognition", *Biomed. Signal Process. Control*, vol. 47, pp. 26–40, 2019.
[45] L. Dutta, C. Talukdar, A. Hazarika, and M. Bhuyan, "A novel lowcost hand-held tea flavor estimation system", *IEEE Trans. Ind. Electron*, vol. 65, no. 6, pp. 4983–4990, 2018.
[46] L. Breiman, "Bagging predictors", *Mach. Learn.*, vol. 24, no. 2, p. 123140, 1996.
[47] R. Fu, H. Wang, and W. Zhao, "Dynamic driver fatigue detection using hidden Markov model in real driving condition", *Expert Syst. Appl.*, vol. 63, p. 397411, 2016.
[48] A. Hazarika, L. Dutta, M. Boro, M. Barthakur, and M. Bhuyan, "An automatic feature extraction and fusion model: application to electromyogram (EMG) signal classification", *Int. J. Multimed. Inf. Retr.*, vol. 7, no. 3, pp. 1–14, 2018.
[49] A. Amarprit Singh Champak Talukdar Hazarika, Manabendra Bhuyan, Lachit Dutta, Mausumi Barthakur and Kishalay Chakraborty, "Microcontroller-based online nerve parameter estimation for diagnosis of healthy subject using real-time NCS signal acquired using voltage controlled neurostimulator", *IEEE Trans. Industr. Inform.*, vol. 21, pp. 193–210, 2021 Jan 11.
[50] A. Hazarika, P. Barman, L. Dutta, C. Talukdar, A. Subasi, and M. Bhuyan, "Real-time implementation of a multi-domain feature fusion model using inherently available large-volume sensor data", *IEEE Trans. Ind. Informat.*, vol. 15, no. 12, pp. 30–38, 2019.

Chapter 11

Feature fusion-based learning algorithm using multi-domain signal features for wearable healthcare devices

B. Patir, A. Saikia, C. J. Kumar, and J. Deka
Cotton University, Guwahati, India

A. Hazarika
Mayang Anchalik College, Marigaon, India

11.1 INTRODUCTION

Real-world experimental analysis requires to assess inherently available diverse sensor measurements for complete understanding of the given phenomena. Each signal carries a different level of information and degree of uncertainty [1]. In such case, the analysis based on a particular measurement will fail in providing logical and reliable inferences. However, analysis of such large sets of data obtained in various experimental analysis in typical ways is often difficult and also fails in provide quantitative information about the process undertaken. Therefore, most recent learning researches focus on developing advanced learning algorithms with special emphasis on the learning framework and decision module.

Use of large sets of signals to extract suitable sets of significant features for developing the support systems that are to be embedded in a portable device setup is essential [2]. With increasing demands of smart sensor technology [3, 4] and requirement of assessment of multiple measurements through advanced signal processing approach, the feature fusion technologies [5–7] in same or different formats get the attention in pattern recognition domains. There are two common feature fusion techniques, namely, feature- and decision-level fusions [8–11]. Feature-level fusion is quite effective while modality data are diverse in nature and model output totally depends how accurately it learns the information gathered from input information space. In multiple modalities data analysis, it is necessary to combine the outputs of various sub-modules that independently handle different modality data in different ways for providing conclusive inferences. In such circumstances, decision-level fusion plays a crucial role. In both cases, feature-level fusion is believed to be more effective, which can be implemented at the initial level of individual sub-model or complete model [12, 13]. This is due

 DOI: 10.1201/9781003474524-11

to the fact that integrity and reliability of the output of complete decision-making model employs decision-level fusion that relies on how accurately a sub-model learns the information from the respective input feature searched space. In such cases, data-driven and feature fusion framework play a crucial role which is widely used in signal processing and other engineering domains.

Principal component analysis (PCA), partial least squares (PLS) [14], locally weighted PLS [15], advanced PLS [16], key performance indicators (KPI) [17], fuzzy positivistic [18], multivariate statistical process monitoring [19], kernel PCA and kernel PLS [20], and others are the commonly used methods. However, they also have inherent limitations for which they are not used as general. That means one method might not ensure its suitability in various real-world applications. For instance, PCA and PLS are not suitable in nonlinear applications due to their linearity assumption [12] but they could be improved according to Zhang et al. [20]. However, such methods often suffer in selecting the kernel function and parameter determination. Additionally, some other methods used KPI with PLS [14] and locally weighted projection [12] from the available resources for detection of abnormality. Despite a reasonable level of improvements in various stages of models, challenges still exist for which it is essential to develop a more robust strategy to improve further. As a result of aforesaid reasons, multi-view learning models (MLM) have gained attention in machine learning–based pattern recognition. This enables interaction of multiple sensor measurement of same or different type and creates high-order input information searched space from which multiple features and their derivatives, termed as multi-view features, can be extracted [21]. The major issue of such strategy is the curse of dimensionality, which could be eliminated by using feature projection, optimization, and feature fusion [22]; for example, canonical correlation analysis (CCA) [13] and its advanced version discriminant correlation analysis (DCA).

This chapter presents multiple view feature fusion–based model learning method using feature projection technique in order to analyze the given process that requires thorough analysis of multiple sensor measurements. The objective of this method includes more signals obtained from the sensor so as to cover more information, and subsequently extracted features would carry more reliable information and energy about the underlying process. In this method, a set of high-dimensional features are created in direct and wavelet domain using discrete wavelet transformation (DWT) for various processes. Then feature spaces are projected to subspace domain and optimization low-order features sets are evaluated. Afterwards, the significant features sets are evaluated using significant statistical test and subjected to various learning models. Finally, the performance of the developed algorithm is evaluated in terms of various markers and compared with various prior methods in order to explore its efficacy.

11.2 METHOD

11.2.1 Mathematical formulation

Consider a particular dataset as mentioned in Hazarika et al. [23], which involves C processes that include diverse nature of real-time signals. Each process partitioned into g subgroups from which sub-feature matrix X is formulated. Now consider X to be a feature matrix that contains m signals of N samples and M is the corresponding DWT version.

$$X = [x_1, x_2, \cdots x_m]^T \epsilon \mathbb{R}^{m \times N}$$
$$Z = [y_1, y_2, \cdots y_m]^T \epsilon \mathbb{R}^{m \times N}$$

$$\mathrm{X_n} \epsilon\, \mathbb{R}^m, \mathrm{y_n} \epsilon\, \mathbb{R}^m \quad n = 1, \ldots\ldots\ldots N.$$

Here, two matrices X and Z are statistically independent. Each element in Z is the low frequency wavelet component (i.e., A_n) of corresponding element in X, which are obtained by performing DWT over each x_n in $\mathbf{X}$. In this analysis the high-frequency components D_ns are eliminated due to fact that maximum real-time signal information contain low-frequency components.

In the next step, feature matrices as mentioned previously are uniformly decomposed into kth sub-features, termed as decomposed-MV (DMV) as follows:

$$X = [X1, X2, \ldots, Xk]T, Xk \in \mathbb{R}^{m \times l} \tag{11.1}$$

$$M = [U1, U2, \ldots, Uk]T, Uk \in \mathbb{R}^{m \times l} \tag{11.2}$$

The terms ratio $l = N/k$ indicates number of samples in each signal sequence of decomposed feature (DF). The DWT processes the signal $x[n]$ through a high- and low-pass filters with impulses $h[n]$ and $g[n]$ respectively. The first stage of decomposition provides $d^1[n]$ and $a^1[n]$ and in the second level or stage a^1 is further down sampled by 2 (See Figure 11.1) and produces an output of $d^2[n]$ and $a^2[n]$ which are expressed as

$$\mathrm{d}^2[\mathrm{n}] = \sum_{k=o}^{N/2-1} a^1[k] h[n-k] \tag{11.3}$$

$$\mathrm{a}^2[\mathrm{n}] = \sum_{k=o}^{N/2-1} a^1[k] g[n-k] \tag{11.4}$$

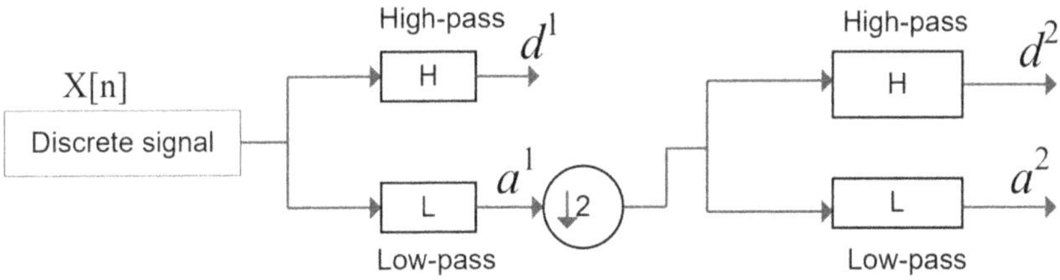

Figure 11.1 DWT two-level decomposition tree of a discrete signal $x[n]$.

It is to be mentioned that DWT method requires proper mother wavelet function that accurately fits the user signal. According to the method mentioned in Kiymik et al. [24], the mother wavelet function, namely, second-order Daubechies, is appropriate for EEG signal classification. The proposed method employs only low-frequency components (i.e., y_n) in evaluating the wavelet DFs. Here DWT is applied over direct feature space to find statistically independent feature space for combining the multi-domain features in fusion mode that are to be discussed later.

11.2.2 CCA and DCA

Let consecutive DMVs X_1, Y_2 be represented by X and Y. Then PCA is employed over input feature matrices; in order to remove redundancy information contents dimension reduction is applied [9, 25, 26]. In order to make the centred data matrices, the mean of each are removed as mentioned in Wang et al. [12]. Then CCA is applied over the input feature matrices for finding two linear transformations, $u = A_x^T X$ and $V = B_y^T Y$ by optimizing Condition 11.11 as follows:

$$\max_{A_x B_y} \rho(u,v) \rightarrow A_x^T \sum_{xy} B_y \tag{11.5}$$

$$s.t. A_x^T \sum_{xx} A_x = B_x^T \sum_{yy} B_y = 1 \tag{11.6}$$

$$\sum_{xx} = XX^T, \sum_{yy} = YY^T, \sum_{xy} = XY^T = \sum_{yx}^T, \tag{11.7}$$

where Σ_{xx} and Σ_{yy} are autocovariance matrices, and Σ_{xy} and Σ_{yx} are cross-covariance matrices of X and Y. The weight vectors A_x R^d and B_y R^d are obtained in projection subspace by solving the following model equations as follows:

$$XY^T\left(YY^T\right)^{-1}YX^T A_x = \rho^2 XX^T A_x, \tag{11.8}$$

$$YX^T\left(XX^T\right)^{-1}XY B_y = \rho^2 YY^T B_y. \tag{11.9}$$

The matrix ρ^2 indicates correlation matrix in terms of non-zero diagonal elements in descending order. The square root of diagonal element represents correction. Each value indicates the canonical correlations between projection vectors $A_x = [A_{1,\ldots\ldots}A_d]$ and $B_x = [B_1,\ldots\ldots\ldots B_d]$. For instance, the first value indicates the correlation of A_1 and B_1.

Unlike CCA–based correlation-based analysis, DCA [26] introduces the class-information into the feature space. Let the matrix X contain n columns, which are divided into C separate groups and n_i columns belong to ith class. It is also considered that $x_{ij}\in X$ represents feature vector corresponding to the jth sample in ith class. The mean values of class and features as well as between-scatter matrix are given below.

$$\bar{x} = \sum x_{ij} / n_i, \mu = \sum n_i \overline{x_i} / n \tag{11.10}$$

$$S_{Bx} = \sum_{i=i}^{C} n_i \nabla_i \nabla_i^T = \Phi\Phi^T \tag{11.11}$$

$$\phi = \left[\sqrt{n_1}\left(\overline{x_1} - \mu\right), \ldots\ldots \sqrt{n_c}\left(\overline{x_c} - \mu\right)\right] \tag{11.12}$$

Where $\nabla_i = \left(\overline{x_i} - \mu\right)$. In order to ensure the well-separation of classes, $\Phi^T \Phi$ is diagonalized as follows:

$$Q^T\left(\Phi^T \Phi\right)Q = \Lambda_r \tag{11.13}$$

Where Q is an orthogonal matrix that consist of r non-zero eigenvectors. The r significant eigenvectors of S_{Bx} is obtained using the mapping $Q \rightarrow \phi Q$ [26].

$$\left(\Phi Q\right)^T S_{Bx}\left(\Phi Q\right) = \Lambda_r \tag{11.14}$$

According to the theory of DCA, the transformation $A_{bx} = \Phi Q \Lambda^{-1/2}$ employs

S_{Bx} and obtains reduced matrix $\bar{X}$ of X such that

$$\bar{X} = A_{bx}^T X \tag{11.15}$$

in the same way $\bar{Y}$ is obtained. Similar to the CCA correlation, in order to achieve the features in one set having non-zero correlation with the corresponding features in other set, the scatter matrix $\Sigma_{\tilde{x}\tilde{y}} = \tilde{X}\,\tilde{Y}^T$ is diagonalized using singular value decomposition.

$$\Sigma_{\overline{xy}} = U\,\Sigma V^T \Rightarrow U\,\Sigma_{\overline{xy}} V^T = \Sigma \tag{11.16}$$

Here Σ is diagonal matrix with non-zero diagonal elements. Let $A_{cx} = U\,\Sigma^{-1/2}$ and $B_{cx} = V\,\Sigma^{-1/2}$ so that

$$\left(U\,\Sigma^{-1/2}\right)^T \Sigma_{\overline{xy}}\left(V\,\Sigma^{-1/2}\right) = I \tag{11.17}$$

Finally, DCA finds two transformed feature sets as follows.

$$\bar{u} = A_{cx}^T \bar{X} = A_{cx}^T A_{bx}^T X = A^T X, \tag{11.18}$$

$$\bar{v} = B_{cx}^T \bar{Y} = B_{cx}^T B_{bx}^T Y = B^T Y \tag{11.19}$$

The covariance matrix $(\tilde{X}\tilde{X}^T)$ and between-scatter matrices of the transformed features are diagonalized to ensure the well-separability of classes. The generalized version that developed herein termed as multiview DCA (MDCA) evaluates all locally features-incorporating class-structure information. It then converts them into single form of vector that contains discriminant vector. Every time, the MDCA finds two transformed sets of features employing direct DMV pair in X and DWT-DMV in M which is applied over all consecutive DMV in X and M.

The summation technique [26] that fuses the features is expressed as follows.

$$z = A^T X + B^T Y = \begin{pmatrix} A \\ B \end{pmatrix}^T \begin{pmatrix} X \\ Y \end{pmatrix} \tag{11.20}$$

The above fusion strategy includes only single transformed feature set. Unlike DCA, the MDCA generalizes the fusion model by combining multi-domain independent subspace features of low order for finding more discriminant vector as follows:

$$Z_{ij} = A^T X + B^T Y + C^T U_1 + D^T U_2 \tag{11.21}$$

Here i, j indicate consecutive features and C and D are weight vectors of DWT-DMV U_1 and U_2. This fusion model as mentioned in Equation 11.21, termed as generalized discriminant features (gDF), includes multi-domain features that improve the quality of discriminant feature [8]. This strategy is applied over all subspace vectors and finally generalized features are evaluated for all $k/2 + 1$ consecutive DMVs, which are later subjected to statistical test and model learning.

11.2.3 Nonlinear medical dataset

In medical analysis, conventionally electrodes are used for acquisition of multiple signals from the given part of body undertaken. As mentioned earlier in Section 11.1, all such signals are highly nonlinear with higher correlation with each other. They are subjected to visual subjective analysis for subjective conclusion as disorders or normal. Usually such medical abnormalities can be further classified into some specific class based on the nature of characteristics and abnormality. Such assessment is also seen in many industrial fault diagnosis models that employ sensor signal [27]. Therefore, for formulation and validation of the proposed method, a EEG dataset is considered herein. These signals are recorded using surface through signal acquisition circuit, known as signal conditioning circuit (SCC), for analysis of brain functional activities. Usually, the amplitude of these signals falls in μV. The SCC contains various submodules including signal processing, signal output, and data storage blocks. In the SCC, signals are usually recorded by using an in-built electrode system which are then further amplified through a signal processing block that includes high-gain differential amplifiers, filters, and an ADC. The data management setup stores the recorded data in a predefined file for future analysis and research.

A publically available online EEG data [28] is considered for model validation and the performance analysis. Signals in the database were collected at the University Hospital in Bonn, Germany. The recording setup contains 128-channel amplifier with filter setting of 0–40 Hz and ADC of 12-bit with sampling rate of 173.61 Hz. It contains five datasets, labelled as A–E. Each set contains 100 single channel signals. Duration and number of samples in each signal are 23.6 s and $173.61 \times 23.6 \approx 4097$. Sets A and B include signals of five healthy subjects in an awake state with an eye open and eye closed respectively. Signals sets C and D are from within the epileptogenic zone and the hippocampal formation of the opposite hemisphere of the brain. These two sets of signals were recorded during seizure-free intervals while in E there was seizure activity. Figure 11.2 shows EEG signal patterns of A, B, and E classes.

11.3 RESULTS AND DISCUSSION

As mentioned in Section 11.2.1, each input process, namely, A–E, is partitioned into $G = 4$ groups with random choice of signals and subsequently DMV features are evaluated using Equations 11.1 and 11.2. Each consecutive

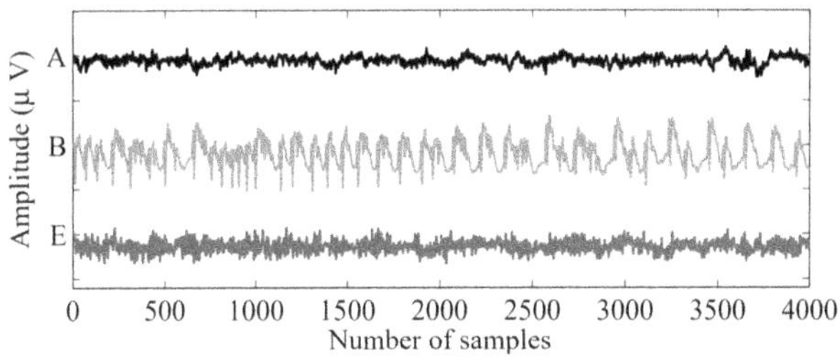

Figure 11.2 Non-linear EEG data patterns of A, B, and E.

pair of DMV feature is subjected to correlation analysis. It is worth mentioning the fact that the proposed MDCA finds low-order subspace feature vectors by maximizing correlation between all possible consecutive DMV pairs and evaluating significant features in two domains independently. It then fuses them to find discriminant feature vectors for diagnosis. In order to apply MDCA over both independent DMV space, DWT–based DMV feature vectors are also evaluated by considering second-level low-frequency DWT coefficients (i.e., a^2s) as shown in Figure 11.3. In order to find the discriminant features from the DMVs feature-searched pace through use of the proposed MDCA method, the correlation analysis is performed first over DMVs, formulating them considering m = 8, 10, 12 number of signals are taken for all the processes. For each value of m, four DMVs (i.e., k = 4) are evaluated. Then correlations analysis is performed in between three consecutive pair–DMV (=$k/2$ + 1 = 2 + 1) of all processes.

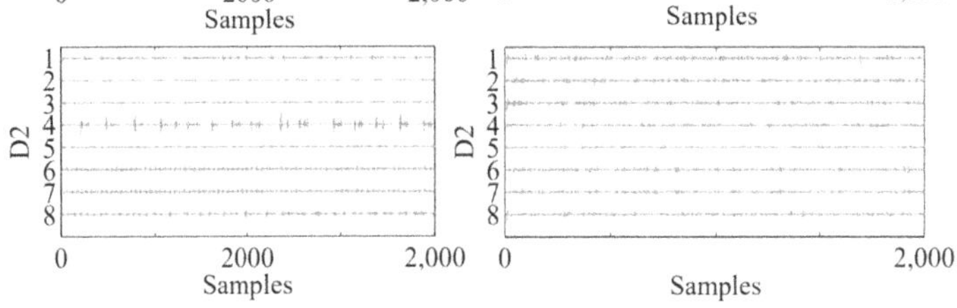

Figure 11.3 Second-level DWT coefficients *A2* (i.e., a^2) and D2 (i.e., d^2) of input processes E and A.

For instance, Figure 11.4 shows the mean correlations for three values m in A. From the analysis over all the process, it is seen that the quantitative value of correlation index between the pair of the subspace features is ≥50% in all cases.

This indicates that obtained features by using projection technique corresponding to higher correlation index are most significant in carrying valuable information of high-order input features. This analysis is also carried out over DWT DMV features in the same way as described. The proposed methods aim to synchronize low-order features of both statistically independent domains that correspond to a higher value of correlation index. In order to improve computational complexity the threshold value of dth is at 8 with m = 10. These features are statistically significant with $p < 0{:}001$. Afterwards three gDFs obtained for three pairs are transformed using $\Phi_i = Z_{ij} + \gamma I$. The term γ indicates class indicator parameter whereas I is unitary matrix. Zero padding is adopted to adjust the dimension of Z_{ij}. To enhance the discriminating ability the values of γ are set at 0 (for A), 3 (for B), and 5 (for E). The mean gDFs are applied over linear discriminant analysis (LDA) for finding more discriminate low-order feature for classification. Finally five significant features are validated using cross-validation strategy. The p-value of each feature is found to be $<1 \times 10^{-4}$ which indicates higher discriminant capability of evaluated features.

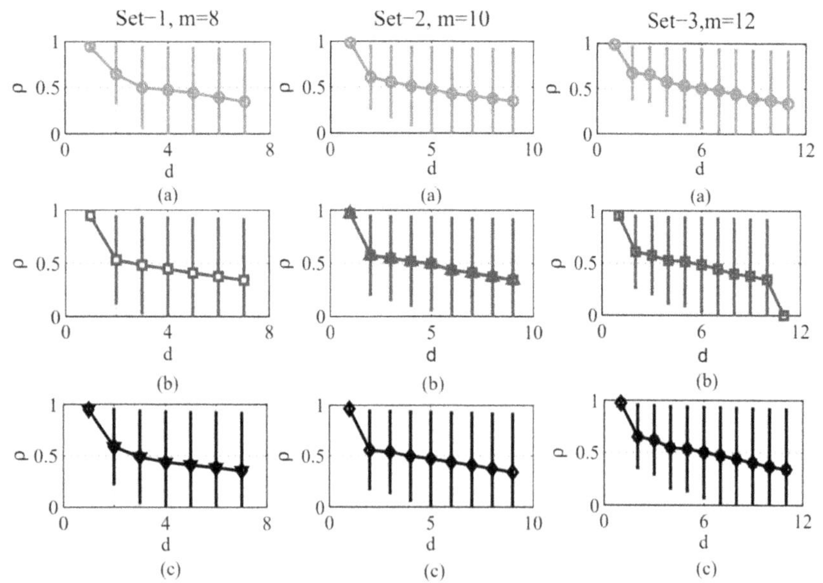

Figure 11.4 Average correlations (with ±σ) of consecutive DMVs, i.e., (1,2), (2,3), (3,4) in A. Sub-figures sets (a)–(c)s indicate three independent measurements for m = 8, 10, 12.

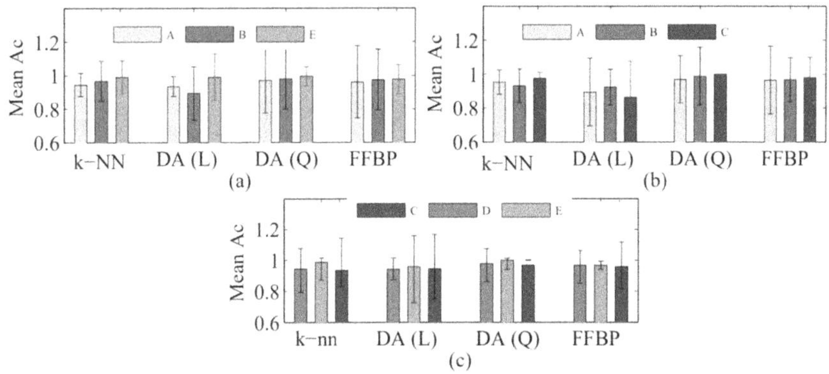

Figure 11.5 Mean classification performances with various AI models.

Now the feature-driven stage is embedded with simple decision-making model for better understanding and suitable implementation in real-world domain. It also establishes the integrity of evaluated feature extraction framework. This method employs 20% of total data for evaluation of features using two-fold cross-validation technique. The model performances are investigated over the remaining dataset. For classification task, the dataset is divided into three separate groups—ABE, ABC, and CDE. The accuracies are investigated by using k-nearest neighbour (k-NN) with $k = 1$. The efficacy of the proposed method is further investigated using neural network (NN) as well as discriminant analysis (DA) with linear and quadratic discriminant functions.

The mean accuracies of various models obtained over five repeated measurements is shown in Figure 11.5. It is observed that the MDCA with DA with quadratic function shows better accuracy, which is $\approx$99%. This is mainly due to wide diversity among the class-specific feature vectors. The network, namely, feedforward back prorogation network, achieves results that are very close to that of the DA with a quadratic function. The network model adopted includes one input, one hidden, and one output layer. The model employs a sigmoid and linear transfer function in hidden and output layers [25]. Eight neurons are used in the hidden layer and three or five neurons in the output layer for three or five classes to be classified. The optimal network parameters are obtained by trial-and-error method. In order to establish the reliability and efficacy of the MDCA–based method, the performances are investigated over combinations ABC, CDE, and ABE and presented in Figure 11.5.

It is observed that the MDCA–based method over CDE shows slightly better results as compared to others. The low accuracies in ABC group is due to similarity among the signals. However, the MDCA+DA with quadratic function obtains mean accuracy of 100%. The method with k-NN provides accuracy of 97:75% in ABC and 98:75% in CDE. The proposed method takes processing time of 15–20 in classifying inputs A, B, and E. It is seen that the proposed method shows a small error in feature biasing due to reduction of redundant information contents from the high-order feature searched space.

11.4 COMPARATIVE ANALYSIS

This section focuses on comparative analysis in order to explain the efficacy of the proposed method. Various prior methods akin to this study are summarized in Table 11.1. Naghsh-Nilchi and Aghashahi [29] used various derivative features including entropy, mean complexity, and so on, and then fed with multiple layer NN (MLNN) to identify epileptic seizures. Nonetheless, this method involves multiple processing steps and use of specific statistical measures derived from specific predefined signals that may not ensure the wide coverage of information of the underlying process. DWT–based methods often provide good results, however, low-order features obtained DWT coefficients may fail in providing important information contents of signals. Nguyen et al. [31] adopted a method that employs the recurrence quantitative measures of heart rate variability data for detection of sleep apnea. The method cited in Kaya et al. [32] introduced a method by employing one-dimensional local binary features for EEG patterns classification. Empirical mode decomposition (EMD) [33] usually suffers mode-mixing

Table 11.1 Comparison analysis of various prior methods with the proposed method

Methods	*Type of feature*	*Classifier*	*Ac (%)*
Naghsh-Nilchi and Aghashahi [29]	Spectral and Time	MLNN	97.50
Orhan and Ozer [30]	DWT	MLNN	95.60
Nguyen *et al.* [31]	Recurrence quantitative	Decision fusion	85.26
Kaya *et al.* [32]	Local binary	BayesNetNN	97.0[a] 95.40[a]
Hassan *et al.* [33]	EMD-statistical	ELM	83.77
Gajic *et al.* [35]	DWT/nonlinear	Quadratic	98.70
Soomro *et al.* [36]	Eigen values	CCA-NN	92.60
Hassan *et al.* [37]	Inverse Gaussian	Adaptive boosting	87.33
Proposed [C3]	**Discriminant**	**DA**	**100.0**
Proposed [C3]	**Discriminant**	**K-NN**	**98.75**
Proposed [C4]	**Discriminant**	**FFBP**	**96.43**[a]

Notes:
a for three and five categories classification performance.
Bold text indicates evolution results.

problem [34]. Gajic *et al.* [35] utilized time-frequency, DWT features, and nonlinear features for classification task.

Although it provided promising results, it involves multiple steps in extracting feature and analysis that limits its utility in real-time domain. Some methods use CCA eigenvalues features and normal inverse Gaussian parameters in wavelet domain respectively to diagnose sleep apnea [36, 37]. In order to get better accuracy as mentioned in [38], it is necessary to select an optimum number of decision tree, which is often tricky. In addition to their inherent theoretical bottlenecks, cited methods did not focus on the use of high-dimensional feature space using fusion technique. However, it is essential to involve large signals in developing feature searched space for wide coverage of information content, and features obtained from that input space using given technique are needed to be fused for better classification task. In contrast, the proposed MDCA employs large data–based input feature searched space. It effectively utilizes large data through discriminant formulation leads to achieve better results as compared to the prior methods that ensure the possible real-time implementation of the MDCA.

11.5 CONCLUSION

The chapter addressed a fusion-based data-driven and classification model using generalized version of DCA using available nonlinear datasets. In this method, first multiple domain feature extraction framework that can handle high-dimensional input searched space are developed. Afterwards, multi-domain feature fusion technique is developed for deriving more discriminant low-order features which are embedded into classification purposes in the subsequent stages. The performance of the proposed method is investigated by integrating the method with various promising decision-making models. Significant classification performances over multiple datasets indicate the efficacy of the proposed method and ensure the possibility of its uses in real-time applications for better diagnosis.

ACKNOWLEDGEMENTS

The authors would like to thank anonymous reviewers for their valuable suggestions that help to improve the chapter. The authors would also like to thank Cotton University fraternity for various supports. This work was supported by Inhouse research project, Cotton University with the Ref: CU/Dean/R & D/2019/05/1995, 4/3/222 Assam, India.

BIBLIOGRAPHY

[1] L. Dana, T. Adali and C. Jutten, "Multimodal data fusion: An overview of methods, challenges, and prospects," *Proc. IEEE*, vol. 103, no. 9, pp. 1449–1477, 2015.

[2] Z. Lv, H. Song, P. Basanta-Val, A. Steed and M. Jo, "Next-generation big data analytics: State of the art, challenges, and future research topics," *IEEE Trans. Ind. Informat.*, vol. 13, no. 4, pp. 1891–1899, 2017.

[3] G. C. Meijer et al., *Smart sensor systems*. Wiley Online Library, 2008.

[4] A. Singh, A. Gaur, A. Kumar, M. K. Singh, K. Kapoor, P. Mahanta, A. Kumar and S. C. Mukhopadhyay, "Sensing technologies for monitoring intelligent buildings: A review,"*IEEE Sensors J.*, vol. 18, no. 12, pp. 4847–4860, 2018.

[5] A. Hazarika, M. Barthakur, L. Dutta and M. Bhuyan, "Multi-view learning for classification of EMG template," *Proc. in Conf. IEEE Signal Proces. and Commun.*, India, 2017, pp. 467–471.

[6] R. C. Luo and C. C. Chang, "Multisensor fusion and integration: A review on approaches and its applications in mechatronics," *IEEE Trans. Ind. Informat.*, vol. 8, no. 1, pp. 49–60, 2012.

[7] R. C. Luo, C. C. Chang and C. C. Lai, "Multisensor fusion and integration: Theories, applications, and its perspectives,"*IEEE Sensors J.*, vol. 11, no. 12, pp. 3122–3138, 2011.

[8] M. E. Sargin, Y. Yemez, E. Erzin and A. M. Tekalp, "Audiovisual synchronization and fusion using canonical correlation analysis," *IEEE Trans. Multimedia*, vol. 9, no. 7, pp. 1396–1403, Oct. 2007.

[9] A. Hazarika, M. Barthakur, L. Dutta and M. Bhuyan, "F-SVD based algorithm for variability and stability measurement of Bio-Signals, feature extraction and fusion for pattern recognition," *Biomed. Signal Process. Control.*, vol. 47, pp. 26–40, 2019.

[10] A. Hazarika, L. Dutta, M. Boro, M. Barthakur and M. Bhuyan, "An automatic feature extraction and fusion model: Application to electromyogram (EMG) signal classification," *Int. J. Multimed. Inf. Retr.*, vol. 7, no. 3, pp. 1–14, 2018.

[11] A. Hazarika, L. Dutta, M. Barthakur and M. Bhuyan, " A multiview discriminant feature fusion-based nonlinear process assessment and diagnosis: Application to medical diagnosis," *IEEE Trans. Instrument. Meas.*, vol. 68, no. 7, pp. 2498–2506, 2018.

[12] G. Wang, S. Yin and O. Kaynak, "An LWPR-based data-driven fault detection approach for nonlinear process monitoring," *IEEE Trans. Ind. Informat.*, vol. 10, no. 4, pp. 2016–2023, 2014.

[13] A. Hazarika, M. Barthakur, L. Dutta and M. Bhuyan, "F-SVD based algorithm for variability and stability measurement of bio-signals, feature extraction and fusion for pattern recognition,"*Biomed. Signal Proces. Control*, vol. 47–60, 596–613, 2019.

[14] S. Yin, S. Ding, A. Haghani, H. Hao and P. Zhang, "A comparison study of basic data-driven fault diagnosis and process monitoring methods on the benchmark tennessee eastman process," *J. Process Control*, vol. 22, no. 9, pp. 15671581, 2012.

[15] S. Yin, X. Xie and W. Sun, "A nonlinear process monitoring approach with locally weighted learning of available data," *IEEE Trans. Ind. Electron.*, vol. 64, no. 2, pp. 1507–1516, 2017.

[16] X. Xie, W. Sun and K. C. Cheung, "An advanced PLS approach for key performance indicator-related prediction and diagnosis in case of outliers," *IEEE Trans. Ind. Electron.*, vol. 63, no. 4, pp. 2587–2594, 2016.

[17] Y. A. W. Shardt, H. Hao and S. X. Ding, "A new soft-sensor- based process monitoring scheme incorporating infrequent KPI measurements," *IEEE Trans. Ind. Electron.*, vol. 62, no. 6, pp. 3843–3851, Jun. 2015.

[18] S. Yin and Z. Huang, "Performance monitoring for vehicle suspension system via fuzzy positivistic C-means clustering based on accelerometer measurements," *IEEE/ASME Trans. Mechatronics*, vol. 20, no. 5, pp. 26132620, Oct. 2015.
[19] A. Haghani, T. Jeinsch and S. X. Ding, "Quality related fault detection in industrial multimode dynamic processes," *IEEE Trans. Ind. Electron.*, vol. 61, no. 11, pp. 64466453, Nov. 2014.
[20] Y. Zhang and C. Ma, "Fault diagnosis of nonlinear processes using multiscale KPCA and multiscale KPLS,"*Chem. Eng. Sci.*, vol. 66, no. 1, pp. 64–72, Jan. 2011.
[21] N. M. Correa, T. Adali, Y.-O. Li and V. D. Calhoun, "Canonical correlation analysis for data fusion and group inferences," *IEEE Signal Process. Mag.*, vol. 27, no. 4, pp. 3950, Jun. 2010.
[22] J. Zhao, X. Xie, X. Xu and S. Sun. "Multi-view learning overview: Recent progress and new challenges,"*Inform. Fusion*, vol. 38, pp. 43–54, 2017.
[23] A. Hazarika, P. Barman, C. Talukdar, L. Dutta, A. Subasi and M. Bhuyan, "Real-time implementation of a multidomain feature fusion model using inherently available large sensor data," *IEEE Trans. Industr. Inform. 15*, no. 12, pp. 6231–6239, 2019.
[24] M. K. Kiymik, M. Akin and A. Subasi, "Automatic recognition of alertness level by using wavelet transform and artificial neural network," *J. Neuroscience Methods*, vol. 139, no. 2, pp. 231–240, 2004.
[25] L. Dutta, C. Talukdar, A. Hazarika and M. Bhuyan, "A novel low cost handheld tea flavor estimation system," *IEEE Trans. Ind. Electron.*, vol. 65, no. 6, pp. 4983–4990, 2017.
[26] M. Haghighat, M. A. Mottaleb and W. Alhalabi, "Discriminant correlation analysis: Real-time feature level fusion for multimodal biometric recognition," *IEEE Trans. Inform. Forensics Security*, vol.11, pp. 1984–1996, 2016.
[27] Z. Gao, S. X. Ding and C. Cecati, "Real-time fault diagnosis and fault-tolerant control," *IEEE Trans. Ind. Electron.*, vol. 62, no. 6, p. 37523756, Jun. 2015.
[28] R. G. Andrzejak, K. Lehnertz, F. Mormann, C. Rieke, P. David and C. E. Elger, "Indications of nonlinear deterministic and finite dimensional structures in time series of brain electrical activity: Dependence on recording region and brain state,"*Phys. Rev. E*, vol. 64, p. 061907, 2001.
[29] A. R. Naghsh-Nilchi and M. Aghashahi, "Epilepsy seizure detection using eigen-system spectral estimation and multiple layer perceptron neural network," *Biomed. Signal Proces. Control*, vol. 5, no. 2, pp. 147–157, 2010.
[30] U. M. H. Orhan and M. Ozer, "EEG signals classification using the K-means clustering and a multilayer perceptron neural network model,"*Expert Syst. Appl.*, vol. 38, no. 10, pp. 13475–13481, 2011.
[31] H. D. Nguyen, B. A. Wilkins, Q. Cheng and B. A. Benjamin, "An online sleep apnea detection method based on recurrence quantification analysis," *IEEE J. Biomed. Health Inform.*, vol. 18, no. 4, pp. 1285–1293, 2014.
[32] Y. Kaya, M. Uyar, R. Tekin and S. Yldrm, "1D-local binary pattern based feature extraction for classification of epileptic EEG signals," *Appl. Math. Comput.*, vol. 243, pp.209–219, 2014.
[33] A. R. Hassan, "Automatic screening of obstructive sleep apnea from single-lead electrocardiogram," *Proc. in Conf. ICEEICT Electrical Eng. and Infor. Commun. Technology*, Dhaka, 2015, pp. 1–6

[34] A. R. Hassan and A. Subasi, "Automatic identification of epileptic seizures from EEG signals using linear programming boosting," *Comput. Methods Programs Biomed.*, vol. 136, pp. 65–77, 2016.
[35] D. Gajic, Z. Djurovic, J. Gligorijevic, S. D. Gennaro and I. Savic-Gajic, "Detection of epileptiform activity in EEG signals based on time-frequency and non-linear analysis," *Front. Comput. Neurosci.*, vol. 9, pp. 38, 2015.
[36] M. H. Soomro, S. H. A. Musavi and B. Pandey, "Canonical correlation analysis and neural network (CCA-NN) based method to detect epileptic seizures from EEG signals," *Int. J. BioSci. Biotechnol.*, vol. 8, no. 4, pp. 11–20, 2016.
[37] A. R. Hassan, "Computer-aided obstructive sleep apnea detection using normal inverse Gaussian parameters and adaptive boosting" *Biomed. Signal Proces. Control*, no. 29, pp. 22–30, 2016.
[38] A. R. Hassan and M. A. Haque, "Computer-aided obstructive sleep apnea screening from single-lead electrocardiogram using statistical and spectral features and bootstrap aggregating," *Biocybern. Biomed.*, vol. 36, no.1, pp. 256–266, 2016.

Chapter 12

The intelligence of WSNs

Nashwan Ghaleb Al-Thobhani
Sana'a Community College, Sana'a, Yemen

Jamil Sultan
University of Modern Sciences, Sana'a, Yemen

12.1 INTRODUCTION AND BACKGROUND

Our world is facing a rapid growth of population accompanied with an increase in the average lifetime of individuals. According to a study by the World Health Organization (WHO), by 2050, the population over 60 will be about 2.1 billion people [2, 20–26, 28]. Ubiquitous healthcare is an emerging technology that promises increases in efficiency, accuracy, and availability of medical treatment due to the recent advances in wireless communication and in electronics offering small and intelligent sensors able to be used on, around, in, or implanted in the human body. Wireless body area networks (WBANs) are an active area of research and development as they offer the potential for significant improvement in healthcare delivery and monitoring [13, 14, 16–19, 27, 29]. WBANs consist of a number of heterogeneous biosensors. These sensors are placed on different parts of the body and can be worn or implanted under the user's skin. Each of them has specific requirements and is used for different tasks. These devices are used to measure changes in a patient's vital signs and detect human emotions or states, such as fear, stress, happiness, and so on. It communicates with a special coordinating node, which is generally power constrained and has greater processing capabilities [8, 9]. It is responsible for sending the patient's biological signals to the medical doctor in order to provide real-time medical diagnostic and allow him to take the right decisions. As shown in Figure 12.1, the overall structure of a WBAN consists of three layers of communication: Intra–BAN communications, Inter–BAN communications, and beyond–BAN communications. Intra–BAN communications denote communications among wireless body sensors and the master node of the WBAN. Inter–BAN communications involve communications between the master node and personal devices such as notebooks, home service robots, and so on. The beyond–BAN tier connects the personal device to the Internet. Communication between different parts is supported by several technologies, such as Bluetooth, IEEE802.15.4 [29]. IEEE802.15.6 was designed especially for WBAN applications while responding to the majority of their requirements [34]. However, it looks lower performing in some

DOI: 10.1201/9781003474524-12

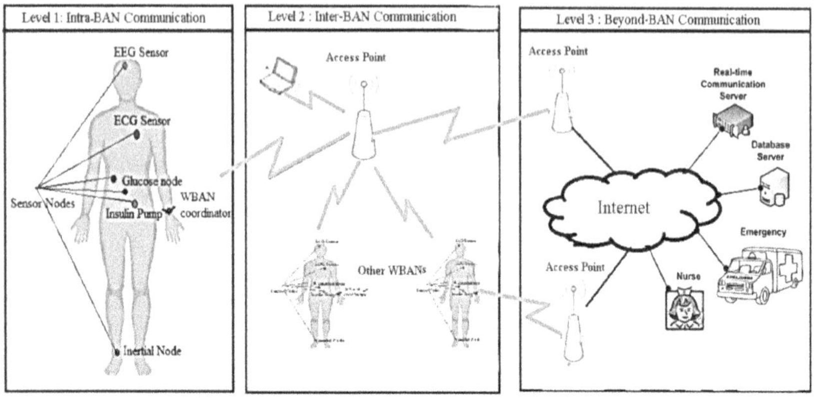

Figure 12.1 Three-level WBAN architecture.

cases in comparison with other technologies supporting WBAN. Wi-Fi, Bluetooth, and mobile networks can be solutions for implementing WBAN applications, since each technology offers specific characteristics, allowing it to meet the constraints of some applications. In fact, WBAN applications cover numerous fields in order to improve the user's quality of life.

According to WHO, cardiovascular disease (CVD) is the primary cause of the deaths in the world. In fact, it is estimated that the number of CVD–related deaths, mainly from heart disease and stroke, will reach up to 23.3 million by 2030 [28]. Besides this, more than 246 million people will suffer from diabetes and the rate of CVD patients or diabetics will increase; similarly, the percentage of individuals in the populace over 60 years old will increase in the upcoming years [29]. In this study, we aim to design and implement a smart healthcare IS with an IoT–based architecture to measure and send heart rate, blood oxygen level, blood pressure, GPS, and continuously monitored body temperature and thermal temperature information to the Internet. In addition, we design an interface for the end user [28]. The developed system can overcome the limitations of fixed health measurement variables and existing health devices, which can be replaced with home-use sensors, and commercial sensors by adding control and communication layers to the sensors without redesigning them or performing mass migration.

12.2 LITERATURE BACKGROUND

Wireless body area network (WBAN) is developed from wireless personal area network (WPAN). In this chapter, we will give more details about WBAN application, architectures, layer, technology, and routing. As companies rely on applications like electronic mail and database management for core business operations, computer networking becomes increasingly more important [1]. Based on designs developed in the 1960s, the Advanced Research Projects Agency Network (ARPANET) was created in 1969 by the U.S. Department of Defense and was based on circuit switching the idea that a single communication line, such as a two-party telephone connection, deserves a dedicated circuit for the duration of the communication. This simple network evolved into the present-day Internet [2].

12.2.1 Network types

The network is divided into two types of connection methods.

12.2.1.1 Wired networking

Wired networking (networking cable) is networking hardware used to connect one network device to other network devices or to connect two or more computers to share printers, scanners, and so on. Different types of network cables, such as coaxial cable, optical fiber cable, and twisted pair cables, are used depending on the network's physical layer, topology, and size. The devices can be separated by a few meters (e.g. via Ethernet) or nearly unlimited distances (e.g. via the interconnections of the Internet) [3].

12.2.1.2 Wireless networking

Wireless networks are networks that use radio waves to connect devices, without the necessity of using cables of any kind. Devices commonly used for wireless networking include portable computers, desktop computers, hand-held computers, personal digital assistants (PDAs), cellular phones, pen-based computers, and pagers. Wireless networks work similar to wired networks, however, wireless networks must convert information signals into a form suitable for transmission through the air medium [4].

12.3 WIRELESS SENSOR NETWORKS (WSNS)

A WSN is a wireless network that contains distributed independent sensor devices that are meant to monitor physical or environmental conditions. A WSN consists of a set of connected tiny sensor nodes, which communicate with each other and exchange information and data [35, 36, 38]. These nodes obtain information on the environment such as temperature, pressure, humidity, or pollutant and send this information to a base station. The latter sends the info to a wired network or activates an alarm or an action, depending on the type and magnitude of data monitored [30]. Typical applications include weather and forest monitoring, battlefield surveillance, physical monitoring of environmental conditions such as pressure, temperature, vibration, or pollutants, or tracing human and animal movement in forests and borders. They use the same transmission medium (air) for wireless transmission as wireless local area networks (WLANs). For nodes in a local area network (LAN) to communicate properly, standard access protocols like Institute of Electrical and Electronic Engineering (IEEE) 802.11

are available [30, 31, 34]. The importance of WSN makes it suitable for application in health, military, education, firefighting and prevention, and psychology. The survival rate in cardiac arrest in the first 720 s (12 min) is 48–75% as reported by the American Heart Association. A detailed description of the WSN application is shown in Figure 12.2.

There are several wireless sensor technologies, which are regarded as an offshoot of WSN, and the commonest is WBAN. WBANs, also referred to as body sensor networks (BSNs) or body area networks (BANs), are designed with thin, lightweight sensors deployed around, on, and in the human body to act as a monitoring device for the body and its immediate environment. WBAN functions as a monitoring, data detection and collection, and wireless data transfer system. Usually, PDAs and smart mobile phones are used to transfer the data to the healthcare professional through a main wireless system. The measured data are either processed or transferred in their raw state through a single gateway or multi-gateway. The sensor nodes are commonly made up of a sensing component, processing component, communicating component, and a power unit. Together they sense, collect, process, and transmit data wirelessly to a central receiver. The following features are important for the most reliable and efficient sensor nodes: low cost, power efficient, wireless capabilities, multi-hop data routing, and decentralized processing [30]. The lifespan of sensor nodes is mostly affected by the quality and duration of the power source, that is, the battery. Previous studies have tried to increase the lifespan of sensor nodes by balancing or duty cycling the load among the sensor node. However, recent studies seek to extend node life by incorporating renewable energy.

Sensor nodes are deployed in large numbers due to their small size and small footprint. Although WBAN is referred to as an offshoot of WSN, there are several differences between these two systems, and these parameters of differences are presented in Table 12.1. Energy efficiency and reliability are

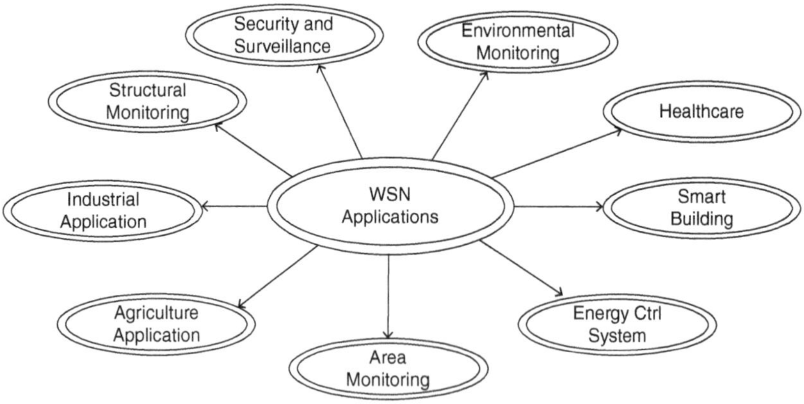

Figure 12.2 WSN application.

Table 12.1 Differences between WSNs and WBANs

Challenges	*WBAN*	*WSN*
node life	the longer the better	months/years
topological	changed	unchanged
node energy	limited and irreplaceable	limited, but replaceable
safety	very high	low
standard	IEEE 802.15.6	IEEE 802.11.4
range	body range (cm/m)	environmental monitoring (m/km)
number of nodes	dozens	hundreds
node size	the smaller the better	no special requirements
node task	many	single or scheduled tasks
data rate	heterogeneous	homogeneous
data loss	intolerable	tolerable
node placement	difficult	easily
biocompatibility	consider	not considering

very important parameters in both systems. The other parameters are more important in WBAN than in WSN; for instance, maximum security is required in WBAN because of the sensitive nature of the data being transmitted [5].

12.4 WIRELESS BODY AREA NETWORK (WBAN)

WBAN was first presented in an article from 1996, but these body networks were named WPAN at the beginning. WBAN is seen as a valuable solution to monitor the human body remotely and fluently. Many works have discussed WBAN definitions, architectures, applications, and so on because various current trends such as growing population have promoted the growth of WBAN. However, most of the existing works focus on theoretical performance enhancement. IEEE 802 has established a Task Group called IEEE 802.15.6 in November 2007 for the standardization of WBAN. The purpose of the group is to establish a communication standard optimized for low-power high-reliability applications for BANS. We can find various annotations for WBAN such as WBSN Wearable Body Sensors Network or WBANS—Wireless Body Area Sensor Networks.

12.4.1 WBAN architectures

Sensors collect physiological data and send them to the concerned entities. Then, diagnosis based on the received information is performed and the right decisions are taken. A three-tier architecture for a WBAN communication system can be proposed as shown in Figure 12.1, the WBAN architectures:

- Intra–WBAN Communication: Communication around the body of sensors between each other's and communication between sensors and Sink node (PDA, phone, sensor, etc.).
- Inter–WBAN Communication: Communication between the Sink node and an Access Point (AP). In inter–WBAN communication, a device coordinator Sink with specific features could be responsible to communicate with the adjacent WBANS. A coordinator is generally considered as a resource-rich device, which can be a multi-standard node to interface with other technologies such as static WSN, Wi-Fi Aps, or broadband cellular networks (Fourth Generation (4G), Long-Term Evolution (LTE), etc.).
- Beyond–WBAN Communication: Connects to the inter– BAN through a gateway. Its functionalities are database and data storage, remote access for medical stuff, and so on. The system performs a real-time analysis of sensors data.

12.4.2 WBAN communication stack layers

The WBAN communication stack layers consist of the following:

- Physical Layer: Three factors: Body path-loss (BPL), receive noise figure (RNF) and dignal-to–noise ratio (SNR) affect the sensor node's transmission power when sending any data wirelessly. SNR is subject to the communication link's quality. RNF is a factor that is device dependent. BPL is influenced by antenna in use and radiation pattern. Some of the main responsibilities of the physical layer include frequency selection, signal detection, modulation, and encryption. For on-body sensors, frequency bands are:1.3–5 MHz, 5–50 MHz, 400 MHz, 600 MHz, 900 MHz, 2.4 GHz, and 3.1 10.6 GHz [32, 33].
- Data-link Layer: This layer is responsible for multiplexing, frame detection, channel access, and reliability. At media access control (MAC) layer, collision occurs when two or more nodes attempt to transmit at the same time. To ensure energy efficiency, MAC protocols propose to synchronize transmission schedule and listening periods to maximize throughput while reducing energy by turning off radios during sleep periods. There is a trade-off between reliability, latency, and energy consumption.
- Network Layer: Commonly, nodes in WBAN are not required to route the packets to other nodes. However, new researches show that multi-hop routing in WBAN is more adequate and is required to guarantee low transmission power, low energy consumption, and efficient data routing by distributing the routing load over the entire network. In

addition, routing is possible when multiple WBANS communicate with each other through their coordinators. WBANS coordinators exploit cooperative and multi-hop body-to-body communication to extend the end-to-end network connectivity [6, 11].

12.4.3 WBAN technologies

WBAN may involve different technologies at different levels. In this section, we present a comprehensive study of the main proposed technologies for WBAN.

- Bluetooth: Bluetooth technology was designed as a short-range wireless communication standard intended to maintain high levels of security. Thanks to this technology, each device can simultaneously communicate with up to seven other devices within a single picante, an ad hoc network including one device acting as a master and up to seven others as slaves for the lifetime of the picante. Slaves have to synchronize by the system clock of the master and follow the hopping pattern, determined by the master. Besides, each device can belong to several picots simultaneously, as they enter radio proximity of other master devices [31, 34].
- Bluetooth Low Energy: A derivative option of the Bluetooth standard is Bluetooth Low Energy (BLE), which has been introduced as a more suitable option for WBAN applications where lower power consumption is possible using a low duty cycle process. Bluetooth LE was designed to wirelessly connect small devices to mobile terminals. Those devices are often too tiny to bear the power consumption as well as cost associated with a standard Bluetooth radio, but are ideal choices for health-monitoring applications [31, 34, 36]. BLE technology is expected to provide a data rate of up to 1 megabit per second. Using fewer channels for pairing devices, synchronization can be done in a few milliseconds compared to Bluetooth seconds. This benefits latency-critical BAN applications, like alarm generation and emergency response and enhances power saving. Its nominal data rate, low latency, and low energy consumption make BLE suitable for communication between the wearable sensor nodes and the AP [31, 34].
- Zigbee and 802.15.4: ZigBee, which is defined by the ZigBee specification, is one of the wireless networking technologies that are widely used in a low-power environment. ZigBee is targeted at radio-frequency applications that require a low data rate, long battery life, and secure networking, thanks to its 128-bit security support to perform authentication and guarantee integrity and privacy of messages. Through the sleep mode, ZigBee-enabled devices are capable of being operational for several years before their batteries need to be replaced.

- IEEE 802.15.6: IEEE 802.15.6 is the first WBAN standard that serves various medical and non-medical applications and supports communications inside and around the human body. IEEE 802.15.6 standard uses different frequency bands for data transmission, including narrowband (NB), which includes the 400, 800, 900 MHz and the 2.3 and 2.4 GHz bands; ultra-wideband (UWB) 4, which uses the 3.1 to 11.2 GHz; and the human body communication (HBC), which uses the frequencies within the range of 1050 MHz. This standard is a step forward in wearable wireless sensor networks as it is designed specifically for use with a wide range of data rates, less energy consumption, low range, ample number of nodes (256) per body area network, and different node priorities according to the application requirements.
- IEEE 802.11: IEEE 802.11 is a set of standards for WLAN. According to the IEEE 802.11 standards, Wi-Fi allows users access to the Internet at broadband speeds when connected to an access point or in ad hoc mode. It is ideally suited for large data transfers by providing high-speed wireless connectivity and allowing videoconferencing, voice calls, and video streaming. An important advantage is that all smartphones, tablets, and laptops have Wi-Fi integrated; however, high-energy consumption is an important drawback [31, 34].
- Other Radio Technologies: Advance Network Technology (ANT) protocol is another emerging standard for wellness and health monitoring applications. ANT is a low speed and low power protocol being supported by several sensor manufacturers. The Zarlink technology is ultra-low power, which makes it suitable for medical implant applications requiring low frequency and low data rates. Rubee active wireless protocol uses Long Wave magnetic signals to send and receive short (128 byte) data packets in a local network. Rubee does not require line of sight communication for its operation. Additionally, Rubee has the advantages of efficient transmission distance, high security level, ultra-low power consumption, stable operation providence, and long battery lifetime, which make it convenient for many WBANs applications such as patient monitoring and mobile healthcare [7, 36].

12.4.4 Types of WBAN devices

As the name indicates, WBAN comprises tiny devices with communication capabilities. Based on their functions and roles, these devices are divided into three classes. This section presents a brief taxonomy of WBAN devices according to their functionality.

12.4.4.1 Wireless sensor node

It comprises four components: transceiver, battery, microprocessor, and the sensor component. WBAN sensor nodes provide wireless monitoring for anybody, anywhere, and anytime. These nodes can be physiological sensors, ambient sensors, or bio kinetic sensors.

Sensors for Wearing: These devices are added to clothing or placed on the body to collect vital signs, such as peripheral oxygen saturation (SpO2) which measures the level of oxygen saturation in a person's blood, which coincides with the cardiac cycle. The electrocardiogram (ECG) sensor investigates the heart function by sampling the heart muscle propagation electric waveform with respect to time. The EEG sensor detects brain electrical activity and the motion detection sensors combine both an accelerometer and a gyroscope to monitor and analyze a person's movements.

Implantable probes: These devices are injected under the skin or into the bloodstream. In Parkinson's disease patients, for example, these sensors are used to send electrical impulses to the brain through neural simulators. Other applications for implantable sensors can be found.

Actuators: Actuators are used to administer medicine to a patient. The required drug is administered directly in a predefined manner when a sensor detects an abnormality or when it is triggered by an external source, according to the doctor's decision. Similar to a sensor node, the actuator consists of a transmitter, receiver, memory battery, and an actuator device that holds and administers the drug. The drugs could be used to control blood pressure, the body's temperature, or to treat many other illnesses. The actuator is activated upon receiving data from the sensors.

Wireless Personal Device (PD): Responsible for establishing communication between sensors, actuators, and a cellular phone in a wireless fashion, its main components are a transceiver, a rich power source, a large processor, and a large memory. This taxonomy is summarized in Table 12.2.

12.4.5 Applications of WBAN

WBAN applications span from the healthcare and entertainment fields to sport and the military among others. WBAN applications are categorized as either medical or non-medical. This section classifies WBAN applications according to their target domain of application. Each application is further classified into medical, non-medical, implanted, and wearable applications. Table 12.1 also gives some examples for each WBAN field.

12.4.5.1 Healthcare

This continual monitoring allows proactive fatal and anomalies detection, which is vital for diagnosing heart and brain activities. Actuators help in automatic drug delivery. Some applications such as cochlear implants, hearing aids, and artificial retinas help enhance the lifestyle of human beings.

Table 12.2 WBAN devices

WBAN devices	*Functionality*	*Examples*	
Sensor node	Samples and communicates physiological attributes and provides a response to the information through wireless communication for anybody, anywhere, and anytime.	Implantable: injected under the skin or in the blood stream.	In Parkinson's disease patients, sensors send electrical impulses to the brain through neural simulators
		Wearable: added to clothes or placed on the body to collect vital signs.	Spo2, ECG, EEG
Personal Device (PD)	Set up communication between a cellular phone, sensors, and actuators wirelessly.	Can be a specialized dedicated unit, PDA, or a smart phone	
Actuators	Administer medicine to a patient when a sensor detects an abnormality according to the doctor's decision.	Control blood pressure, the body's temperature, and treat many other illnesses	

Additionally, given that medical accidents can and do happen, WBAN applications help to reduce them and increase public safety by using profiles of previous medical accidents to alert medical personnel before similar accidents occur. Consequently, WBAN is expected to improve the management of illnesses and reaction to crises, which will increase the efficiency of healthcare systems. WBAN healthcare applications can be further classified as follows:

Medical Applications

WBAN medical applications enable the continual monitoring of physiological parameters such as heartbeat, body temperature, and blood pressure. The data collected can be sent through a cell phone, which acts as a gateway, to a remote location such as an emergency center so that the relevant action can be taken. WBAN is considered key to the early detection and treatment of patients with serious cases such as diabetes and hypertension. Medical applications of WBAN can be further divided according to the position of the medical sensors as follows [36]:

(a) Wearable Applications: Wearable medical healthcare applications include temperature monitoring, blood pressure monitoring, glucose monitoring, ECG, EEG, EMG, SpO2, and drug delivery.

(b) Implant Applications: These applications comprise nodes implanted either under the skin or in the bloodstream such as in diabetes control systems, cardiovascular diseases, and cancer detection [29].

Non-Medical Applications

These applications are considered to fall within the wearable sensor class of applications and include two applications, which are:

(a) Motion Detection: This application is used to detect, capture, recognize, and identify body gestures and motions and send alerts to the owner of the application. For example, fear increases heartbeat, which leads to sweating and other symptoms. Thus, emotional status can be measured and monitored.
(b) Secure Authentication: This is a very promising WBAN application because it is the core of both multimodal biometrics and EEG. This application harnesses physiological and behavioral human body biometrics such as fingerprints and facial patterns.

Sports: Sport activities and fitness can be improved by keeping a log of vital physiological data such as temperature, heartbeat, and blood pressure. The data can be used to avoid sport accidents and injuries and to plan for future training .WBAN sport applications are considered medical wearable applications. Such applications enhance professional and amateur sport training especially for athletes.

Entertainment: Entertainment is also a very promising field for WBAN. The film industry, for example, makes use of motion capturing and post-production mechanisms to produce films in which actors play their roles using smart (things) object(s). Using the on-body accelerometers and gyroscopes for capturing motion facilitates the possibility of tracking the different positions of body parts.

12.4.5.2 Military and defense

WBAN provides new capabilities to improve performance of individuals and teams of soldiers in military situations. To avoid threats at the individual tier, a group of sensors sample important information on the surrounding emerging actions and environment. At the team level, the taken information enables the commander to coordinate team tasks efficiently. Inter–WBAN communications and security play a key role in preventing critical data from being hacked by enemies. WBAN applications can be considered as either medical wearable or non-medical wearable as follows:

Military Medical WBAN Applications: These types of applications are used to assess soldier fatigue, prepare for battle, and to protect uniformed personnel. For example, sensors surrounding soldiers, firefighters, or police officers can foresee a life-threatening situation by monitoring the level of air toxins.

Table 12.3 Fields and applications of WBAN

WBAN fields	*Application types*		*Examples of applications*
Healthcare	Medical	Wearable	ECG, EEG, EMG, SPO2, temperature, blood pressure, drugs delivery
		Implant	Diabetes control
	Non-medical	Motion detection	
		Secure authentication	
Sports	Medical wearable	Heartbeat, temperature, blood pressure, motion sensor	
Entertainment	Non-medical wearable	Gaming purposes, virtual reality, ambient, intelligence areas, personal item tracking, and social networking	
Military and defense	Medical wearable	Assess solider fatigue, detect life-threatening situations	
	Non-medical wearable	Fire, poisonous gas detection	

Non-Medical Military WBAN Applications: Such applications involve off-body sensors (on buildings) that are used for emergencies. Such sensors are capable of, for example, detecting a fire in the home or a poisonous gas and must directly send this information to on- and in-body devices to notify the wearer of the emergency situation. Table 12.3 gives some examples for each WBAN field.

12.4.6 IOT with WBAN

The Internet of Things (IoT) is a platform that provides enormous access to the network communication the world has emerged into, an evolution that affected and enhanced the way human beings live. Throughout decades and centuries, these information technologies have continuously interchanged positively to ultimately benefit the world with endless solutions and improvement of technologies, which has also led to numerous employment opportunities [38].

Furthermore, IoT has affected human life in many ways, from the way a driver navigates a car to a person who easily can shop and purchase online within seconds, and smart watches with a lot of services that monitor and display life and health activities; on the other hand, modern houses have remarkably data-collecting smart appliances such as smart room temperature adjusting along with air conditioners, smart washers, and smart televisions. In addition, the IoT is embedded in any item that consists of the necessary factors, which are electronics, software, and sensors. Where IoT acts as data storage, that data can be stored, managed, remolded, and shared to the benefit of the consumer and increasing data efficiency. The data can be collected from the sensors, and the work can be done afterward. A perfect example that can simplify the work of an IoT is with an air conditioner, which has a built-in sensor with health and temperature information that can be sent to an IoT platform, where feedback services alert the consumer in the event of a replacement or emergency. Wire body area networks work with the Internet of Things, according to the connection of data or information from sensors embedded in electronic and other devices such as sensors, monitors, mobile phones, and so on to the Internet or exchange information from one device or sensor to another. IoT facilitates WBAN technology to connect to the Internet, store and process data, collect information, control the sensors and nodes and connect to mobile communication to provide more reliability and synchronized data to ensure expedient arrival of the patient's health information to save his life in opportune time. Figure 1.1 shows WSNs for IoT [3, 36–38].

12.4.7 WBAN systems for healthcare

Healthcare is one of the most important processes that define the economic and social growth of a country. All developed and developing countries are striving to increase the usage of technology in bringing excellent healthcare to their population. However, existing medical resources cannot satisfy the future healthcare demands of different types of patients (older or younger). The resources are quite limited, and it is impossible for most patients to stay a long time in the hospital because of economic restrictions, work, and other personal reasons, even though their health status must be monitored in real time or frequently. As a result, wireless monitoring medical systems will become part of mobile healthcare centres with real-time monitoring in the future. WBANs supporting healthcare applications offer different contributions at monitoring, diagnosis, and therapeutic levels. Figure 12.1 shows typical WBAN. They cover real-time medical information obtained from different types of sensors with secure data communication and low power consumption.

Electronic health (E-Health) and telemedicine are two areas that are leveraging current wireless communication technologies to provide emergency medical services, enable outpatient monitoring and treatment, facilitate patient recovery, and directly connect doctors and nursing staff with patients. WBAN healthcare applications can offer valuable contributions to improve patient healthcare, including diagnosis and therapeutic monitoring. Patients, while performing their activities comfortably at home or outdoors, can be monitored by the medical staff. In this field, data reliability and energy consumption (considering 24/7 monitoring) are fundamental characteristics to consider when choosing appropriate WBAN sensor nodes. These nodes operate in close proximity to the human body, collecting data for various medical and non-medical applications. Medical bands used in WBAN provide physiological data from sensor nodes. They are chosen in such a way that it reduces interference and thus increases the coexistence of sensor node devices with other network devices available at medical centres. The collected data is sent to stations using medical gateway wireless boards [8, 9, 12–15].

12.4.8 WBAN scenarios for the healthcare system

Following are the scenarios where WBAN can be used for wireless health monitoring.

(a) Telemonitoring of Patients with Cardiac Arrhythmia: Cardiac arrhythmia is very common and in many cases, is related to coronary heart disease. It's estimated around 200 million people are living with coronary heart disease [39]. In patients suffering from arrhythmia, ECG measurements have to be taken regularly to monitor the efficacy of drug therapy. To save time and reduce costs, the patient can transmit ECG and blood pressure via General Packet Radio Service (GPRS) from home or elsewhere to the health call center, where a cardiologist monitors the vital signs. The intention is that irregular patterns will be detected quickly and appropriate intervention can be initiated. This scenario will evaluate how the patients and the cardiologist can gain time and reduce the related costs.

(b) Integrated Homecare for Women with High-Risk Pregnancies: Women with high-risk pregnancies are often admitted to the hospital for longer periods because of possible pregnancy-related complications. Admission is necessary for the intensive monitoring of the patient and the unborn child. Homecare with continuous monitoring is desirable and can postpone hospitalization and reduce costs, as well as offering more security for the mother and unborn child. In this scenario, patients are monitored from home using BAN and bio signs (maternal and fetal) are transmitted to the hospital. An additional objective of the scenario is to evaluate if such a solution postpones hospitalization and reduces costs. The scenario will use both GPRS and Universal Mobile Telecommunication System (UMTS) networks.

(c) Teletrauma Team: The trauma patient BAN will measure vital signs, which will be transmitted from the scene to the members of the trauma team located at the hospital. The paramedics wear trauma team BANs that incorporate an audio system and a wireless communication link to the hospital. The purpose of this scenario is to evaluate whether use of mobile communications can improve quality of care and decrease lag time between the accident and the intervention. When using telemetry technology, time can be saved and thus treatment and chances for patient recovery improved.

(d) Support of Home-Based Healthcare Services: This scenario involves use of GPRS for supporting remote assistance and home-based care for elderly and chronically ill patients suffering from co-morbidities. The wireless health monitoring nurse BAN will be used to perform patient measurements during nurse home visits and the wireless health monitoring patient BAN will be used for continuous monitoring during patient rehabilitation at home, or even outdoors. It is very important to facilitate patients' access to healthcare professionals without saturating the available resources, and this is one of main expected outcomes of the wireless health remote monitoring approach. Parameters to be measured are oxygen saturation, ECG, spirometer, temperature, glucose, and blood pressure.
(e) Outdoor Patient Rehabilitation: The patients involved in this scenario are chronic respiratory patients who are expected to benefit from rehabilitation programs to improve their functional status. The physiotherapist will receive online information on the patient's exercise performance and will provide feedback and advice. It is expected that by enabling patients to perform physical training in their own local settings, the benefits in terms of cost and social acceptance can be significant. Parameters to be measured are pulse oximetry, ECG, and mobility with audio communication between patient and remote supervising physiotherapist.
(f) Physical Activity and Impediments to Activity for Women with Rheumatoid Arthritis (RA): This scenario will centre women with rheumatoid arthritis. The use of the BAN together with the mobile communications will enable collection of a completely new kind of research data, which will enhance the understanding of the difficulties and limitations these patients face. The objective is to offer solutions that will make their lives easier. Through this collection of data, the scarce knowledge about what factors impede normal life will be supplemented and the quality of life of RA patients may thereby be improved. By use of wireless health monitoring BANs, the activity of the patients will be continually monitored. Parameters measured include heart rate, activity level, and walking distance and stride length.
(g) Monitoring of Vital Parameters in Patients with Respiratory Insufficiency: The group of patients involved in this scenario suffers from respiratory insufficiency due to chronic pulmonary diseases. These people need to be under constant medical supervision in case they suffer an aggravation of their condition. Besides needing regular checkups, they are also dependent on oxygen therapy at home, which means oxygen delivery and close supervision. The use of wireless health monitoring BANs is designed to enable the early detection of this group of diseases but also to support home care for diagnosed patients by detecting situations where the patient requires intervention. The expected benefits are a reduction of the number of checkups and

hospitalizations needed, thus saving both time and money. Parameters measured are pulse rate, oxygen saturation, and signals from a motion sensor (accelerometer).

(h) Home-Care Services and the Possibility to Monitor Health Conditions Remotely:

The way patient care is provided has changed. If suitable, home-based services are provided and patients do not need to be in hospital, for example they are recovering from an intervention, then by investing in home care, hospitals have been able to significantly reduce pressure on beds and on staff time dedicated to this kind of patient.

This scenario tests transmission of clinical patient data by means of portable GPRS/UMTS equipment to a physician or a registered district nurse (RDN) from patients living in a rural, low population density area. The expected benefit is that this solution will reduce the number of cases where the patient is supposed to visit a hospital for consultation unnecessarily.

(i) Ambient Assisted Living: The aging population, the increasing cost of formal health care, and the importance that the individuals place on living independently all motivate the development of innovative-assisted living technologies for safe and independent aging. Applications in this field improve quality of life to maintain a more independent lifestyle using home automation. In fact, assisted living facilities have emerged as an alternative housing facility for people with disabilities and elderly who are not considered independent but do not need around-the-clock medical care, as in nursing or retirement homes. An ambient sensor network can sense and control the parameters of the living environment and then deliver the body data to a central station, thanks to continuous cognitive and physical monitoring. The health condition of these people can be estimated from their heartbeat rate, blood pressure, and accelerometer data. The system may be connected to a healthcare center for observation and emergency assistance, in case of strong changes in the observed parameters or deviations from the normal range [10].

12.5 CONCLUSION

In this study, we implemented an integrated system of equipment consisting of Arduino Nano and the necessary sensors, as well as software to control and measure blood pressure, body temperature, and other necessary measurements and obtain information via SMS and locate the patient, as well as via cloud computing via mail. Possible electronic means of communication (computer networking technologies) were used, and one of the new sections in the field of networking, namely WBAN and WSN, was implemented to help save patients' lives as quickly as possible. We carried out the study on cardiac patients from the period 01/02/2022 until 30/07/2022 in an ongoing study to monitor the designed technology, its devices, its application, its layers, and its relationship to IOT, how it can help in the healthcare system, and important healthcare applications that WBAN technology helps in. We expect this technology will reduce the death rate.

REFERENCES

[1] Sahu, M. L., Atulkar, M., Ahirwal, M. K. and Ahamad, A. Vital sign monitoring system for healthcare through IoT based personal service application. *Wireless Personal Communications*, 122, 129–156, 2021.

[2] Al-Thobhani, Nashwan Saeed Ghaleb Implementation wearable WBANs for e-healthcare. 2022. doi: 10.13140/RG.2.2.15354.57286/1, https://www.researchgate.net/publication/364107439_Implementation_Wearable_WBANs_for_e-Healthcare [accessed January 12 2023]

[3] Al-Thobhani, N.G., Aotoban, H., Al-Handali, K., Al-Sharafi, Z. and Al-Hamadi, Z. Design and Implementation of a SWBAN for Diabetes Management in Healthcare (No. 9124). EasyChair, 2022. doi: 10.13140/RG.2.2.15626.54728/1. https://www.researchgate.net/publication/364331489_Design_and_implementation_of_a_SWBAN_for_diabetes_management_in_healthcare. [accessed January 14 2023]

[4] Salazar, Jordi. *Wireless Networks*, 1st Edition. Czech Technical University of Prague Faculty of electrical engineering, 2017. https://upcommons.upc.edc/bitstream/handle/2117/110811/LM01_F_EN.pdf. [accessed January 15 2023]

[5] Al-Turjman, Fadi. *Wireless Medical Sensor Networks for IoT-based eHealth*. Institution of Engineering and Technology, 2020. https://ar.1lib.limited. [accessed January 17 2023]

[6] Beji, Wafa Badreddine. *Communication Protocols in Wireless Body Area Network (WBAN)*. HAL open science, 2018. https://ar.1lib.limited

[7] Al-Thobhani, Nashwan Ghaleb. Wireless Body Area Networks for Healthcare (No. 8874). EasyChair. 2022. https://www.researchgate.net/publication/363417862_Wireless_Body_Area_Networks_for_Healthcare_System [accessed January 18 2023]

[8] Movassaghi, Samaneh and Lipman, Justin Wireless body area networks: A survey. *IEEE Communications Surveys & Tutorials*, 2014. www.researchgate.net [accessed January 20 2023]

[9] Negra, Rim, Jemili, Imen and Belghith, Abdelfettah. *Wireless Body Area Networks: Applications and technologies*. Elsevier, 2016. https://cyberleninka.org/article/n/633432.pdf [accessed February 01 2023]

[10] Al-Thobhani, Nashwan Ghaleb. IoSTs in e-healthcare services. 2022. doi: 10.13140/RG.2.2.34927.71841/3, https://www.researchgate.net/publication/366186751_IoSTs_in_e-healthcare_services [accessed February 03 2023]

[11] Yuce, Mehmet R. and Khan, Jmil Y. *Wireless Body Area Network Technology, Implementation, and Applications*. Pan Stanford Publishing, 2012. https://ar.1lib.limited [accessed February 01 2023]

[12] Salayma, Marwa, Al-Dubai, Ahmed and Romdhani, Imed and Nasser, Youssef Wireless body area network (WBAN): A survey on reliability, fault tolerance,

and technologies coexistence. *ACM Computing Surveys, New York*, 2016, doi: 10.1145/3041956

[13] Ghamari, M., Janko, B., Sherratt, R.S. Harwin, W., Piechockic, R. and Soltanpur, C. A survey on wireless body area networks for healthcare systems in residential environments. *Sensors*, 16, 831, 2016. doi: 10.3390/s16060831 [accessed February 01 2023]

[14] Cavallari, R., Martelli, F., Rosini, R., Buratti, C. and Verdone, R. A survey on wireless body area networks: Technologies and design challenges. *IEEE Communications Surveys & Tutorials*, 16, 1635–1657, 2014. doi: 10.1109/SURV.2014.012214.00007 [accessed February 03 2023]

[15] Hasan, Khalid, Biswas, Kamanashis, Ahmed, Khandakar, Nafi, Nazmus S. and Islam, Md Saiful. A comprehensive review of wireless body area network. 2019. https://www.sciencedirect.com/science/article/pii/S1084804519302218 [accessed February 03 2023]

[16] Maheswar, Kanagachidambaresan, Jayaparvathy and Thampi, Sabu M. *Body Area Network Challenges and Solutions*. EAI/Springer Innovations in Communication and Computing, 2019.

[17] Wahane, Varsha and Ingole, P.V. A Survey: Wireless body area network for health monitoring. *American Scientific Research Journal for Engineering, Technology and Sciences (ASRJETS)*, 2017. https://core.ac.uk/download/pdf/235050277.pdf [accessed April 03 2023]

[18] Send receive SMS and call with SIM800L module and Arduino. https://electronicsmith.com/send-receive-sms-call-with-sim800l-gsm-module-arduino/ [accessed February 10 2023]

[19] Interface ublox NEO-6M GPS module with Arduino. https://lastminuteengineers.com/neo6m-gps-arduino-tutorial/ [accessed March 17 2023]

[20] Interfacing Max30100 pulse oxmeter and heart rate sensor with Arduino. https://lastminuteenginee-rs.com/max30100-pulse-oximeter-heart-rate-sensor-Arduino [accessed March 19 2023]

[21] Introduction to Arduino nano. https://www.theengineeringprojects.com/2018/06/introduction-to-arduino-nano.html [accessed March 20 2023]

[22] Lin, J., Yu, W., Zhang, N., Yang, X., Zhang, H. and Zhao, W. A survey on Internet of Things: Architecture, enabling technologies, security and privacy, and applications. *IEEE Internet of Things Journal*, 99, 1–1, 2017.

[23] Al-Thobhani, Nashwan Ghaleb. Smart wireless sensor network. 2023, https://www.researchgate.net/publication/367545388_Smart_Wireless_Sensor_Networks [accessed April 04 2023]

[24] Lin, J., Yu, W., Zhang, N., Yang, X. and Ge, L. On data integrity attacks against route guidance in transportation-based cyber-physical systems. In *Proceedings of the 14th IEEE Annual Confernece in Consumer Communications and Networking Conference (CCNC 2017)*, 2017.

[25] Sahu, M. L., Atulkar, M., Ahirwal, M. K. and Ahamad, A. Vital sign monitoring system for healthcare through IoT based personal service application. *Wireless Personal Communications*, 122, 129–156, 2021.

[26] Nasser, N., Emad-ul-Haq, Q., Imran, M., Ali, A., Razzak, I. and Al-Helali, A. A smart healthcare framework for detection and monitoring of COVID-19 using IoT and cloud computing. *Neural Computing & Applications*, 35, 13775–13789, 2023.

[27] Nashwan, Saeed M. and Al-Thobhani, Ghaleb. Using the technology of wireless body area networks. 2022, https://www.researchgate.net/publication/366356191_Using_the_Technology_of_Wireless_Body_Area_Networks [accessed May 01 2023]
[28] Taleb, Houssein, Nasser, Abbass Andrieux, Guillaume, Charara, Nour and Cruz, Eduardo Motta. Wireless technologies, medical applications and future challenges in WBAN: A survey. *Wireless Networks*, 27, 5271–5295, 2021. doi: 10.1007/s11276-021-02780-2
[29] Jin, Z., Oresko, J., Huang, S. and Cheng, A. C. HeartToGo: A personalized medicine technology for cardiovascular disease prevention and detection. In *Life Science Systems and Applications Workshop, 2009*. LiSSA 2009. IEEE/NIH (pp. 80–83). IEEE, 2009, April.
[30] Latha, R. and Vetrivelan, P. Wireless body area network (WBAN)-based telemedicine for emergency care. 2020. doi:10.3390/s20072153. www.mdpi.com/journal/sensors
[31] Filipe, Luis, Riverola, Florentino Fdez, Costa, Nuno and Pereira, António. Wireless body area networks for healthcare applications: Protocol stack review. *International Journal of Distributed Sensor Networks*, 2015, 213705, 23 pages, 2015. doi: 10.1155/2015/213705
[32] Mkongwa, Kefa, Liu, Qingling, Zhang, Chaozhu and Siddiqui, Faizan A. Reliability and quality of service issues in wireless body area networks: A survey. *International Journal of Signal Processing Systems*, 2019, March. doi: 10.18178/ijsps.7.1.26-31
[33] Ghamari, Mohammad, Janko, Balazs, Sherratt, R. Simon, Harwin, William, Piechockic, Robert and Soltanpur, Cinna. A survey on wireless body area networks for eHealthcare systems in residential environments. *Sensors*, 16, 831, 2016. doi:10.3390/s16060831
[34] Yuan, D., Zheng, G., Ma, H., Shang, J. and Li, J. An adaptive MAC protocol based on IEEE802.15.6 for wireless body area networks. *Wireless Communications and Mobile Computing*, 2019. doi:10.1155/2019/3681631.s
[35] Gardašević, G., Katzis, K., Bajić, D. and Berbakov, L. Emerging wireless sensor networks and internet of things technologies-foundations of smart healthcare. *Sensors*, 20(13), 3619, 2020.
[36] Al-Thobhani, Nashwan Ghaleb. Using WBANs in healthcare services. 2022. https://www.researchgate.net/publication/363840865_Using_WBANs_in_Healthcare_Services [accessed May 04 2023]
[37] Lin, C. T., Wang, C. Y., Huang, K. C., Horng, S. J. and Liao, L. D. Wearable, multimodal, biosignal acquisition system for potential critical and emergency applications. *Emergency Medicine International*, 2021, 9954669, 10 pages, 2021.
[38] Raza, M., Awais, M., Singh, N., Imran, M. and Hussain, S. Intelligent IoT framework for indoor healthcare monitoring of Parkinson's disease patient. *IEEE Journal on Selected Areas in Communications*, 39(2), 593–602, 2021, Feb.
[39] BHF Global Heart & Circulatory Diseases Factsheet January 2024. https://www.bhf.org.uk/-/media/files/for-professionals/research/heart-statistics/bhf-cvd-statistics-global-factsheet.pdf?

Chapter 13

Applications of wireless sensor networks in IoT

Radhika Sreedharan
Presidency University, Bangalore, India

13.1 INTRODUCTION TO WIRELESS SENSOR NETWORKS

A network where each node of every sensing element connects wirelessly and has the capacity to compute, compact data, and aggregate and analyze the data is known as a wireless sensor network (WSN). Every node has the capacities of communication and networking. A WSN incorporates structurally distributed autonomous gadgets (sensing elements) and varieties of nodes of sensing elements having the capability to sense, compute, and communicate wirelessly. WSN technology is utilized for controlling and supervising the surroundings and soil specification in agronomical areas. WSN is utilized as a part of agronomy for purposes such as giving high clarification, growing crop production, less power exhaustion, and collecting distributed data. Productive supervision of water presupposes a significant role in agronomy. Monitoring and controlling units supervise and manage numerous agronomical specifications such as soil dampness, air temperature and humidity, and others. WSNs have captivated much awareness in numerous areas of investigation such as health-care supervision, supervision of surroundings, and supervision of structural health. Lately WSN is widely utilized for providing solutions on exact agronomy for overcoming numerous issues in actuality (field). The industries that have lately begin to utilize WSN for growing efficiency and systematizing agronomical demand at an inexpensive price are agronomy and cultivations made up of smart objects that are wirelessly connected and are known as nodes. The actuality that the framework is not there for contemplating with WSNs is definitely a substantial benefit for adaptable utilizations, however there are variations of model barriers to contemplate with these wirelessly linked smart objects.

The important restrictions of smart objects in WSNs are given as follows:

- Finite computing power
- Finite store
- Lossless transmission
- Restricted transference speeds
- Finite potential

 DOI: 10.1201/9781003474524-13

These restrictions exceedingly affect how WSNs are created, utilized, and deployed. The actuality that individual sensing element nodes are normally so restricted is a reason that they are generally utilized in very huge numbers. When the cost of nodes of sensing elements gets lower, the capacity to utilize most unnecessary sensor elements becomes increasingly suitable. As a lot of sensing elements are very affordable and subsequently incorrect, the capacity for utilizing smart objects excessively permits for greater correctness.

More numbers of sensing elements allow the establishment of a pyramid of smart objects. Such a pyramid issues, among other supervising benefits, the capability of aggregating comparable sensing element readings through sensing element nodes that are in closeness to one another. Smart objects having restricted computing, store potential, and so on are known as constrained nodes.

Smart objects that are wirelessly connected normally have one of the following two communication patterns:

(1) Event-driven: Transference of receptive information is activated only when a smart object discovers a specific event or predetermined threshold.
(2) Periodic: Transference of receptive information occurs only at periodic intervals.

The conclusion of which of these communication plans is utilized relies considerably on the particular application. For instance, in certain medical scenarios, sensing elements intermittently transmit post-operative necessities, like readings of temperature and blood pressure. In other medical scenarios, the same readings of temperature or blood pressure are activated to be transmitted only when particular crucially low or high readings are estimated. When WSNs expand to very large number of smart objects, there is a tendency toward ever-growing levels of self-sufficiency. For instance, non-automatic positioning of possibly thousands of smart objects is impracticable and unmanageable, hence smart objects in a WSN are particularly auto-configured or self-activated by an IoT supervision policy in the environment. In addition, further levels of self-determining tasks are needed for developing coordinated communication between the numerous nodes of extensive WSNs which are frequently impromptu utilizations by not considering consistent distribution of node and/or denseness. For instance, there is a growing tendency with regard to "smart dust" applications, where very small sensing elements nodes are distributed across a territorial region for detecting temperature, vibrations, humidity, and so on. This technique has essentially infinite capacities, like military (e.g., detection of opponent soldiers' activity), meteorological (e.g., detection of earth tremors or forest fires), and industrial (instance.g., detection of manufacturing deviations, tracing of properties). A certain level of self-organization is needed for networking plenty of wireless smart objects in a way that these nodes individually come in conjunction for forming a true network with a typical motive. This capacity to self–co-ordinate is a capability for adjusting and developing the logical topology of a WSN for optimizing communication (between nodes as well as to centralized wireless controllers), clarify the foundation of current smart objects, and enhance dependability and approach to services.

Further benefits of utilizing more wireless low-cost smart objects are the fundamental capability for providing error toleration, dependability, and the capacity to increase the existence of a WSN, mostly in situations in which the smart objects have finite battery life. Self-determining techniques, like self-healing, self-protection, and self-optimization are frequently performing these tasks in the interests of an all-inclusive WSN system. IoT applications are commonly operation crucial, and in extensive WSNs, the all-inclusive system can't be unsuccessful when the surroundings varies all of a sudden, wireless communication is momentarily lost, or a finite number of nodes have a shortage of battery power or perform inappropriately.

13.2 WSN HISTORY

Inspection and tracing systems based on sound-waves-for enemy subaqueous were utilized during the 1950s [7]. Networked radars based on wireless are utilized at this moment. Investigation into the distributed sensor networks (DSN) making use of network communication began prior to 1980. From that time onwards, DSNs are utilized. A lot of structurally distributed sensing nodes integrate and network self-sufficiently based upon best attempt. Exploration and approaches of low-power wireless integrated micro-sensors (LWIM) utilization in DSNs was in progress before 1998. WSNs have many IoT applications, illustrated by smart cities and smart homes. The junctions of WSN are able to communicate data, making use of the Internet, in a wireless manner through a distant locality, like manufacturing business unit appliances, plantations, reservoirs, gas, and oil pipelines, that might not be sometimes effortlessly available.

13.2.1 Node operation on the basis of conditions

A node of WSN can adjust, modify, or perform another work utilizing a sensing element and linked circuit, calculations, networking abilities, and conditions at the node.

Whenever nodes can adjust their functioning in reciprocation to variations in surroundings, it is known as surroundings-reliant sensing, computing, and networking. The programs of the application layer assist the junction to distinguish the functions to be managed on varied surroundings. The choice of data, power, memory, and supervision of routine path, the routing protocol, users, gadgets, and application interactions in the system of WSN are set to work according to the environment and contemplate the surroundings at the time of computations and networking. Surroundings can be an organizational, physical, or terrestrial condition.

Subsequent circumstances might signify the need for reorganization of the measures of the WSN nodes:

- Previous and recent circumstances of environment.
- Measures like the network.
- Local gadgets of systems.
- Variations in the condition of the linking network.
- Physical specifications like the current time of the day.
- Convenient connectivity presently accessible.
- Previous succession of the measures of the gadget user.
- Previous succession of applications.
- Data records cached earlier.
- Leftover storage and battery power in recent scenario.

A WSN reprogramming can be over-the-air (OTA), which implies codes can be wirelessly altered in non-volatile storage using an entry control point by the gateway, application, or services.

13.2.2 WSN protocols

Network protocols have objectives of creation:

- Restrict requirements of calculation,
- Restrict battery power utilization and this bandwidth restricts functioning in self-created impromptu configuration mode, and
- Restrict memory needs of network protocol.

The physical layer does flexible radio frequency (RF) capacity management at the receiver; it decreases capacity in instance of close-by junction, enlarges the potential when the closest junction is at a considerable extent, and utilizes CMOS low-power ASIC circuits and potency-productive nodes.

Data-link layer media access control (MAC) protocol: Sr-MAC (Sensor-MAC) protocol can be utilized in the data-link layer. Nodes of S-MAC get to sleep for a continuous term. They are required to concur at regular meantime. S-MAC protocol permits usage of the potency-productive transmission without collision and continually manages the functioning. Transmission without collision is the consequence of channel scheduling. The channel is allotted to every node. Channels are re-used such that transmission without collision happens and retransmission is not utilized to.

Routing protocols: Multi-hop determination, potency-productive routing, caching of route, and managed data distribution are utilized by the network layer.

Routing protocols can be proactive or receptive. Proactive protocols store and find the route prior. Receptive protocols find the path on request. Routing protocols are table-operated when a routing information base escorts the routes accessible. The request drives the routing protocols when an origin requests the path and it escorts the accessible routes. Arranged wireless sensing elements networks can utilize CGSR (cluster-head gateway switch routing) that escorts the paths utilizing explanatory routing procedures.

13.2.3 Accomplishment of WSN mechanism

Whenever a WSN infrastructure is designed, the following measures require consideration:

- Sensing elements with linked CMOS low power ASIC circuits, their potency-productive coding and radio spans.
- Choosing mechanism, (i) arranged connecting mechanisms of junctions, relays, coordinators, routers, gateways or (ii) movable impromptu network of movable WSNs having restricted or undetermined movability area.
- Topology and architecture of network as per the services and applications, multi-cluster architecture or wireless multi-bounce mechanism network architecture [2].
- Network junctions self-contemplating, self-arrangement, and self-error correction protocols, location, movability scope, protection, link quality indicator (LQI) (packet reception ratio), data-link and routing protocols, exploration scope, and quality of service (QoS) needed as per the services and applications.
- Cluster, gateways of cluster, cluster-heads and hierarchy of clusters.
- Aggregation of data, routing, compression, synthesis and uninterrupted distribution.
- Times synchronization at random meantime as time is utilized as source in identical distance approximation, considering for postponement, localization, services of location, and ranging.

Numerous methods of integration for WSN: WSN requires consolidated methods for:

- Creation of nodes and supply of resources
- Localization of nodes
- Movability of nodes
- Connecting architecture of sensing elements
- Networking architectures of sensing elements

- Data broadcasting protocols
- Protection protocols
- Routing protocols and data link layer
- Incorporation with sensing elements data, instead of wireless sensing elements data, for IoT services and applications.

Quality of service: QoS is a mean weighted measure over the lifespan of a network. Various metrics are as follows:

- Mean hold-up: Estimation of the duration taken to generate data of a sensing element and its transmission till the destination.
- Lifespan: Time till where the performance of a WSN's productivity or given potency expedients of the methods conclude.
- Production: Bytes per second transmitted till destination. Low throughout means high hold-ups. It is related to the high frequency of the network as well.
- Link quality indicator (LQI): Implies transmitted or delivered packets through the junctions.

Actual time applications of communication over the WSNs require allocation of assured requirements for highest hold-up, lowest capability, or alternative QoS specifications.

High QoS ultimatums are:

- Routing via junctions having high throughput, less delay and paths that utilize resources with energy.
- Connection.
- Perpetuating the product of delay and preference, which implies lower delay paths taken over by higher-preference packets whereas higher delay paths are taken over by lower-preference packets.

The analysis of sensing element network is the metric. The analysis relies on the density and positions of the junctions in the area, range of transmission and junctions' sensitiveness. Another criterion is the percentage of times the network covers incidents of an event in circumstances of observation systems, dangerous chemicals, or fire-identification systems.

Configuration: Ultimatums to configure considering inadequate resources with the nodes dynamically or statically or spontaneously are as follows:

- Locations and movability scope of the junctions of WSN
- Clusters
- Gateways
- Cluster heads
- Rate of sampling of the sensed specifications
- Their collection, compression, and combination.

Requirements need to be met having restricted battery power, storage, capacity of calculation, frequency, and adaptability restrict one of the junctions.

13.2.4 Categorization of wireless sensor networks

The categorization of WSNs is done on the basis of application, however its features generally vary on the basis of the sort. Normally, WSNs are grouped into distinct types such as the following.

- Static & Mobile
- Deterministic & Nondeterministic
- Single Base Station & Multi-Base Station
- Static Base Station & Mobile Base Station
- Single-Hop & Multi-Hop WSN
- Self-Reconfigurable & Non–Self-Configurable
- Homogeneous & Heterogeneous
- Static & Mobile WSN

Every sensing element node in a lot of applications can be placed in the absence of motion, hence these networks are static WSNs. Predominantly certain applications such as biologic systems utilize moving sensing elements nodes known as mobile networks. The foremost instance of a mobile network is the animals' supervision.

- *Deterministic & Nondeterministic WSN*: The arrangement of sensing element nodes can be permanent and computed in a deterministic kind of network. This sensing element node's predetermined functioning can be feasible in certain applications. The locality of sensing elements nodes cannot be set on due to various characteristics such as unfriendly working conditions and raucous surroundings in many applications. Hence, these networks that adjure a complicated control system are known as nondeterministic.
- *Single Base Station & Multi-Base Station*: One base station is utilized and it can be organized very near to the area of the sensing element node in a single base station network. The activity among sensing element nodes is done with the base station. Many base stations are utilized and a sensing element node is utilized for moving data close to a convenient base station in a multi-base station type network.
- *Static Base Station & Mobile Base Station*: Base stations are either static or mobile comparable to sensing element nodes. The static-type base station incorporates a firm locale usually near to the sensing region while the mobile base station moves in the sensing element's region so that the load of sensing element nodes can be equitable.
- *Single-Hop & Multi-Hop WSN*: The organization of sensing element nodes is done straight near the base station in a single-hop type

network. Both the peer nodes and cluster heads are employed for transmitting the data for decreasing the depletion of energy in a multi-hop network.

- *Self-Reconfigurable & Non–Self-Configurable*: The organization of sensing element networks is not done within the network and relies on a control unit to gather data in a non–self-configurable network. The sensing element nodes organize and protect the network and jointly function by utilizing remaining sensing element nodes for achieving the task in wireless sensing element networks.
- *Homogeneous & Heterogeneous*: Every sensing element node primarily comprises energy usage, storage capacities, and computational power alike in a homogeneous wireless sensor network. Certain sensing element nodes incorporate high computational power and energy prerequisites in contrast to others in a heterogeneous network. The communication and processing functions are apart in consequence.

13.2.5 Kinds of movability in wireless sensing elements networks

Movability is a fundamental characteristic of every node in impromptu networks. Movability prevails normally to disconnect the network components and more explicitly it relies on the application in networks of wireless sensing elements. Applications of wireless sensing elements networks have been necessitated in various areas; however, in a lot of areas, association of movability is not there. Hence movability takes on a main part in which networks of wireless sensing elements are utilized. We can characterize three various kinds of movability in WSNs as follows.

- Movability of nodes of sensing elements
- Movability of scattered notes
- Movability of supervised occurrence or object

The first kind of movability, such as movability of nodes of sensing elements, mostly takes place each time the sensing element node's smallest component is movable. The greatest illustrations of this kind of movability are when nodes of sensing elements vanish and openly proceed inside the supervised region. These are arranged on animals to supervise and track the animals. The second kind of movability pertains to a state in which scattered nodes are capable of independently moving inside the supervised region to gather information through the sensing element network. Finally, the third kind of movability normally happens when a network of wireless sensing elements is utilized to track/supervise motives and responsibilities beneath an incident-maintained data model. Similarly, when the network of wireless sensing elements is utilized to track a goal, the activity of goal modeling is exceedingly applicable to predict the pattern and quantity of processed information inside the network all around tracing the goal.

13.2.6 Protected communication of nodes of WSN

Networks of sensing elements require protected communication for secrecy and coherence of data. Authentication assures data only through the transmitting node, preserves coherence of data, and disables the message's transmission through unauthenticated origins. Secrecy assures privacy of data and intruding (joining in the absence of approval).

SPINS (Security Protocols in Network of Sensors) was recommended by Berkeley laboratories. SPINS utilize the most relevant cryptographic rules due to the causes mentioned—an unsymmetrical cryptographic process having high overhead with regard to memory and calculations for computerized signatures, created and collaboration of key. It also has high memory needs and communicates a higher number of bytes in contrast to the symmetric method.

- SPINS is a collection of protection rules for the networks of sensing elements that are:
  - Secure Network encryption protocol (SNEP) and
  - Micro-Tesla (m-Tesla)

Functions of block cipher encryption are utilized by SPINS. Secure point-to-point communication is enabled by SNEP, which assures secrecy and coherence of data. It protects communication once the authentication procedure is done. It does not need replay of message; hence, the message stays new. The entry control point issues session keys to A and B making use of SNEP. Two nodes, A and B, share six keys. The cipher-block chaining message validation keys, K_{AB} and K_{BA}; encryption keys, K_{AB} and $K_{BA;}$ and counter keys, C_A, C_B are the shared keys. MAC assures secrecy and coherence of message. A lightweight version of TESLA is micro-tesla. It permits validated transmission. The validation is micro-scheduled and productive. It causes stream-loss–tolerant protected validation. An access point validates utilizing micro-tesla. Initially a packet is heard and contemplated as a parent. Then, it gets validated. This causes protected validation stream-loss tolerance. Networks of distributed sensing elements utilize management of the key program, which can be (i) on the basis of sharing or (ii) sharing of random predistribution key.

Needs for wireless sensing element networks: The current establishments in communication, engineering, and networking have given to design of new sensing elements, wireless systems, and information technologies [6]. Such progressive sensing elements can be utilized as a link between the physical world and the digital world. Sensing elements are utilized with diverse gadgets, industries, machineries, and the environment and assist to avoid failures in foundation, mishaps, protecting natural resources, safeguarding fauna, growing production, and issue protection. The utilization of a distributed sensing element network or system has also come up by the scientific approaches in MEMS, VLSI and wireless communication. Hence, by the assistance of present-day semiconductor techniques, you can establish stronger microprocessors that are importantly of smaller size when contrasted to the products of earlier generation. This reduction of processing, computing, and sensing technologies has given rise to minute, insufficiently powered, and inexpensive sensing elements, actuators, and controllers.

13.2.7 Operating systems for WSN

OS for WSN are like embedded system OS.

- TinyOS is an operating system deliberately created for WSN. TinyOS software consists of event handlers and tasks with run-to-completion semantics on the basis of an event-driven programming model.
- LiteOS is another recently established OS for WSN. It has a Unix-like environment and utilizes C programming language.
- Contiki is an OS which keeps up futuristic techniques like 6LoWPAN and Prothreads.
- RIOT is a modern, real-time OS having the same utility as Contiki.
- PreonVM OS keeps up Java programming language deliberately created for WSN and issues 6LpWPAN on the basis of Contiki.

13.2.8 Categories of WSN services

- **Supervising:** WSNs can be established to obtain information regarding surroundings for supervising motives. As an illustration, a WSN established for precise agronomics can supervise climatological states like humidity, temperature, wind speed, and so on and transmit this information every hour. Correspondingly, a WSN for information on quality of water in a water body can make estimations of numerous specifications like turbidness, dissolved oxygen, and so forth and transmit this information twice a day. Those supervising illustrations are sufficient in numerous fields and WSNs have demonstrated the ability to issue correct, constant, and punctual information that can be utilized for decision-making purposes.
- **Notifying:** A WSN can be utilized in distant or reachable lands or regions that are susceptible to calamities. The nodes of WSN can be supported for sending notifications once a uniform condition crosses specific threshold, for illustration, the level of water more than 10 meters for water level sensor nodes in flood. In the same way, notifications can be problems from a tsunami-supervising WSN that consist of

nodes of sensing elements on buoys deployed on the oceans. A WSN employed to monitor landslides can send advance notifications for avoiding accidents and loss of life and property.

- **Information on request:** The network can be inquired regarding the actual values of certain features of interest, and it responds by providing information accordingly. For example, soil moisture values can be queried at a particular location or during the last two days. Further, sub-setting of WSN archived data based on the current time instant, arithmetic, Boolean, logical operators can also be performed.
- **Activating:** On the basis of surrounding factors where the WSN is supervising, the nodes of sensing elements can transmit a signal and transform the actions of an outward system. Specific illustrations of activations are (a) climatic conditions felt by a WSN node can activate a node of sensing element for capturing the GPS location of the mobile entity where it is utilized, or (b) activating a sensing element like a camera for acquiring data for particular supervising motives.

13.3 WSN COMMUNICATION PROTOCOLS

The main objective of communication is to improve the network for working in an energy-efficient manner so that WSN lifetime is lengthened. The physical layer of WSN is regarding the communication hardware. The physical layer is answerable to detect signal, generation of career frequency, modulation, choosing frequency, and encryption of data. MAC protocols are utilized in the data link layer. The network layer manages data routing coming through the transport layer, which is answerable to the flow of data to the WSN application. Numerous types of applications and usefulness are supported by the application layer relying on the type of sensing needed.

Single channel MAC protocol: It is split into two types:

- Synchronous single channel MAC protocol
- Asynchronous single channel MAC protocol

The synchronized single channel method is based on concurrency among nodes. Many synchronization MAC protocols are proposed like S-MAC, T-MAC, RMAC, and DW-MAC (demand wakeup MAC). In these protocols, wireless sensor nodes are synchronized to schedule the active and sleeping periods of nodes to make the WSN energy efficient. The data exchange occurs only in the active time of the sensor node. This method saves energy by decreasing the idle listening time of each node. The overhead of implementing synchronous protocol is substantial.

In an asynchronous approach, each sensor node wakes up independently as per its duty cycle. A sender transmits a preamble for a period of sleeping of the receiver node before transmitting data. If the receiver receives the preamble, it remains active and becomes ready to receive the data. A receiver initiates a transmission by sending a control message in asynchronous MAC protocol initiated by receiver. Thus, energy efficiency is achieved in asynchronous MAC protocol without synchronization.

- *Asynchronous single channel MAC protocol*: Every sensing element node in a network frequently checks the availability of the wireless means as per its own program. If the channel is unavailable, it holds back further until it becomes inactive or a data packet arrives from the sender. The sender assumes that in such a case the receiver is not ready to obtain the data even after obtaining the long preamble, because of network errors. This problem consumes energy; WiseMAC protocol utilizes the preamble sampling method to overcome this problem. In this method, dynamic length preambles are utilized to reduce inactive listening time. In dynamic length preamble, the sensor nodes keep a track of sleep-live schedule of their neighbors. The drawback of WiseMAC is that buffering time of the packet increases in a sender buffer. While broadcasting, the sender sends the packets to all its neighbors and that causes waste of energy.
- *Pseudorandom Asynchronous MAC protocol*: This protocol utilizes a hash function to determine the succeeding wakeup time. The next wakeup time decided by the hash function is non-periodic and this way the pseudorandom asynchronous MAC protocol saves energy wasted in idle listening and even collision is reduced.
- *Medium Reservation MAC (MRMAC)*: In this protocol, additional information about subsequent packet arrival time and medium reservation information is there in the beacon message to reduce end-to-end suspension in delivery of message. Due to this additional information in the beacon message, every node in WSN knows when the wireless channel is idle, so every node decides its own flexible schedule of transmission and reception, hence idle listening and collision is reduced. MRMAC shows better performance than RI-MAC if the network has periodic traffic pattern.
- *Multi-channel MAC protocols*: As single channel-based communication utilizes only one requirement, capability and efficiency of the network is low. The multi-channel radio utilizes a lot of orthogonal channels and divides the bandwidth into multiple channels. Adjacent nodes utilize various channels for sending the packets. This multi-channel MAC protocol enhances efficiency of the network and decreases collision. The multi-channel MAC protocol is effectively created for WSN by considering the following points. Multi-channel networks do not interrupt one another and are orthogonal. The junctions cannot communicate with one another as they belong to different channels; the same channel should be assigned for nodes to communicate.

There are three methods for channel assignment: fixed, semi-dynamic, and dynamic. The approach is based on cluster approach in the fixed assignment. The nodes of sensing elements are split into several clusters. The cluster communicates through the assigned channel. This method obstructs interruption among clusters. Every node is assigned to a certain channel for transmission and reception in approach of semi-dynamic assignment. Every node is allotted a different channel after every wakeup schedule of the node in the approach of dynamic channel assignment. The primary benefit of multi-channel protocol is that there is low interruption and collision while transmitting and receiving the data. But, in semi-dynamic and dynamic approaches, because of switching among channels, there is energy overhead.

- *Flooding*: This is a vigorous algorithm that can deliver a data packet through source till destination in a network. The WSN node that receives the message broadcasts it to all its neighbors in flooding. That causes unwanted message retransmission and increased collisions, causing waste of limited battery power of sensing elements. Thus, flooding algorithms are not beneficial for dense WSN.
- *Gossiping*: Gossip protocol handles certain crucial issues of WSN. The primary objective of gossip protocol is to decrease transmission. Certain nodes throw away messages and do not forward every message it receives in gossip protocols. Gossip protocols are of two types: nondeterministic and probabilistic.

In nondeterministic behavior, any probabilistic distribution is not followed for scheduling the wait time before retransmitting the packet if the sink node is not active. On the contrary, in probabilistic behavior, based on pre-specific gossip probability p_{gsp}, the packets are forwarded by a node. When a node receives a message, rather than forwarding it immediately as in flooding, it gives the decision control to p_{gsp} whether to forward the packet or not. The main advantage of gossiping protocol is that it is easy to implement and reduces energy loss overhead due to unnecessary retransmission. Gossiping protocol suffers from latency, and propagation delays. Hence, based on specific application requirements, a suitable choice of protocol is used.

13.4 COMMONLY USED SENSORS

(1) *Magnetometer*: It measures the strength, direction, or relative change of a magnetic field. Some applications of magnetometers include mobile devices using it as a compass, detecting metal objects under water (for example submarines), various surveys for minerals and oil exploration, locating underground buried objects, equipment that need to maintain precise direction, and so on.

(2) *Gyroscope*: It consists of a rotor (routing disk) mounted at the centre of a larger spinning axis. Angular velocity (variation in rotational angle per unit time and expressed in degrees per second) is sensed by gyro-sensors. A gyroscope device uses earth's gravity to find out direction based on the principles of angular momentum. There are three fundamental sorts of gyroscopes, namely rotary, vibrating structure, and optical. Various kinds of gyroscopes can measure the rotational velocity in three directions and also tilt and lateral orientation. Gyroscopes are generally implemented with a three-axis accelerometer for issuing complete six degrees of freedom, which is useful for motion tracking systems that are used in aircraft, autonomous vehicles, robots, and others.

(3) *Proximity sensor*: This emits a beam of electromagnetic radiation and any change in the field results in a return signal. These sensing elements have the capacity to identify the existence of objects in their vicinity in the absence of physical contact.

Proximity sensors are of four types:

- Capacitive proximity sensing element
- Inductive proximity sensing element
- Photo-electric or opto-electronic sensor
- Ultrasonic proximity sensor

Some of the common application areas include identification, points, investigation, and computing on automatic machines and in manufacturing systems.

Any variation in the di-electric medium encompassing the active face is responded to by the capacitive sensor mostly without making physical contact. They can be configured to sense nearly any material. Capacitive sensing elements can also feel any object beyond a layer of thin carbon, glass, or plastic.

Inductive proximity sensors can sense and react to ferrous and non-ferrous metal objects as well. These sensors can discover metal covered in a layer of non-metallic material. Inductive proximity sensing elements are made of a coil wound throughout a soft iron core. When a ferrous object is nearby, the change in inductance of the sensor takes place. The change in inductance is then converted to voltage to operate a switch. This concept is used in mobile phones to discover unintentional touch when held close to the ear.

(4) *Light-sensing element*: Light-sensing elements that transform light energy (photons) into electronic signals are also called photo-electric gadgets or photo-sensing elements. The light-sensing element is a non-resistant gadget. In a photo-sensing element, a beam of light is utilized for identifying existence or non-appearance of an object. A visible or infrared light beam is emitted through its light-emitting element. For

detecting the light beam reflected through a target, a reflective-type photo-electric sensing element is utilized. The emitted beam of light from the light-emitting element is received by the light-emitting element. In a single housing, both the light-emitting and light-receiving elements are installed. Some common types of light intensity sensing elements are photo-resistors, photo-transistors, and photo-diodes. When this interfaced circuit is balanced with potentiometer, the variation in light intensity is interpreted as voltage variation. Bright light permits a larger amount of current flow. These sensing elements are widely used to measure light intensity as they are simple, cheap, and reliable.

(5) *Radio frequency identification (RFID)*: RFID technology uses an RFID tag (a small chip with an antenna) to embed data in it with an RFID reader. This technology is similar to bar code technology. Contrary to a conventional bar code, a line-of-sight communication is not required among the tag and the reader and the RFID tag can be identified from a distance. The range of RFID can be varied with the frequency variation and can be extended up to hundreds of meters.

Two types of RFID technologies are used often, namely near and far. In a near RFID reader, there is a coil from where alternating current is passed and a magnetic field is generated. The ambient change in the force field is registered by the tag with a small coil, which generates a potential difference. This voltage is then integrated with a capacitor, which actuates the tag chip. The charge is accumulated by producing a small magnetic field which encodes the signal to be transmitted and the RFID reader reads it.

RFID tags are also classified as active and passive. An external power/energy source is linked with active tags; passive tags do not need any such external power source. Passive tags can sense the electromagnetic waves that the reader emits without any external power source and are hence a cheaper option with a long lifetime. Many accessible user-level tools can process the data gathered by specific RFID readers and the raw data from RFID tags is then processed and managed. This processed data can be further analyzed, and inferences are drawn from this analysis for some control section. A dipole antenna is present in a far RFID reader, which is utilized to propagate electromagnetic waves and to sense the alternating potential difference. The object to be tracked is attached with the RFID tag and its existence is discovered by the reader when the object passes across it. Thus, object motion can be followed with RFID technology to send smart things.

An authorized object with RFID tag is beneficial for access control. For illustration, small RFID chips are attached to the front of vehicles and when the car reaches a toll plaza on which an RFID is present the tag data is read for car authorization. If authenticated successfully, it opens automatically. RFID technology is utilized in many other applications like tracking of objects, supply chain management, identity authentication, and access control.

(6) *Accelerometers*: There are two common types of accelerometers, namely piezoelectric accelerometers and seismic mass type accelerometers. For high frequency applications, the piezoelectric accelerometer is preferred due to its compact size. The seismic mass type accelerometer is based on the relative motion between a mass and sources used for low-to-medium-frequency applications.

An accelerometer is an electromechanical device. Forces are measured by an accelerometer. These forces may be static such as the constant force of gravity pulling some object or they could be dynamic, which is caused by moving or vibrating the accelerometer. An accelerometer produces digital or analog outputs. Output of an analog accelerometer is a continuous voltage that is proportionate to acceleration with some specific calibration. Digital accelerometers commonly use pulse width modulation for output. In some accelerometers, the piezoelectric effect is used where accelerative forces stress microscopic crystal structures, and it leads to voltage generation. The other way is by sensing variations in capacitance. Two micro-structures side by side have some capacitance among them. If the accelerative force moves one of the structures, then the capacitance will vary. Using an accelerometer, this variation in capacitance is measured to convert from capacitance to voltage.

13.5 SURVEY OF TECHNICAL PAPERS

13.5.1 Understanding of accurate agronomy supervising system on the basis of network of wireless sensing elements

This exploration interpolates the concepts of the supervision system and explains the characteristics of software and hardware plans of the formulated constituents, network communication protocol, network topology, and the existing requirements [1]. Demonstrations prove that the junction can attain agronomical surrounding information gathering and broadcasting. The system has the characteristics of being small in structure, insubstantial, stable in accomplishment, and smooth in functioning.

It substantially enhances the agronomical output and spontaneous measure exceedingly. The creation and execution of an original wireless sensing element network for supervising agronomical surroundings estimated the vulnerability, strength, and durability of the network in the field. In contrast to conventional agronomy, the system exceedingly enhances agronomical productiveness.

13.5.2 Preciseness agronomy supervising framework on the basis of WSN

This section presents the iFarm framework system, a convenient and expandable agronomical monitoring solution for improving land efficiency by more finely controlling water, enhancing the socio-economic characteristics of agronomists and their perceptions, and planning and forecasting the yield production.

The iFarm system recommends WSNs as a favorable procedure to agronomical supplies expansion, making of conclusions, and supervising of land. WSNs make it feasible for knowing at any time information regarding crop and land conditions so that agriculturists can be helped with numerous messages and advice at the time of cultivation tasks. It describes the benefits of the precise agronomy method to assist making priceless decisions which could not only enhance the efficiency of land but also improve resources utilization. It delivers an explanation of the precise agronomy supervising method, which issues significant assistance to agriculturists. The utilization of wireless sensing elements networks in precise agronomy has been established by propounding the architecture and creation of a precise agronomical supervising system.

The iFarm system is explained through the view of agronomical efficiency, architecture, characteristics, and major serviceability.

The iFarm issues a set of services for agronomists that include hydration and the supervision of water, control of disease and pests, planning and forecasting of crop production, and the expansion of facilities. The iFarm system issues alerts making use of SMS and website only; however at present it is performing on the translation of text to speech and voice messages.

13.5.3 Preciseness agronomy utilizing networks of wireless sensing elements: Chances and ultimatums

The work analysis on the WSN system established in this scheme is for utilization in precise agronomical applications, in which actual data of atmospherical and other ecological characteristics are recognized and authoritative conclusions are taken on that basis to improve them. Agriculturists rely deliberately on the rainfall as they go without an approach to means of hydration. Their production of crops is extremely undependable because of differences in both quantity and spread of raindrops. As well, these agriculturists rely deliberately on the forecasting values of numerous characteristics like climate, water, soil, and so on.

Agronomy comes across a lot of issues, like variation in climate, scarcity of water, scarcity of workers because of an aging population, and enlarged communal interests regarding problems like animal well-being, food harmlessness, and atmospheric influence. Sensing elements networks and alternative agronomic techniques may assist for storing and utilizing the downpour, growing their crop production, decreasing horticulture cost, and utilizing actual values in place of relying just on forecasting. The fundamental goal is to shift the Indian agronomist from forecasting to the definite values that are advantageous for their farmsteads.

13.5.4 Wireless sensing element networks in precise agronomy

We perform the preparatory arrangement of the Lofar Agro project that focuses on supervising atmospheric conditions in a harvest region. Additional to the agrarian demonstration, Lofar Agro focuses on collecting reports on the wireless sensing element network itself. These reports will formulate the source for algorithm simulations in wireless sensing elements networks and will be allocated.

13.5.5 Energy-effective wireless sensing element networks for agronomy accuracy: An analysis

This analysis defines the current implementations of WSNs in agronomy investigation as well as categorizes and differentiates numerous wireless communication protocols, the systematization of energy-effective and energy-gleaning techniques for WSNs that can be utilized for agrarian supervision systems, and differentiation among previous investigation tasks on WSNs based on agronomy. The requirements and restrictions of WSNs in the agronomical field are surveyed, and various power depletion and agronomical direction techniques for continuing supervision are accentuated.

13.5.6 Utilization of wireless sensing element networks for greenhouse requirements control in agronomic accuracy

Here, the utilization of Programmable System on Chip Technology (PSoC) as a part of WSNs for monitoring and controlling numerous parameters of greenhouses are suggested and examined. A large number of the specifications have to be controlled as the crop differences are more in greenhouse technology.

Every day they are growing due to the establishment of agronomy technology. In this circumstance, the wireless sensing element network having extra hardware and software is an effective resolution for greenhouse control.

Analytically it is demonstrated that the hardware established by Cypress Inc. is the foremost resolution that functions with less power having low complicatedness and more dependability for greenhouse control.

13.5.7 Wireless sensing element network in precise agronomy application

This procedure issues real-time information regarding crops and fields, which can assist agriculturists make correct decisions. Using the fundamentals of WSN and Internet technology, IoT technology–based precise agronomy systems are described particularly on the network architecture, software process control, and hardware architecture of the precise irrigating system. The software supervises data through sensing elements in a feedback loop that triggers the control gadgets on the basis of threshold value. WSN application in precise agronomy will improve utilization of water fertilizer and also increase crops production.

13.5.8 Agronomics field supervision utilizing wireless sensing elements networks for enhancing crop yield

The motive of this study is to create and establish an agronomical supervising system utilizing wireless sensing element network for increasing the efficiency and standard of farming by not monitoring it at all times manually. Levels of water, temperature, and humidity are the foremost significant characteristics for the efficiency, enhancement, and standard of plants in agronomy. The water, temperature, and humidity level sensing elements are utilized to collect the values of humidity and temperature. The sensing element has to broadcast the collected information using the wireless communication network to the data server (cloud).

The IOT gateway manages communication among the distant-control serial gadgets and central control system. Agriculturists can notice the estimations through the web at the same time. With the continual supervision of many surrounding factors, the grower can analyze the optimal environmental states for achieving the highest crop efficiency, for better production, and for achieving considerable energy resources. This study presents a crop supervision system on the basis of wireless sensing element network.

IoT has importance to promote agronomical information. The ARM 7 processor combines with the sensing elements (humidity, temperature, and water level) utilized for agronomy supervision and productivity of crops. Relying on the threshold value, the model is supervised spontaneously. The supervised details of crops are transmitted to the cloud using IoT gateway. Thus, the agriculturists can effortlessly access and control the agronomical productivity, while economizing the base materials and enhancing efficiency and aptness in farming production system.

13.5.9 Case study 1: Monitoring snow and ice on UK highways during winters

Winters are perhaps the most pleasant seasons of all seasons. However, when the roads get blocked by thick layers of ice, then you can't say the same thing. Every year thick deposits of ice cover up the highways in the United Kingdom. To tackle this problem, the UK councils had instructed about 1.3 million tonnes of salt before the beginning of winter in 2017. This amount has almost doubled the orders of 2016.

Mayflower Smart Control is a smart company that facilitates solutions for exterior lightening and linked brightened devices. The Amey and Hampshire County Council has been approached by the company to debate about the use of smart technology and minimize the excessive use of gritting of roads during the winters. They have deployed the project in various locations with their first priority being the Winchester area (Hampshire, UK). Mayflower Smart Control has used a smart sensor platform in the pioneering project that will help the highway maintenance team to take a crucial gritting decision on the basis of climatic conditions.

13.5.10 Weather station to assist in decision-making

Every decision linked with the gritting method is normally based on the weather prediction issued by various sources such as Met Office. But, this weather prediction relies on a large geographic area which could issue incorrect readings. The primary issue is that the temperature of the road surfaces may vary as much as 15^0 C. This could result in a day and night difference in the readings. The installed wireless sensing elements by Mayflower Smart Control efficiently supervise the road surface temperatures and especially the weather on the gritting routes. This naturally assists the gritting service teams to make conversant decisions on the basis of correct data. The reading devices have been installed on the street light poles that utilize the Mayflower Smart Control Street lighting network to communicate to the main database for analysis of the readings. The installed sensing elements have the capacity to measure atmospheric pressure, temperature, humidity, pluviometer, anemometer, and wind direction.

Mayflower Smart Control is primarily based on the Zigbee communication platform. The information gathered through smart sensing elements is sent via ZigBee to Mayflower gateway. The data present on the gateway is transmitted to the cloud services by utilizing 2G or 3G technology to get envisaged in the Sentilo Platform which is based on Amazon web services. This collected information is very beneficial to take real-time decisions and also for the weather prediction and to talk about the conditions of the road to the public.

13.5.10.1 Saving money and improving the environment

The fundamental aim of this project was to show instant operational advantage and savings. Now the authorities are capable of making a decision in marginal situations. Amey of Hampshire is receiving up-to-date real-time data with correctness and localized information on temperatures of air and surface to help in the gritter area decision taking.

This exciting innovative project enables the Hampshire County Council to conserve their natural resources. It could also change our view of engaging on the problem of when to grit the road. This project is not only a great money saver but also decreases the carbon footprint of the local authority.

13.5.11 Case study 2: Forecasting volcanic eruptions in Masaya with smart wireless sensing elements

Volcanoes formation is the outcome of natural phenomenons. A volcano occurs when the soft molten rocks and gases trapped under Earth's mantle slicea path through the crust. There are various types of volcanoes across the planet. Some are explosive in nature, while some stay non-explosive. Often a volcanic eruption can happen all of a sudden; it is very hard to forecast regarding their eruption in advance. Many scientists around the world are trying to study volcanoes for forecasting future eruption to save countless lives. A volcanic eruption could destroy entire human and animal settlements with hot lava flows. This could endanger the surroundings and human health; the gas emanations and ash falls have catastrophic effects on our surroundings.

Experts are also considering to work with the current technique for supervising the volcanoes in real-time to learn what is exactly occurring inside and outside the craters. This might greatly assist them to forecast the eruptions. Qwake is an up-and-coming global brand with astounding scientific researchers along with state-of-the-art technology for bringing positive variations. They have been utilizing a smart platform for developing a wireless sensor network. This project took place in the Masaya volcano in Nicaragua.

Masaya is one of the most active Latin American volcanoes. The volcano threw ash and steam rising to a height of 2.1 km in 2008. At the time of this eruption, a lava lake was created in one of the craters with the huge size of 600 m^2. This gave a view of the behavior of magma.

13.5.12 Real-time monitoring at Masaya volcano with 80 sensors

The Qwake team along with Sam Cosman, a filmmaker and explorer, had worked together on this project to make it the world's first online volcano. Their mission took place in July and August 2016. The Qwake team was determined to find a wireless monitoring platform capable of collecting, transmitting, and storing data in real-time. For this reason, they chose a smart sensing element platform for the project. David Gascon, a member of the expedition and also the first person to encounter the impressive landscape around the Masaya volcano, said that they are pleased to help in a project that could possibly save a million lives not just in Nicaragua but also in other worldwide projects to monitor other volcanoes.

To reach the Santiago Centre and the open-air lava lake safely, their team built a zip-line system that would allow the personnel to descend efficiently with gear. They had installed the smart sensing elements just next to the crater to obtain information in some of the most demanding and extreme environments.

The sensor platforms in Masaya volcano were Smart Environment and Ambient Control sensors used as their data repeaters. More than 80 different sensors were installed in order to monitor H_2S, CO_2, temperature, atmospheric pressure, and humidity. These sensors were vacuum sealed to protect them from extreme heat coming from the volcano. The temperature was around 65° C where the sensors were placed. However, the temperature could be as high as 537° C on some parts of the active volcano.

The collected data from the sensors would directly send information to a gateway. The IoT gateway obtained the data and sent it by using 3G communications to the database. The data was later visualized by using Predix, a cloud software platform GE built for the industrial internet.

13.5.12.1 Saving lines by opening data

The principal objective of the project was to build a digital early warning system to forecast volcanic explosions in advance. The data obtained from the sensors will be utilized by scientists and data experts to digitally simulate what is happening inside the volcano. The project provides the public service of giving them access to the general population and decision-makers to encounter the volcano in real time.

13.5.13 Case study 3: Protection of beluga whales in Alaska utilizing flexible sensor platform

The oceans consist of vast numbers of aquatic creatures. It is estimated that we humans have only discovered a tiny percentage of what is actually present under the water. But, all the aquatic species are not doing well under the water. Beluga whales are considered to be an endangered aquatic species. They commonly live in the cold Arctic Ocean and near the coastal areas of Alaska, Canada, Greenland, and Russia. Their population is estimated to be 150,000 in the world, with only 400 said to inhabit Alaska.

These whales have a very fixed behavioral pattern. They commonly live in the open ocean in winters and migrate to the warmer areas in summers. However, they remain close to the continents and sometimes they even travel to rivers by several kilometers. The rivers give them temporary refuge and facilitate the seasonal renovation for their epidermal layer. Sadly, the environmental and acoustic pollution resulting from human activities place a real danger to these species. The pollutants caused by humans can have a catastrophic effect on the animals, leading to an increase in cases of cancer, deterioration of the immune system, and reproductive pathologies. Many studies have already proven that cancer frequency is higher in the determined areas.

A natural gas leak took place in 2017, approximately 6 kms in Nikiski, Alaska. People reported the leakage to the Alaskan authorities, which took the prospect of monitoring the area due to its being a censorious place for the already endangered beluga whales.

Aridea Solutions, a company from West Virginia, was given the task to design a solution to monitor any kind of leaks. The company has already been developing monitoring technologies to tackle environmental challenges. The Aridea Solutions team had the goal to develop something that continuously monitors the environmental parameters in real-time and can communicate with people and machines. After some intense hours of critical research, the company decided that a buoy would be the most suitable platform for the required application. Aridea Solutions got a smart sensor platform to permit the interfacing of the different industrial protocols needed to interface the sensors. The smart platform they had chosen would easily allow them aggregation and transmission of data to any shop located several miles away from the buoy by using a 900 MHz communication protocol.

Aridea deployed the wireless sensing element platform for supervising air and water close to the leak-affected area where the beluga whales and other aquatic organisms inhabit. The project aimed to monitor methane, CO_2, and oxygen levels above the water surface. Due to the urgency of the problem, the project had a very hectic and tight timeline. Thus, the engineers at Aridea worked together for 2 weeks to design, build, and deploy the buoy-based monitoring system to the affected location. Once the system was deployed and training was given to the operators, sensors calibration also took place for superior accuracy. The system worked fine as it was meant to and gave back data to the scientists for further analysis and to quantify the environmental impact in the area.

The parameters measured in the air are:

- Temperature, relative humidity and pressure
- Carbon dioxide (CO_2)
- Oxygen (O_2)
- Methane (CH_4) and
- Volatile organic compounds (VOC) and lower explosive limit (LEL) Parameters measured in water are:
- Dissolved methane Pro Oceanus Mini CH_4
- Dissolved oxygen
- pH levels
- Conductivity

The deployed platform effectively collected and transmitted important environmental data into a gateway on a nearby ship. Then the data was gathered by the gateway and sent to Aridea's Terralytix platform by using a 4G cellular connection. This platform helped the scientists to view and analyze the data in real time. Until the leakage was repaired, the buoy system was used many times to monitor the environmental conditions in the affected area.

The sensors deployed for the job were well designed to outperform the extreme conditions. The main reason behind the success was perhaps the limited time frame, which led the engineers to work too hard to tackle the environmental conditions. The deployment process was itself a big job as the buoy had to drift through the affected area within reasonable times.

13.6 FEATURES OF WIRELESS SENSING ELEMENTS NETWORK

- *Energy productive*: In WSN, energy is utilized for various tasks like sensing, communication, computation, and storage. If the nodes of sensing elements run out of power they cannot be recharged at a remote location, and as a consequence, fail to sense the surrounding physical environment. Thus, to reduce consumption of power of a node of sensing elements, efficient protocols and algorithms are utilized when creating a WSN.
- *Low cost*: Normally, to measure the physical surroundings, a large number of nodes of sensing elements are employed in a WSN. It is desirable to have the cost of individual sensor nodes as low as possible and as a consequence decrease the overall cost of the network. Further, deployment cost also needs to be taken into account while using a large WSN.
- *Computational power*: Typically, the sensor node has limited computational capabilities and is usually decided by cost, size, and energy of the node.
- *Communication capabilities*: Generally, radio waves are used for communication in WSNs. The communication channel can be unidirectional or bidirectional. Sometimes, WSN is used in a remote and inaccessible terrain environment; hence, it must be robust and resilient enough to quickly recover in case of failure of some of the nodes of WSN.
- *Cross-layer design*: This type of design is recently emerging in wireless communication as it improves WSN performance in terms of energy efficiency, data rate, QoS, and so on. The traditional layered approach faces problems such as:
  - In a traditional layered approach, network optimality cannot be ensured as there is no data communication to share information among different layers leading to incomplete information at each layer.

  - Dynamically changing environment cannot be adapted by the traditional layered approach of WSN.
  - Other limitations include interference between different users, access conflicts, fading, and others.

- *Distributed sensing and processing*: To enable robustness and resilience, the sensor nodes in a WSN are distributed randomly and uniformly. A sensor node sends the data to a sink node that is capable of collecting, processing, sorting, aggregating, and sending the data.
- *Security and privacy*: Sufficient security must be provided to each sensor node to protect from unauthorized access, unintentional damage, and malicious attacks to damage the information inside the node.
- *Multi-hop communication*: In case a node requires to communicate with another node or a base station beyond its radio frequency range, then to reach to the sink node, a multi-hop route is needed.
- *Robust operation*: WSN nodes are left unattended most of the time and are prone to physical damage, battery drainage, communication failures, and impacts of harsh environments. Therefore, it is necessary to have sensor nodes with capabilities of fault and error which is a measure of its reliability. The sensor nodes must have the capability of self-testing, self-configuring, and self-repairing for robust use.
- *Dynamic network topology*: A sensor node can fail due to battery exhaustion or physical or sensor data tampering. To overcome such issues, WSN nodes must have the capabilities to dynamically reconfigure and adjust themselves under various conditions such as addition of new nodes or replacement of existing ones.
- *Application oriented*: WSN is highly dependent on the specific application (e.g., military, health, environmental, etc.) for which it is designed.

Multimedia wireless sensor networks: Multimedia WSN [8] is used in situations where tracking and monitoring of the events is performed using multimedia (images, audio, and video). These WSNs consist of low-cost sensor nodes with devices such as camera and microphone. The sensor nodes are connected to form a network using wireless connections to collaborate with each other for performing data preprocessing operations such as data retrieval, correction, and compression. However, these are challenging to perform as they require high bandwidth and energy.

Movable wireless sensing elements networks: An accumulation of nodes of sensing elements that can proceed by itself and meet with the physical surroundings forms a movable WSN. The movable sensing elements nodes have abilities of calculating and communication. As the nodes of sensing elements can proceed, movable WSNs have broad adaptable applications compared to static sensing elements networks because these WSNs have enhanced exploration of the surroundings and better energy productivity and channel capability.

Undersea wireless sensing element network: Though water covers greater than 70% of the earth, its supervision is not as up to date compared to earthly land mass. That is because of the vast requirements included in utilizing sensing elements undersea. A WSN in which nodes of sensing elements are utilized undersea is known as an undersea WSN. Particular tests of undersea WSN are prolonged detaining in propagation, restriction in bandwidth, and failure of nodes of sensing elements. To design and develop undersea WSN, the considerable problem is conservation of energy. The data is collected through these nodes of sensing elements by utilizing self-determining undersea vehicles. For instance, in the ocean observing systems, both drifting and protected buoys are utilized for measuring various marine, biological, and water parameters.

Underground wireless sensor network: Underground WSN is specifically designed for subsurface region. Potential applications include monitoring precision agriculture, landslides, earthquakes, environmental, infrastructure, and so on. The cost of underground WSN is high due to the need for special infrastructure for developing underground WSN such as equipment, cost, careful planning, and maintenance. For data transfer from underground sensing elements nodes to the base station, deployment of added sink nodes is required over ground, which also secures resilience of the WSN. Since the sensor nodes are underground, it is difficult to recharge them; also the communication signal from the sensor nodes gets attenuated.

Climate supervisions: Climate supervision systems based on IoT can gather data through a number of sensing elements attached (like pressure, degree or intensity of heat, wetness, and so on) and transmit the data to applications based on cloud and back-ends of storage. The gathered data in the cloud can then be studied and conceptualized by cloud-based applications. Climate notification can be sent to the supported users through such applications. AirPi is a climate and air feature supervision tool efficient to record and upload information regarding air pressure, humidity, temperature, extent of light, extent of UV, nitrogen dioxide, carbon monoxide, and extent of smoke to the Internet. A prevalent climate supervision system is integrated with buses for measuring climatic variables such as temperature, humidity, and quality of air at the time of bus path.

Structural health supervision: Structural health supervision systems utilize a network of sensing elements for supervising the levels of vibration in structures like buildings and bridges. The data gathered through these sensing elements is examined to estimate the health of the structures. By data analyzation, it is feasible to identify cracks and mechanical breakdowns, locate the damages to a structure, and also calculate the remaining life of the structure. Warnings can be given in advance in the case of imminent failure of the structure by utilizing those systems.

Since structural health supervision systems utilize large numbers of wireless sensing element nodes that are powered by traditional batteries, researchers are exploring energy-harvesting technologies to harvest ambient energy, like mechanical vibrations, sunlight, and wind.

Smart Grids: A data communications network consolidated using electrical matrix that gathers and analyzes data caught in near real time regarding transmission, distribution, and utilization of power is a smart grid. Smart grid technology issues foretell information and directions to resources, their providers, and their consumers on how best to control power. Smart grids gather data about generation of electricity (centralized or distributed), consumption (immediate or foretelling), storage (or energy transformation into other forms), distribution, and health data of equipment. Smart grids utilize rapid, incorporated, co-operative communication technologies for real-time information and exchange of power. The health of the device and the integrity of the grid can be found utilizing sensing and measurement technologies based on IoT. Smart meters can capture nearly real-time utilization, distantly control the consumption of electricity, and switch off the supply remotely when needed. Smart metering can be utilized to prevent power thefts. Smart grids can enhance productivity all over the electric system by analyzing the data on generation of power, transmission, and utilization. Storage accumulation and smart grids data analysis in the cloud can assist in dynamic enhancement of system operations, prolongation, and organizing. Cloud-based supervision of smart grids data can enhance energy utilization levels through energy feedback to users integrated with concurrent pricing information. Real-time demand response and management strategies are utilized to lower peak demand and overall load through appliance control and mechanisms of energy storage. Condition supervision data gathered through power generation and transmission systems can assist to detect faults and forecast disruptions.

13.7 HOW CAN WIRELESS SENSING ELEMENTS NETWORKS CREATE A SECURED AND PRODUCTIVE MINING INDUSTRY?

Mines that utilize wireless sensing elements are now called "smart mines," and they are coming up due to innumerable courses of action that can be utilized in a functioning mine.

In the absence of farther problems, there are five major ways where wireless sensing elements networks assist the mining industry [5].

- *Keep workmen safe*: Lacking uncertainty, the most significant reason wireless sensing elements are utilized in mines is to make certain the manpower is secure. In spite of latest security gadget, elegant mining items, and the best health and security means, there is always a possibility of a mishap when operating over the diggings. Utilizing wireless sensing elements networks takes away the requirement for human staff to monitor likely threatening readings in regions that are unsteady or hard to access, which in turn decreases the accident's possibilities. The sensing elements can discover possible risks such as barrage cracking, certainly one of the worst catastrophes in mining. Workmen can also wear wireless gadgets for tracing purposes. This means it can be systematically pinpointed by superiors where they are and can advise them of any possible mishaps in the region. Sequentially, this assists to stop mishaps.

- *Decrease costs*: Wireless sensing elements can importantly decrease the costs of running a mining operation. With the sensing elements taking on many of the tasks that manual labor habitually made up, economizing money is unavoidable. They also stop the requirement for certain high-priced devices such as cables and costs of perpetuation. Sensing elements that work on anticipating capability can decrease costs farther, discovering variations, malfunctions and several other issues using machinery before they become an actual issue, which means downtime is notably decreased and prices on current devices and repairs are few.
- *Economize duration*: Given that wireless sensing elements can speedily and correctly collect and outline data, it's given that the duration required for getting something done is less compared to a group of workers doing so. While they are crucial to save time when the mine is functional, they are just as beneficial at emerging a new mining operation. The wireless sensing elements, linked to the Internet of Things, can assist to ensure the preliminary mining is accurate, a method that would be more laborious when left to a purely manual worker.
- *Automatize with simplicity*: Mining workers in smart mines having wireless sensing elements can more clearly automatize their instruments compared to those functioning in mines in their absence. They can form their previously occurring wireless network of sensing elements by putting mechanization software in place, which assists them to distantly manage certain forms of devices. For example, automatic trucks are becoming progressively in demand in the mining industry. These trucks work using wireless sensing elements, permitting them to function independently. These vehicles mean various functions that consistently utilized human labour are now being accomplished through machines with wireless sensing elements.
- *Secure organization data*: The security and recording of organization data is predominant in the industry of mining. Correctly taking down data with purely human labour is vital, particularly in the instance of a mishap, an issue, or a major mistake. On top of that, always there is a possibility that if this data is digital, it can be stolen or hacked.

Luckily, appropriate wireless sensing elements take down data in their field of operation, which in this situation can be mining. This is a protected measure to acquire data because the sensing elements are only attainable locally instead of in the cloud, where they can be hacked, manipulated, or stolen. Cybercriminals will not be able to access the network, however data is still given to the mining staff nevertheless.

- This sort of technology that is demonstrating itself to be crucial in the mining industry is provided at Sensor-Works.

13.7.1 Design of a wireless sensing element network

The WSN design mostly consists of numerous topologies utilized for network of radio transmissions such as a mesh, star, and hybrid star. These topologies are explained as follows [3]:

Star Network: When the base station can only send or receive a message approaching remote nodes, communication topology such as a star network can be utilized. There are a lot of junctions accessible that are not permitted for sending messages to each other. The advantages of this junction mostly have clarity, the ability to keep the power usage of distant junctions to a minimal.

It also allows communications with low dormancy between the base station and a distant node. The fundamental disadvantage of this network is that the base station should be in the radio range for every unconnected node. Compared to other networks, it is not strong as it relies on a single node for networking handling.

Mesh Network: This type of network allows data transmission among two nodes in the network that are in radio transmission scope. If a message transmission needs to happen between two nodes and it is away from radio communications scope, then it can use a node like an intermediate for message transmission approaching the favored node. The primary advantage of a mesh network is flexibility. Whenever a single node discontinues functioning, a distant node can confabulate with any other node type in the scope within, then send the message approaching the favored position. Furthermore, the scope of the network is not spontaneously restricted by means of scope among single nodes; it can expand directly by prepending many nodes to the system.

The primary drawback of this type of network is capacity usage of the network nodes that carry out communications in a way appropriate to multi-hop is normally higher compared to other nodes that don't have this power to restrict battery life regularly. Furthermore, whenever the number of communication hops becomes higher approaching the target, then the time taken for transmitting the message will be higher as well, especially when the process with low capacity of the nodes is a need.

Hybrid Star–Mesh Network: A hybrid among the two networks like star and mesh issues a robust and adjustable communications network when it maintains the lowest utilization of capacity of wireless sensor nodes. The nodes of sensing elements having low capacity are not permitted for transmitting the messages in the network topology of this type. This allows it to maintain usage of minimum capacity. However, remaining network node elements are permitted with the power of multi-hop by permitting them for transmitting messages between two nodes. Normally, the nodes having multi-hop capability have more power and are regularly plugged into the mains line. This is the accomplished topology with the imminent quality mesh networking called ZigBee.

13.7.2 Design of a wireless sensing element node

The constituents utilized for making a wireless sensing element node are contrasting parts such as processing, sensing, power, and transceiver. It also incorporates extra elements that rely on an application such as an electric generator, a locality detection system, and a mobilizer. Normally, sensing parts incorporate two subunits, specifically analog to digital converters (ADCs) and sensing elements. Sensing elements produce analog signals that can be converted to digital signals using ADC, which is later passed on to the processing unit. Usually, this part is linked across a very small storage part for handling the activities for making the sensing element node function with remaining nodes for obtaining the allotted sensing functions. The sensing element node can be linked to the network using a transceiver part. The power units are assisted using parts of power rummage such as photovoltaic cells while the remaining subparts rely on the application.

In the functional block diagram of wireless sensing nodes, the modules provide an adaptable policy for dealing with needs of broad applications. For example, on the basis of sensing elements to be organized, the substitution of a signal controlling block can be done. This allows usage of various sensing elements along with the wireless sensing node. Moreover, the radio link can be swapped for a particular application.

13.7.3 Advantages of using a wireless sensing elements network

WSNs have a lot of benefits [4] compared to standard supervision systems based on cable. WSNs are convenient, profitable, and dependable. A WSN comprises structurally allocated self-sufficient gadgets that utilize sensing elements for detecting sound, temperature, and other frameworks in various applications. In the beginning, WSNs were established for military purposes. Nowadays WSNs are utilized for mainly civilian purposes, like supervising conditions, health care, and traffic control. Moreover, wireless sensing element nodes are utilized for detecting tenancy of vehicles in parking lots. Magnetometers are utilized for detecting the presence of a vehicle in the hardware node. Magnetometers and micro-radars are also utilized for automotive tracking.

WSNs comprise of a large number of supplies, mainly having sensing elements nodes with less cost, for establishing a densely utilized network with the help of a wireless communication constituent enabled on the nodes. Every sensing element node is enabled with various sensing elements, computation units, and storage gadgets for sensing, processing, and transmitting all types of supervised information. Information related to freeway traffic can be gathered using video cameras and well-organized bends in the road. Certain features of WSNs are identifying correctness, coverage area, error tolerance, connectivity, minimum human interaction, feasibility in unpleasantly rough surroundings, and organizing of effective sensing elements.

13.8 CONCLUSION

Currently, in the creation and establishment of WSNs, various programming approaches are estimated, of which importance are commonly on problems of low-level systems. However, as mentioned before, for the streamlining of the design and establishment of WSNs and summary from technological link load balancer particulars, certain hardware load balancers (HLB) methods have been predicted and developed for its purposes. As per BenSaleh et al., [9] putting in the model-driven engineering technique is becoming a favorable resolution in particular and these HLB methods would be of great help to ease the establishment and design and also lessen certain ultimatums of WSNs.

BIBLIOGRAPHY

[1] Nifasath Piyar, S., & Baulkani, D.S. (2022). A Review of Wireless Sensor Networks in Agriculture. *International Journal for Research in Applied Science and Engineering Technology*, 10(XII), ISSN: 2321–9653; IC Value: 45.98; SJ Impact Factor: 7.538. https://www.ijraset.com/research-paper/a-review-of-wireless-sensor-networks-in-agriculture

[2] Agarwal, T. (2021, January 25). Wireless Sensor Network architecture: Types, working & its applications. ElProCus-Electronic Projects for Engineering Students. https://www.elprocus.com/architecture-of-wireless-sensor-network-and-applications/

[3] Ukhurebor, K. E., Odesanya, I., Tyokighir, S. S., Kerry, R. G., Olayinka, A. S., & Bobadoye, A. (2021). Wireless Sensor Networks: Applications and challenges. *IntechOpen eBooks*. https://doi.org/10.5772/intechopen.93660; https://www.intechopen.com/chapters/73287

[4] Rodrigues, J. J. P., Compte, S. S., & De La Torra Diez, I. (2016d). Body area networks. In *Elsevier eBooks* (pp. 97–121). https://doi.org/10.1016/b978-1-78548-091-1.50006-3; https://www.sciencedirect.com/topics/engineering/wireless-sensor-network
[5] Aram, & Aram. (2021, October 15). How wireless sensor networks are creating a safer, more efficient mining industry. *Sensor Works Condition Monitoring Technology*. https://www.sensor-works.com/wireless-sensors-for-mining-safety/
[6] M. Monisha Macharla, Basics of Wireless Sensor Networks, Topologies and Application. https://iot4beginners.com/basics-of-wireless-sensors-networks-topologies-and-application/
[7] Surya Durbha, Jyoti Joglekar. *Internet of Things*. Oxford University Press, 2021; ISBN 0190121092, 9780190121099
[8] Types of Wireless Sensor Networks: Attacks & Their Applications ElProCus; https://www.elprocus.com/introduction-to-wireless-sensor-networks-types-and-applications
[9] M.S. BenSaleh, R. Saida, Y.H. Kacem, M. Abid. (2020). Wireless sensor network design methodologies: A survey. *Journal of Sensors*, 2020, 9592836.

Chapter 14

Security considerations in IoT using machine learning and deep learning

S. Hemalatha and K. Jothimani
Kongu Engineering College, Tamil Nadu, India

S. Anbukkarasi
SRM University, Tamil Nadu, India

14.1 INTRODUCTION

As the world becomes increasingly connected, more and more appliances are being integrated with machine learning (ML) techniques and deep learning (DL) capabilities. While these smart appliances offer a variety of advantages including increased efficiency, convenience, and personalization, they also introduce new security considerations that must be taken into account [1]. One major concern is that smart appliances are connected to the Internet, which is vulnerable to attacks. Hackers can also easily gain access to these devices and steal some secret information like client data and use them as a gateway to institute attacks on other devices within the same network. Another issue is that the ML and DL algorithms used by these appliances may be trained on sensitive data, such as user behaviour or personal information. If these algorithms are not properly secured, they can be manipulated or exploited by malicious actors to compromise user privacy. To mitigate these risks, manufacturers of smart appliances must ensure that their devices are designed with security in mind. This includes implementing robust authentication mechanisms to prevent unauthorized access, encrypting all data transmitted between the device and the cloud, and using secure software development practices to prevent vulnerabilities. Additionally, it is important for consumers to take steps to secure their smart appliances. This includes changing default passwords, keeping software up to date, and avoiding connecting their devices to unsecured public networks. In conclusion, while smart appliances offer a range of benefits, they also introduce new security considerations that must be addressed. By implementing strong security measures and practicing good cybersecurity hygiene, both manufacturers and consumers can help ensure that these devices remain safe and secure. In this book chapter, we have discussed the security consideration in Internet of Things (IoT) using machine learning and deep learning algorithms. The major contribution of this chapter is deep study of the security consideration and the effect of machine learning and deep learning on them. This chapter discusses various topics such that recent technologies, architecture, ML and DL algorithms on security of IoT and its significance [2]. Hence this chapter provides an overall idea on security consideration in IoT domain.

DOI: 10.1201/9781003474524-14

14.2 DEFINITION

Smart appliances are household or commercial appliances that are designed with the ability to connect to the Internet, interact with other devices, and use ML and DL algorithms to perform various functions. These appliances can be controlled remotely using a smartphone or other internet-enabled device, and can often be programmed to perform tasks automatically or adapt to user preferences. Examples of smart appliances include smart refrigerators, ovens, thermostats, and security systems [3]. These appliances are meshed with Amazon Alexa, which allows users to interact with them by using voice commands. Smart appliances are designed to improve efficiency, convenience, and personalization for users.

14.3 SMART APPLIANCES IN VARIOUS FIELDS

Smart appliances have been introduced in various fields and industries and have become increasingly popular due to their ability to improve efficiency, convenience, and accuracy. Here are some examples of smart appliances in different fields:

Home Appliances: Smart home appliances that include ovens, smart refrigerators, and lighting systems are becoming increasingly common. These appliances can be controlled remotely using a smartphone app, and can often be programmed to perform tasks automatically or adapt to user preferences.

Healthcare: Smart medical devices, such as wearable blood pressure monitors, fitness trackers, and blood glucose monitors are becoming increasingly popular. These devices can be used to monitor health and provide real-time feedback to users and their healthcare providers.

Agriculture: Smart agriculture appliances, such as soil sensors, weather stations, and automated irrigation systems are being used to improve crop yields and reduce water usage. These appliances can collect data and use ML algorithms to optimize crop growth and reduce waste.

Transportation: Smart transportation appliances, such as self-driving cars and traffic management systems, are being developed to improve safety, reduce traffic congestion, and enhance efficiency. These appliances use DL algorithms to make decisions and adapt to changing road conditions.

Energy: Smart energy appliances, such as smart meters and energy management systems, are used to minimize energy usage and electricity waste [4]. These appliances can collect data on energy usage and use ML approaches to optimize energy consumption. Overall, smart appliances are being integrated into various fields and industries, and are expected to continue to grow in popularity as technology continues to advance.

14.4 FEATURES

14.4.1 Significant features of ML and DL

Machine learning and deep learning techniques are two important subsets of AI that are transforming various industries by enabling machines to learn and make decisions based on data. Here are some significant features:

Automated Learning: Both techniques have the ability to learn from large datasets without being explicitly programmed. They can identify patterns and make predictions based on the data they have learned from.
Adaptive Learning: It has the ability to adapt and improve performance over time as they are exposed to new data. This enables them to make better predictions and decisions as they learn from more data.
Non-Linearity: Deep learning algorithms are able to learn complex relationships between inputs and outputs, including non-linear relationships. This means that they are able to make more accurate predictions and decisions than traditional linear models.
Feature Extraction: Both algorithms are able to automatically extract relevant features from complex data. This enables them to make accurate predictions and decisions based on the most important data points.
Scalability: AI is able to scale large datasets and complex problems. This enables them to be used in a variety of industries and applications. These are powerful tools for making predictions, identifying patterns, and making decisions based on data [5]. Their ability to learn and adapt from data makes them particularly valuable for industries such as healthcare, finance, and transportation, where making accurate predictions and decisions is critical.

These two techniques have overlapping but distinct use cases. Here's a general guideline on when to use each approach.

14.4.2 Machine learning

Limited Data: When you have a moderate amount of labeled data available, and the task at hand can be solved using traditional statistical and algorithmic techniques, machine learning is often a good choice [6]. It involves training models on features extracted from the dataset, by using these models to make decisions based on those features.

Structured Data: Machine learning is particularly effective when dealing with structured data, such as tabular data with well-defined features and relationships. Tasks like classification, regression, and anomaly detection can be efficiently tackled with machine learning algorithms.

Interpretable Models: If the interpretability of the model is crucial for your application, machine learning methods tend to provide more transparent and explainable results. Decision trees, linear regression, and support vector machines are examples of interpretable models.

14.4.3 Deep learning

Large-Scale Data: Deep learning excels when dealing with vast amounts of unlabelled or labelled data. It can automatically learn hierarchical representations from raw or high-dimensional input, making it suitable for tasks where manual feature engineering would be difficult or time consuming.

Unstructured Data: Deep learning is particularly effective for unstructured data types, such as audio, text images, and video [7]. Convolutional neural networks are commonly used for image analysis; recurrent neural networks (RNNs) for sequential data; and transformers for natural language processing tasks [8].

High Complexity: If the problem is highly complex and requires capturing intricate patterns or dependencies, deep learning models with their multiple layers of neurons and non-linear activations can capture and model complex relationships more effectively than traditional machine learning methods.

It's important to note that these guidelines are not strict rules, and there can be cases where the boundary between machine learning and deep learning is blurred. Additionally, the choice of approach also depends on the available computational resources, time constraints, and expertise of the practitioners.

14.4.4 Machine learning in security considerations smart appliances

Machine learning is a critical tool for addressing security considerations in smart appliances. It can be used to identify patterns and anomalies in data, enabling them to detect and prevent potential security breaches. Here are some ways in which ML can be used to improve the security of smart appliances.

Anomaly Detection: It will be trained to identify abnormal behaviour on a smart appliance, such as unauthorized access or unusual usage patterns. This can help to detect potential security breaches before they can cause harm.

Predictive Maintenance: It will be used to identify potential security vulnerabilities in a smart appliance before they become a problem. This can enable proactive maintenance and patching of security vulnerabilities before they can be exploited.

User Authentication: Used to improve the accuracy of user authentication mechanisms, such as facial recognition or fingerprint scanning [9]. This can reduce the risk of unauthorized access to a smart appliance.

Threat Detection: It will be trained to identify potential security threats, such as malware or cyber-attacks, on a smart appliance. This can help to prevent or minimize the damage caused by these threats.

Data Encryption: It will encrypt sensitive data transmitted between a smart appliance and the cloud, reducing the risk of data breaches and unauthorized access.

Overall, ML can be a valuable tool for improving the security of smart appliances by detecting and preventing potential security breaches and help to ensure that these devices remain safe and secure for users.

14.4.5 Taxonomy of the survey: Deep learning in security considerations smart appliances

Deep learning is another critical tool for addressing security considerations in smart appliances. DL algorithms can be used to analyse large amounts of data and identify complex patterns that may be difficult to detect using traditional security methods. Here are some ways in which DL can be used to improve the security of smart appliances:

Intrusion Detection: It can be trained to identify patterns of behaviour that are indicative of a potential security breach or intrusion. This can help to detect and prevent unauthorized access to a smart appliance [10].

Malware Detection: It can be trained to identify patterns of behaviour that are indicative of malware or other malicious software on a smart appliance. This can help to prevent or mitigate the damage caused by these threats.

Threat Intelligence: It can be trained to analyse large amounts of security data to identify emerging threats and vulnerabilities in smart appliances. This can help to ensure that security measures are up to date and effective.

Fraud Detection: It can be trained to detect patterns of behaviour that are indicative of fraudulent activity, such as credit card fraud or identity theft [11]. This can help to prevent financial losses and protect user data.

Natural Language Processing: Deep learning algorithms can be used to improve the accuracy of voice authentication systems, such as those used by smart speakers or smart assistants. This can reduce the risk of unauthorized access to a smart appliance.

Overall, DL can be a powerful tool for improving the security of smart appliances by analysing large amounts of data and identifying complex patterns and helping to ensure that these devices remain safe and secure for users.

14.5 MACHINE LEARNING AND DEEP LEARNING METHODS IN NEW INFORMATION TECHNOLOGIES

14.5.1 IoT

The term Internet-of-Things refers to a wildly diverse network made up of smart gadgets that are connected to the Internet. IoT devices generate huge amount of data, which are handled using ML and DL techniques. ML and DL algorithms can be crucial in tackling the problems with resource management in large-scale IoT networks [12]. Older IoT networks used intrusion detection systems (IDS) such as behavior based, knowledge based, and intrusion prevention systems. But now IoT uses artificial intelligence among other things, and is more complex and flexible. IoT networks have only recently been employed with these techniques. This chapter discusses the potential security risks and intrusion detection using ML and DL methods.

14.5.2 Cloud computing

Machine learning as a service cloud platforms have gained popularity because of advancements in ML and DL approaches and the power of cloud

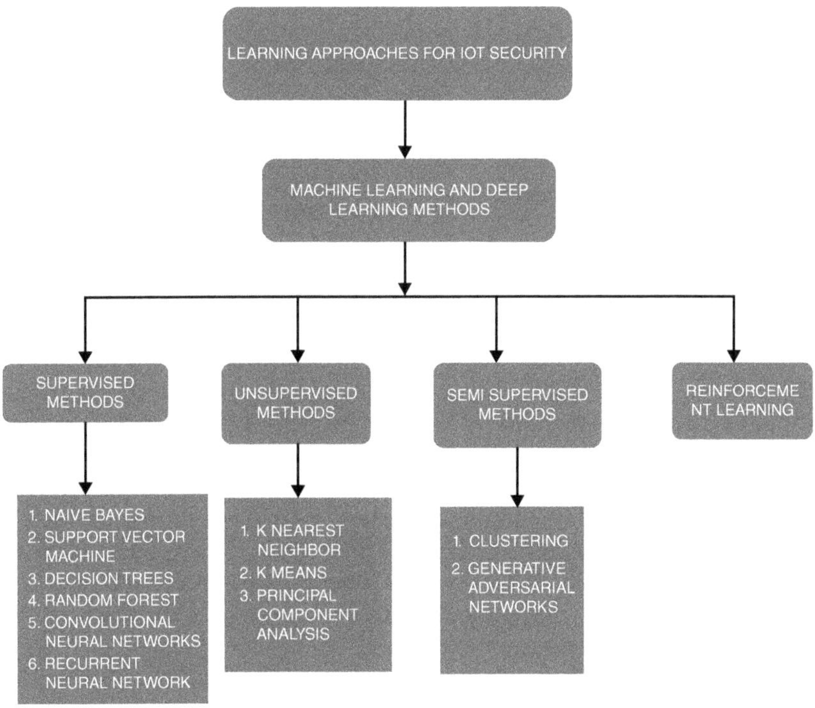

Figure 14.1 Machine learning and deep learning methods for Internet of Things.

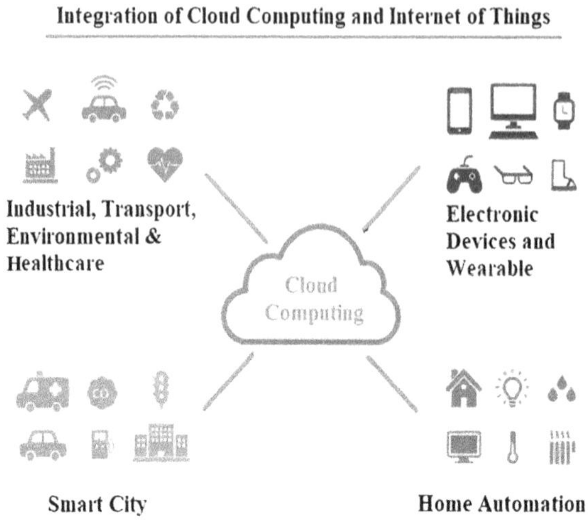

Figure 14.2 Cloud computing in IoT.

Source: https://iotdesignpro.com/articles/iot-and-cloud-computing.

computing to deliver services quickly and affordably. In addition, third-party cloud services are becoming more widely used to outsource the training of DL models, which requires high-performance graphics processing units (GPUs). Because cloud-hosted ML/DL services are widely used, a variety of attack surfaces are now available for adversaries to use the system for their own evil purposes (Figure 14.2). The literature on cloud-hosted models is evaluated systematically in this study along both the critical security dimensions of assaults and defences.

14.5.3 Edge computing

Massive volumes of data are being produced by smart devices and familiar sensors in businesses and communities. Processing power is pushing the centre of computation and services for the network connection [13] and the development and deployment of AI such as deep learning in IoT applications. These apps and services are widely transforming how people live their lives, from facial recognition to aspiring smart factories and cities. Previous techniques such as cloud computing make it difficult to realise the goal of providing artificial intelligence for every individual and every company all-around due to latency and efficiency. As a result, it has become desirable to launch DL services by taking the resources at the edge of the network that close towards the datasets.

Edge intelligence has therefore attracted a lot of attention as a means of easing the implementation of DL services using edge computing. Moreover, the AI representative technique can be used in edge computing frameworks to create intelligent edges for adaptive, dynamic edge management. According to their technical architecture, three categories may be made for emerging edge AI hardware: 1) Hardware based on graphics processing units (GPUs), which often have good performance and compatibility but typically use more energy; 2) Hardware based on field programmable gate arrays (FPGAs), which save energy and require fewer computational power but have less programming flexibility and compatibility than GPUs; 3) Hardware based on application specific circuit designs such as Google's TPU, which generally have a custom design and also provide power consumption. The discriminator is in charge of accurately identifying if the data set is coming out from genuine data or the generator, while the generator tries to learn as much as possible about the true data distribution by actively introducing information at the back-fed input cell [14]. In the adversarial process, these two participants must keep improving their ability to create and differentiate until a suitable solution is found.

Edge computing and DL are likely to complement each other as important AI techniques. The various possible scenarios and fundamental supporting methods for edge intelligence and the intelligent edge have been thoroughly introduced and discussed in this survey. In conclusion, the main challenge of moving deep learning to the cloud to the network's edge is how to design and build edge computing architecture under various constraints of networking, interaction, energy consumption, and computing power to achieve the best performance of inference and training. Edge intelligence will become more ubiquitous as the edge computing capability rises and smart edge computing will play a crucial supporting role in helping edge intelligence function better. Using this survey, we intend to intensify debates and research projects.

14.5.4 Hybrid models

Recurrent Neural Network (RNN): RNNs are capable of processing sequential data produced by data sources. Each neuron in an RNN layer receives data from the channel before it as well as the output from the layer above. RNNs are typically excellent options for making predictions about the future or filling in gaps in sequential data. The gradient explosion is a significant issue with RNNs. This problem can be resolved by LSTM, which enhances RNN by including a gate structure and a clearly defined memory cell, by controlling (allowing or preventing) the flow of information.

14.5.5 Fog computing

By moving resource-intensive tasks including computing, analytics, communication, and storage closer to the clients, fog computing (FC), a new architecture, intends to reduce network stresses throughout the core network and cloud computing. In order to overcome the issue of excessive energy usage with electricity for IoT apps that require speed, FC systems can employ intelligence features to profit from data that is easily accessible with computational resources. It produces vast amounts of data, which leads to the development of an increasing variety of FC services and applications. Additionally, the vital discipline of DL has achieved substantial advancements in a number of study fields, including robotics, facial recognition, decision-making, neuromorphic computing, and computer vision.

Speech synthesis with computer graphics along everal investigations have been suggested to investigate the use of DL to address FC problems [15]. These days, D L is more frequently used to enhance FC applications and offer fog services including cost, delay, data processing, accuracy, and energy reduction as well as resource management, security, and traffic modelling.

14.5.6 Internet of Drones

Internet of Drones (IoD) provides high flexibility over a large area of difficult outlines and it has recently acquired step. Unpiloted aerial vehicles can be exploited practically in a group of applications, together with search and rescue operations, agriculture, mission-critical services, surveillance systems, and so on. This is achieved by other technological and real advantages, such as high flexibility or the ability to reach locations and the ability to extend wireless connectivity areas that are inaccessible to humans. As well, the use of drones shows potential in enhancing a number of network design performance criteria, including dependability, throughput, connectivity, and delay. However, the usage of drone networks raises a variety of issues relating to the dependability of the wireless medium, high degree of mobility, the life of batteries, and which might also result in frequent topological changes. Moreover, security and privacy issues must be investigated thoroughly. This explains the remarkable volume of recent literature on IoD–related topics that has recently been generated. The objective of the current work, in comparison to previous surveys on IoD–related issues, is to describe the various features of IoD by bringing out a classification technique of the IoD environment that develops along two primary directions. On a macroscopic level, it follows the Internet protocol stack's hierarchy, starting at the physical layer and moving up to the top layers while considering cross-layer and optimization methods into considerations. Each layer of the stack's most major works are further categorised at a finer level in accordance with the different issues that are particular to that layer.

14.5.7 Internet of Vehicles

In the automobile sector, mobile ad hoc networking has been applied extensively through the vehicular ad hoc network (VANET). The IoT, a cutting-edge technology in the 5G/B5G era, is gradually transforming the present Internet into such a fully integrated internet architecture. Additionally, it will stimulate the advancement of current research areas in new directions, including smart communities, smart homes, and smart health, and when it comes to meeting the required specifications of intelligent transportation, vehicle automated control, smart transport systems, and intelligent road internet, the VANET must increase the speed of technology advancement. The Internet of Vehicles (IoV), which intends to achieve the information exchange between vehicles, has been developed in light of this background [16].

The objectives of IoV are lessened road congestion, reduced accidents, and other information services. IoV is presently getting a lot of attention from both business and academics to design a new network architecture for the future network with higher security, lower latency, higher data throughput, and huge connectivity in order to facilitate relevant research. This article also offers an extensive literature study of the basic details of IoV, including fundamental VANET technology, various network designs, and common IoV applications.

- IoV's smart application servers are divided into four types: service subscriptions, entertainment, traffic safety, and traffic management. For smart servers, both internal and exterior processing engines are taken into consideration. The internal engine comprises a big data unit, a big data processing unit, and a big data processing component analysis unit. These three units' tasks are carried out in full using the fundamental cloud solutions offered by the cloud platforms. The outside engine comprises an information distribution unit in charge of delivering end-to-end services to client applications and a data gathering unit in responsibility of gathering information from internal sources.
- Each car would have a worldwide unique number on the internet, which is a requirement for web presence of vehicles. Despite flaws, some government groups have started implementing GPS–based identification to improve safety in public transport services. The GPS–based vehicle registration approach would not need to be used thanks to the ubiquitous Internet ID.
- The incorporation of RFID and GPS would substantially improve current working parameters. This is due to the fact that new application domains for ITS are accessible and old application domains could be improved for efficiency and service quality.
- IoV is founded on the growing idea of connected travelling in smart transportation. Automating numerous vehicle security and efficiency features in addition to popularising vehicular transport are the IoV's primary objectives.

14.6 IOT WITH MACHINE LEARNING/DEEP LEARNING

A new class of social applications is emerging as a result of the development based on IoT infrastructure and indeed the increasing prevalence of smart handheld devices. The majority of these apps are consumer focused and data intensive. Such applications are successful in creating a more interconnected society fueled by DL and ML approaches. The recent COVID-19 pandemic highlights the importance of such applications across a wide range of industries, including fitness tracking, entertainment, and many others, in addition to the healthcare sector. The cloud infrastructure is frequently utilised by these applications to deliver advanced user experiences [17]. This collection of social applications and research issues related to it, as well as the implementation elements, will all be covered in detail throughout the chapter.

14.6.1 IoT architecture

The IT profession is already inundated with innovative terms like data science, analytics, AI, and the IoT, but what do these terms actually mean? Internet of Things refers to the connection between various devices and users. The devices collect data from various devices and processes those data, which is then sent to a server. Finally the data centres perform additional analysis, enabling automation and actions. However, there is still a substantial and largely unseen architectural structure that depends on different components and interactions between your command and the fulfilment of tasks. IoT technology has been more well liked recently and has a wide range of uses. IoT applications work in accordance with how they were created depending on the many application domains. There is not a set of standards-defined architecture of work, nevertheless, that is rigidly followed worldwide. Depending on the specific business job at hand, different architectural layers and levels of complexity are used. The most common and accepted architecture is a four-layer one.

Different Layers of IoT Architecture:

Sensing / Perception Layer: The first layer of IoT system is the physical layer. Here all the devices are connected with each other for providing the connectivity between real word and digital. The physical layer, which contains sensors and actuators capable of collecting, accepting, and processing data across a network, is referred to as perception (Figure 14.3). There are two types of connections available; wired and wireless connections can be used to connect sensors and actuators. The components' range and location are not restricted by the design.

System Layer: The sensor is used to capture real time data by the sensing layer. This data is passed to system layer. In this layer are Network Gateways and Data Acquisition Systems (DAS). DAS is responsible for aggregating the data that is captured by the sensing layer. (Sensing, collecting, arranging, and aggregating data from sensors, then converting these analogue data into digital data). Finally, these data assembled by the sensor devices must be conducted and processed. The system layer performs that function. It provides connection between these gadgets

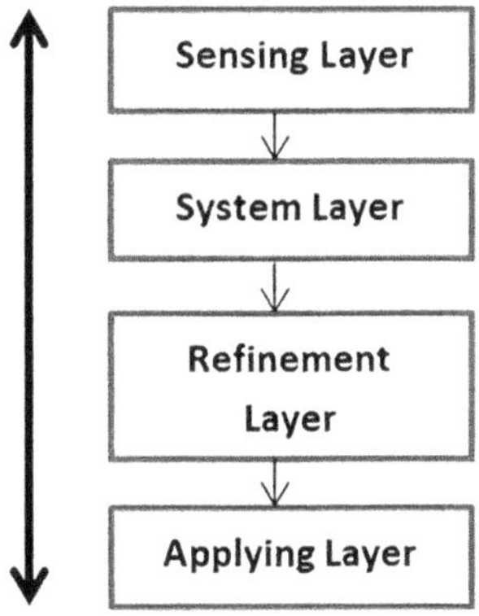

Figure 14.3 Different layers of IoT architecture.

to servers and cloud and network gadgets. Furthermore, it controls each device's data transmission.

Refinement Layer: The refinement layer works similarly to processing layer functions of the human brain. The data before transfer to the data centre is being evaluated, pre-processed (i.e. removing noisy data), analysed, and stored. Future actions are retrieved by software applications that handle the data and prepare it. This is done by edge analytics or edge IT.

Applying Layer: The applying layer, which provides the user interface of IoT solutions, is useful for administrators for managing IoT device and creating new rules for operations of the IoT. A dashboard that displays the status of the gadgets in a system or a smart home application where users may turn on a coffee maker by touching a button in applications are two examples. There are various ways to use the Internet of Things.

14.6.2 Deep learning methods for IoT

Image identification and time-series interpretation for IoT applications have seen widespread use of deep learning techniques, such as neural nets, convolutional neural networks, reinforcement learning, and prolonged term memory [18]. For example, advanced driver assistance systems and autonomous vehicles that undertake forward collision warning, blind pinpoint monitoring, driver assistance, traffic safety, infrastructure, and overcrowding, among many other functions have been established utilising deep computing and machine learning techniques (Figure 14.4). Through vehicular communication systems, such as dedicated short-range communication systems, VANETs, long term evolvement, and fifth generation mobile networks, autonomous vehicles can cooperate with each other by sharing information they have detected, such as road signs, collision events, and so on.

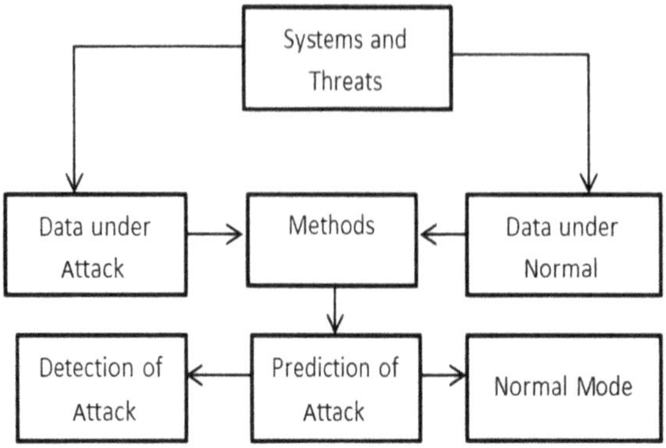

Figure 14.4 DL methods in IoT.

14.6.3 Behaviour modelling and analysis of IoT using deep learning

Deep learning is credited as having given rise to modern artificial intelligence. Many additional application disciplines including robotics, computer vision, and speech recognition have made substantial use of DL. In comparison to traditional ML techniques, DL has certain important advantages (Figure 14.5). Deep learning uses a variety of hidden layers inside a neural network design, allowing it to tolerate complex nonlinear connections

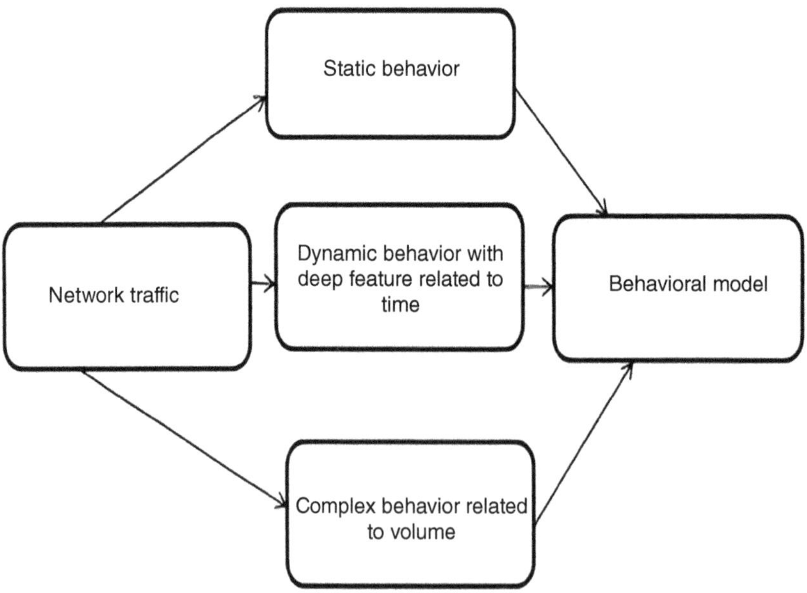

Figure 14.5 Behaviour modelling and analysis of IoT using deep learning.

between attributes. Popular topologies including neural network convolution and longer short attention span networks have the ability to extract and identify relevant features directly from raw data as opposed to depending on manually constructed statistical characteristics as is done in traditional ML (e.g., auto encoders). DL is particularly important for tackling big data challenges [19].

IoT applied in several fields including smart energy, smart homes, Industrial 4.0, and healthcare, has many advantages such as increased agility and mobility and higher productivity and efficiency, but it also adds a lot of new risks. The issue is data privacy challenges and security and interoperability challenges. IoT networks are ad hoc in nature, as shown in Figure 14.4. The management of IoT now faces substantial issues in the areas of security and privacy [16]. Current research has shown that DL algorithms are quite effective and have numerous benefits over previous approaches for doing security assessments of IoT devices.

14.6.4 ML/DL in IoT security

IoT device monitoring capabilities can intelligently offer a defence against fresh or zero day threats. With DL and ML, by using effective data exploration techniques, it is possible to identify normal and abnormal behaviour as shown in Figure 14.3 based on how IoT devices and components interact with one another. The required information of each component of both the IoT system may be gathered and examined to establish typical patterns of interaction and allowing for the early detection of malicious behaviour. Furthermore, because techniques may intelligently anticipate upcoming unknown assaults by learning from past examples they may be crucial in anticipating attack patterns, which are frequently mutations of earlier attempts. IoT systems must make the switch from just enabling secure connectivity between devices to access control [20].

14.6.4.1 Safety and security threats in IoT

Through an internet connection, IoT combines physical items and their environment. Most of the time, the gadgets operate in an undesired and unwelcome internet environment. As a result, hackers and attackers may use the weak IoT (Figure 14.6).

Devices to disclose personal data and login passwords from sensors. Threats may be divided into two categories: passive threats and aggressive threats. Threats that are passive try to access the system's data and information but do not interfere with its resources, like eavesdropping.

Active threats occur when an attacker or hacker tries to change the information and take over the hardware. Threats that are currently active include Sybil, distributed denial of service (DDoS), denial of service (DoS), Trojans, spoofing, phishing, and smishing. Authorization, authentication,

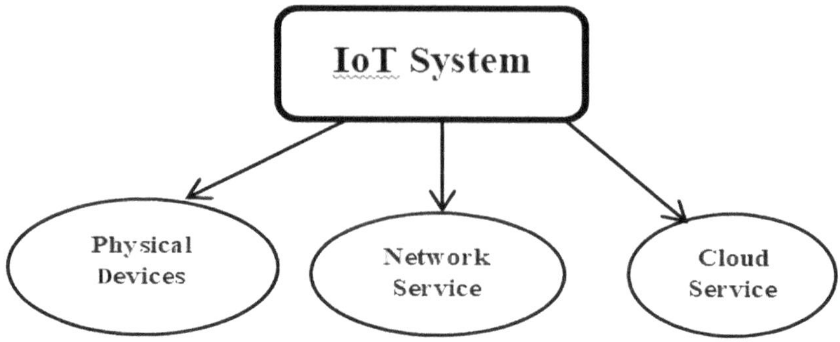

Figure 14.6 Safety and security threats in IoT.

confidentiality, availability, integrity, and quasi are examples of potential security assaults that might have an impact on the security requirements.

14.6.5 Attack surface for IoT (s)

This section talks about potential attack surfaces and dangers. The network system, physical device, and cloud application services are shown in Figure 14.5, the categories that make up the attack vectors of IoT.

14.6.5.1 Physical devices are the attack surface

The main vector for cyber-physical dangers is the physical surfaces. There are sensors, RFIDs, and actuator on this IoT surface. Many kinds of data are gathered from the IoT ecosystem using the sensors. RFIDs are crucial for wireless network connection because they automatically identify objects using a unique identifier. This surface is easily accessible and is more at risk from physical dangers. The majority of physical equipment are vulnerable to assault since they have important information and limited resources. For instance, access requests from physical devices might monitor device information, which could lead to threats like DoS, DDoS, or other cyber-attacks [21].

14.6.5.2 Network service at attack surface

An IoT systems network service is made up of two key components, namely RFID and WSN. Both components are susceptible to online threats. Possible attacks against RFID at the session layer include blackmail, synchronisation attacks, and reply attacks. At the network and application layers, potential risks include fake routing and eavesdropping. The Sybil attack, jamming attacks, and replay attacks are among the potential assaults on the network level of WSN. The IoT system faces these security risks when it is directly connected with traditional networks since those networks systems are no longer safe. Another possible point of vulnerability for the network infrastructure service, which may be subject to major security risks, is the routing protocol. A secure forwarding is therefore required for a protected IoT system. Attackers can also target open ports to gather data such as MAC addresses, router IP addresses, and network gateways.

14.6.5.3 Cloud service at attack surface

By laying a solid cybersecurity foundation, an organisation may successfully manage cloud-related IoT risks. This requires using a sensible strategy that includes determining what assets and data to safeguard, doing an extensive risk analysis across all elements of the IoT–cloud ecosystem, and making sure that security precautions are not just in place but are continuously followed. As long as businesses embrace a proactive security posture and as many best practices as they can, they'll profit from the IoT–cloud convergence.

14.6.5.4 Emerging ML techniques in IoT security

With the capability to send data across a network, the Internet of Things is becoming more and more commonplace and converting inanimate objects into intelligent ones. IoT is present in a variety of businesses and is not just found in homes or utilities. By linking the physical and digital worlds, IoT is fast making the world smarter. It is predicted that by 2024, more than 20 billion gadgets will likely be linked. It provides possibilities but also a variety of threats. Concerns regarding securing the IoT environment and how ML methods could assist to solve these security challenges are centred on how to keep billions of devices secure and what can be done to assure the security of the networks these devices operate on.

14.7 ARCHITECTURE (SMART HOME)

The development of the widely used worldwide network known as the Internet, along with the use of mobile devices and cloud computing in smart objects, which provides new convenience for the development of creative solutions to various troubles in daily life, popularized the idea of the Internet of Things. The idea of the IoT creates a network of things that can interact, communicate, and work together to accomplish a single objective [22]. Once each IoT gadget ceases to function as a standalone unit and joins a larger, fully linked system, they can improve our daily life. It gives us access to the collected data, which we can use to make smarter decisions, monitor our businesses, and keep a check on our properties even when we are far away. As the IoT archetype develops it permeates every part of our existence. A wider range of applications, such as electronic medical solutions and the smart city, makes life easier. The idea of a smart city intends to increase resource efficiency and enhance the level of amenities provided to inhabitants and reduce the cost of public administrations. The major focus of this project is home automation, which is another application. The system's freshly constructed mock-up sends notifications whenever any type of human activity is detected close to the house's entryway [23].

14.7.1 The MufHAS architecture (smart home)

Automation is playing a more and more crucial component in both daily life and the world economy. In order to build complex systems for a rapidly growing variety of uses and human activities, engineers attempt to combine automated equipment with numerical and organizational tools. Home automation is a strategy that has existed since the late 1970s. However, technology and smart systems have advanced people's expectations and have changed significantly over time regarding how well a conventional house can be transformed into a smart home (Figure 14.7). There are now different ideas about what a home should accomplish or how services should be offered and accessed at home to become a smart home and this has affected the concept of home automation systems.

A system for home automation entails removing the responsibility for managing and handling electric appliances from the end users. If society examines various home automation systems across time, it will see that they have always provided efficient, practical, and secure means for residents to enter their houses. The emergence of various existent, recognised home automation systems rely on data connection, such as Microcontroller-based and Raspberry Pi–based automation systems, regardless of shifts in users' expectations and evolving technology or passage of time. This is not a concern until the system is set up and well programmed during the actual construction of the structure. The cost of construction is fairly expensive for buildings that are already standing, regrettably.

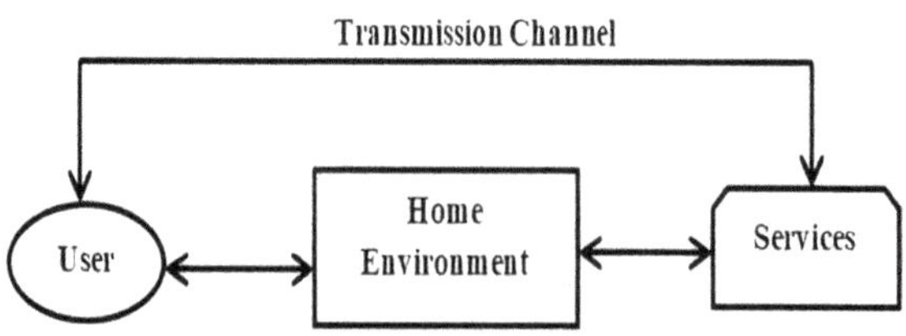

Figure 14.7 The MufHAS architecture.

Recent techniques:

- Building Automation Using IoS Software Platform.
- Smartphone-Based Home Automation System.
- Zigbee Inspired Connected Home System.
- Area Network Inspired Connected Home System.

14.7.1.1 System architecture

IoT will now have a major effect on people's existence in poor nations. A smart home is one that combines smart items with various purposes. The implementation of an intelligent house is intended to increase security and efficiency and we may manage home equipment without the assistance of people by employing several sensors to monitor activity within the house (Figure 14.8). A laptop or intelligent phone device that is linked to the Internet can access testing methods and automated controllable devices.

A network that is controlled by a single network owner is considered to be a local area network. This network is built as a vehicle local area network group, which separates the process efficiently and improves network management. Smart objects are referred to as Internet of Everything (IoE) devices despite the fact that the term IoT just was recently coined. The devices and activities that have been recorded on the IoE server as an administrative assistant home internet gateway were just what we considered to be the creative stuff for this investigation. A suggested method for making innovative home networks is called home automation network design, which combines the IoE with networking technologies to carry out specific tasks on network systems [24].

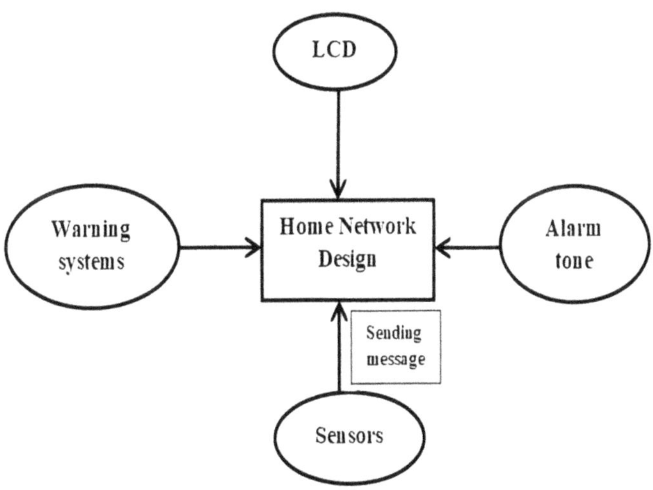

Figure 14.8 System architecture.

In addition to offering security, home automation may provide an array of features, such as programmable security using various warning system, LCDs, alarm sounds, and sending texts to approved consumers when a sensor detects security problems as shown in Figure 14.1 Home computerization refers to the use of a PC or microcontroller invention to manage and monitor household items. Computerization is commonplace because it makes processes efficient, safe, and lucrative. By bringing up numerous sensors for house automated processes, smart home assumes client commitment in managing home settings and running home appliances. IoT and Internet of Services are advanced technologies that depend on connected sensors to improve the quality of life. A hypothesis known as the IoE networks expands on the importance put on machine-to-machine communication via the IoT to reflect a more comprehensive framework that also incorporates people and processes. IoT is clever with the connected individual who comprehends strategy, specifics, and group relations.

The program Cisco Packet Tracer Simulator was used to simulate the construction of an intelligent home network. The smart home may offer a variety of benefits in addition to security, such as planned protection utilizing numerous warning systems, LCDs, and alarm tones, as well as sending messages to approved clients when a sensor finds security difficulties. The connections can be done wirelessly or through wires. At our house, the servers, PC, and routers are all linked through wired connections because they are all closer together. The very last IoE devices, such as household appliances, are connected wirelessly via DHCP protocols.

14.8 INTRUSION DETECTION IN IOT

14.8.1 Anomaly/intrusion detection

IDS are either hardware or software systems that automate the process of keeping track of analyzing networking or computer system activity for indications of security issues. Intrusion detection systems are now a required component of the security infrastructure of the majority of companies as a result of the rise in frequency and severity of network attacks over the past few years [25]. The guide was created as an introduction to intrusion identification for those who need to know what protection objectives their mechanisms support, where to pick and configure intrusion detection systems for the unique networks and systems environments, how to oversee the outcome of intrusion detection systems, and how to integrate intrusion prevention operations with the rest of the organizations safety facilities.

14.8.2 Intrusion detection systems for IoT

IoT is a group of several technologies that collaborate with one another. Sensors, actuators, CPUs, and transceivers are built into IoT devices. Sensors are tools that gather information about the outside world. Temperature, camera, pressure, UV, and some other sensors are common [26]. On the network's edge or even on distant servers, data from sensors is stored and processed. Actuators are tools that alter the environment based on information that has been processed. IoT devices need wireless communication because they are located in geographically dispersed locations. The back-end services, Internet, local network (gateway), and devices try to compensate for the IoT architecture core infrastructure. Enhancing the comfort and effectiveness of human life is one of the objectives of smart environments. A technique for creating smart surroundings has appeared recently from the IoT concept. In any real smart home environment built on the IoT model, privacy and security are considered to be key concerns [27, 28]. Applications for intelligent devices are challenged by security because of security flaws in IoT–based systems. IoT systems for intrusion detection are essential for preventing security attacks on IoT that take advantage of some of these security flaws. It's possible that traditional IDSs may not be a suitable option for IoT due to the constrained CPU, ways described of IoT devices, and the particular protocols used.

14.9 APPLICATIONS AND SECURITY CHALLENGES

IoT devices collect data and transfer it to a centralized data server, where it is processed, compiled, distilled, and used to facilitate a variety of operations [29]. The advantages of IoT are utilized by the corporate sector, the government, organizations, and the average consumer. Mobile phones, computers, coffee makers, microwaves, Google, and Apple watches are just a few examples of specific IoT–enabled items. Any device that has a sensor and Internet access can be connected to the IoT. IoT and Big Data are frequently brought up together since the former produces the same kinds of enormous amounts of data that the latter does.

Key cybersecurity problems and obstacles include

Neglecting to evaluate and create: It is difficult to get gadgets on the market; several IoT manufacturers neglected to consider security. Gadget security issues could have gone unnoticed during development, and after release, there might not have been any security upgrades. Yet as IoT security has become better known, so has device security.

Default passwords that promote brute-force attacks: The default passwords that come with many IoT devices are frequently insecure. The buyers of them might not be aware that they may be changed [30]. IoT devices are susceptible to brute-force attacks and password cracking when they use strong passwords and access data.

IoT malware and ransomware: Since there have been an increasing number of IoT, more have linked IoT devices recently, and this pattern is projected to persist, there is now a greater chance that malware and ransomware may try to take advantage of these devices. One of the most often seen malware types is IoT botnet malware [31, 32].

Increased cyber-attacks: Coordinated denial of services assaults can exploit infected IoT devices. Here, hacked devices serve as an offensive platform to infect more computers or hide harmful activities. DDoS assaults on IoT devices might strike home automation even though they mostly target businesses [33].

14.10 RESEARCH AND FUTURE DIRECTION OF SMART APPLIANCES

Research in smart appliances is a rapidly evolving field, with new advances and innovations being made every day. Here are some current and future directions of research in smart appliances.

Interoperability and Integration: One of the major challenges in smart appliances is interoperability between different devices and platforms. Future research will focus on developing standards and protocols that enable seamless integration and interoperability between different smart appliances and platforms. **Personalization:** As smart appliances become more advanced, there is a growing trend towards personalization. Future research will focus on developing algorithms and models that can personalize the user experience based on individual preferences and behaviors [34].

Energy Efficiency: Smart appliances can help to reduce energy consumption and carbon emissions. Future research will focus on developing algorithms and models that optimize energy consumption and reduce waste, while maintaining user comfort and convenience.

Security and Privacy: As smart appliances become more prevalent, there is a growing need for improved security and privacy measures. Future research will focus on developing more advanced security and privacy mechanisms that can protect user data and prevent unauthorized access.

Artificial Intelligence: Advances in AI are driving innovation in smart appliances. Future research will focus on developing more advanced AI algorithms and models that can enable smart appliances to make more intelligent and informed decisions based on user data and preferences [35, 36].

14.11 CONCLUSION

Overall, research in smart appliances is a rapidly evolving field, with new advances and innovations being made every day. As technology continues to advance, the potential applications and benefits of smart appliances will continue to expand, improving our lives in countless ways. The future of security in smart appliances is likely to be characterized by increasing emphasis on cybersecurity measures to protect against potential threats. As smart appliances become more interconnected and capable of transmitting data, there is a greater risk of them being targeted by hackers who could gain access to sensitive information or even take control of the appliances themselves. To mitigate these risks, smart appliance manufacturers are likely to adopt a range of security measures, such as encryption, secure communication protocols, and multi-factor authentication. They may also implement regular software updates to address any vulnerabilities that are discovered, as well as collaborate with security researchers to identify and fix any issues [37]. In addition, regulations and standards are likely to be put in place to ensure that smart appliances meet certain security requirements, and that manufacturers are held accountable for any breaches that occur. While there are undoubtedly risks associated with smart appliances, the industry is likely to take significant steps to address these risks and ensure that consumers can enjoy the benefits of these devices without compromising their security or privacy.

REFERENCES

1 Ioan Fitigau, Gavril Toderean. 2013. Network Performance Evaluation for RIP, OSPF and EIGRP Routing Protocols. *International Conference on Electronics, Computers and Artificial Intelligence (ECAI)*, pp. 1–4.
2 R. Rajakumar, Amudhavel. 2017. GWO-LPWSN: Grey Wolf Optimization Algorithm for Node Localization Problem in Wireless Sensor Networks. *Journal of Computer Networks and Communications*, 7(1), 01–10. DOI: 10.1155/2017/7348141
3 I. Fiţigă, G. Toderea. 2013. Network Performance Evaluation for RIP, OSPF and EIGRP Routing Protocols. *Proceedings of the International Conference on Electronics, Computers and Artificial Intelligence - ECAI*, Pitesti, 2013, pp. 1–4. DOI: 10.1109/ECAI.2013.6636217
4 Sai M. Krishna. 2019. A Hybrid Approach for Enhancing Security in IOT using RSA Algorithm. *HELIX*, 9(1), 4758–4762, 10.29042/2019-4758-4762
5 Anuka Pradhan, Biswaraj Sen. 2018. A Brief Study on Contention Based Multi-Channel MAC Protocol for MANETs. *International Journal of Emerging Trends in Engineering Research (IJETER)*, 6(12), 74–78. https://doi.org/10.30534/ijeter/2018/016122018
6 Salil S. Kanhere, Sherali Zeadally, Mehmet A. Orgun. 2021. Machine Learning Techniques for IoT Security: A Comprehensive Survey. *IEEE Communications Surveys & Tutorials*, 15, 9231–9253.
7 Muhammad Usama, Sheraz Ahmed, Khaled Salah. 2021. Intrusion Detection Systems for IoT: A Comprehensive Survey. *IEEE Communications Surveys & Tutorials*, 15, 8845–8872.

8 Krishna P. Gopi, Ravi K. Srinivasa, S. Adluri, D.S. Devineni 2017. Implementation of Bi Directional Blue-Fi for Multipurpose Applications in IoT Using MQTT Protocol. *International Journal of Applied Engineering Research*, 12(Special Issue 1), 692–700.

9 Lakshmana Phaneendra Maguluri. 2018. Efficient Smart Emergency Response System for Fire Hazards using IoT. *International Journal of Advanced Computer Science and Applications*, 9(1), 314–320.

10 Tong Li, Jiawei Chen, Guanbo Bao. 2020. Machine Learning for Security and Privacy in the Internet of Things: A Survey. *IEEE Internet of Things Journal*, 13, 287–312.

11 V. Dankan Gowda, S. B. Sridhara, K. B. Naveen, M. Ramesha, G. Naveena Pai. 2020. Internet of Things: Internet Revolution, Impact, Technology Road Map and Features. *Advances in Mathematics: Scientific Journal*, 9(7), 4405–4414 ISSN:1857-8365.

12 Abdelrahman I. Saad, Yasser M. K. Omar, Fahima A. Maghraby. 09 October 2019. *Predicting Drug Interaction With Adenosine Receptors Using Machine Learning and SMOTE Techniques*, IEEE Access, pp. 146953–146963.

13 Mahsa Rakhsha, Mohammad Reza Keyvanpour, Seyed Vahab Shojaedini. 2021. Detecting Adverse Drug Reactions from Social Media Based on Multichannel Convolutional Neural Networks Modified by Support Vector Machine. *2021 7th International Conference on Web Research (ICWR)*, IEEE, INSPEC Accession Number: 20839679. DOI: 10.1109/ICWR51868.2021.9443128, 02 June.

14 B.V. Anjali, G.K. Ravi Kumar 08 June 2022. A Broad Review on Adverse Drug Reaction Detection Using Social Media Data. In *2022 6th International Conference on Intelligent Computing and Control Systems (ICICCS)* (pp. 1852–1856). IEEE. DOI: 10.1109/ICICCS53718.2022.9788381

15 A. Tripathy, N. Joshi, H. Kale, M. Durando, L. Carvalho. 2015. Detection of Adverse Drug Events Through Data Mining Techniques. *2015 International Conference on Technologies for Sustainable Development (ICTSD)*, Mumbai, India, pp. 01–06.

16 Yawei Yue, Shancang Li, Phil Legg, Fuzhong Li. 2021. Deep Learning-Based Security Behaviour Analysis in IoT Environments: A Survey. *Security and Communication Networks*, 2021, 8873195, 13 pages. DOI: 10.1155/2021/8873195

17 Fatima Hussain, Rasheed Hussain, Syed Ali Hassan, Ekram Hossain. 2020. Machine Learning in IoT Security: Current Solutions and Future Challenges. *IEEE Communications Surveys & Tutorials*, 22(3), 386–410.

18 Naercio Magaia, Ramon Fonseca, Khan Muhammad, Afonso H. Fontes N. Segundo, Aloísio Vieira Lira Neto, Victor Hugo C. De Albuquerque. April 15, 2021. Industrial Internet-of-Things Security Enhanced With Deep Learning Approaches for Smart Cities. *IEEE Internet of Things Journal*, 8(8), 119–132.

19 Hao Lin, Wenwen Zhang, Ruikai Miao. 2020. Deep Learning for IoT, Big Data and Streaming Analytics: A Survey. *IEEE Communications Surveys & Tutorials*, 16, 982–1015.

20 Mohammed Ali Al-Garadi, Amr Mohamed, Abdulla Khalid Al-Ali, Xiaojiang Du, Ihsan Ali, Mohsen Guizani. 2020. A Survey of Machine and Deep Learning Methods for Internet of Things (IoT) Security. *IEEE Communications Surveys & Tutorials*, 22(3), 652–683.

21 Ghulam Abbas, Amjad Mehmood, Maple Carsten, Gregory Epiphaniou, Jaime Lloret. 2022. Safety, Security and Privacy in Machine Learning Based Internet of Things. *Journal of Sensor and Actuator Networks*, 11, 38. DOI: 10.3390/jsan11030038

22 Yiwen Zhang, Youhuizi Li, Mingkui Wei, Jianxin Li, Laurence T. Yang, Albert Y. Zomaya. 2020. Edge Computing: A Survey on Recent Developments and Future Directions. *ACM Computing Surveys*, 11, 8852–8873.

23 Olutosin Taiwo, Absalom E. Ezugwu, Olaide N. Oyelade, Mubarak S. Almutairi, Enhanced Intelligent Smart Home Control and Security System Based on Deep Learning Model. *Hindawi Wireless Communications and Mobile Computing*, 2022, 9307961, 22. DOI: 10.1155/2022/9307961

24 Khan Muhammad, Afonso H. Fontes N. Segundo, Aloísio Vieira Lira Neto, Victor Hugo C. de Albuquerque. April 15, 2021. Industrial Internet-of-Things Security Enhanced With Deep Learning Approaches for Smart Cities. *IEEE Internet of Things Journal*, 8(8), 95–114.

25 L. Duan, M. Khoshneshin, W. Street, M. Liu. Adverse Drug Effect Detection. *IEEE Journal of Biomedical and Health Informatics*, 17(2), 305–311, 2013.

26 Daemin Shin, Keon Yun, Jiyoon Kim, Philip Virgil Astillo, Jeong-Nyeo Kim, Ilsun You. A Security Protocol for Route Optimization in DMM-Based Smart Home IoT Networks. Special Section On Security And Privacy In Emerging Decentralized Communication Environments, 7, 142531–142550. DOI: 10.1109/ACCESS.2019.2943929

27 Jhonattan J. Barriga, Sang Guun Yoo. Securing End-Node to Gateway Communication in LoRaWAN With a Lightweight Security Protocol. *IEEE Access*, 2022. DOI: 10.1109/ACCESS.2022.3204005

28 T. Huynh, Y. He, A. Willis, S. Rüger. 2016. Adverse Drug Reaction Classification With Deep Neural Networks. *Proceedings of COLING 2016, the 26th International Conference on Computational Linguistics: Technical Papers*, Osaka, Japan, pp. 877–887.

29 Y. Liu, U. Aickelin. 2012. Detect Adverse Drug Reactions for Drug Pioglitazone. *2012 IEEE 11th International Conference on Signal Processing.*

30 L. Yang, G. Li, Z. Wu et al. 2021. Robust Truncated L2-Norm Twin Support Vector Machine. *International Journal of Machine Learning and Cybernetics*, 12, 3415–3436.

31 Olutosin Taiwo, Absalom E. Ezugwu, Olaide N. Oyelade, Mubarak S. Almutairi. Enhanced Intelligent Smart Home Control and Security System Based on Deep Learning Model. *Hindawi Wireless Communications and Mobile Computing*, 2022, Article ID 9307961. DOI: 10.1155/2022/9307961

32 Daemin Shin, Keon Yun, Jiyoon Kim 1, Philip Virgil Astillo, Jeong-Nyeo Kim, Ilsun You. A Security Protocol for Route Optimization in DMM-Based Smart Home IoT Networks. Special Section On Security And Privacy In Emerging Decentralized Communication Environments. DOI: 10.1109/ACCESS.2019.2943929

33 Xin Shen, Lingfeng Niu, Zhiquan Q, Yingjie Tian. March 2017. Support Vector Machine Classifier with Truncated Pinball Loss. *Pattern Recognition*, 68. DOI: 10.1016/j.patcog 2017.03.011

34 Arash Heidari, Nima Jafari Navimipour, Mehmet Unal. 2022. Applications of ML/DL in the Management of Smart Cities and Societies Based on New Trends in Information Technologies: A Systematic Literature Review. *Sustainable Cities and Society*, 85, 104089.

35 K. Jothimani, N. Gowsalya, S. Logeshwaran, E. Monika, M. Vinothini et al. 2022. A Deep Learning Approach For LSTM Based Covid-19 Forecasting System. *Journal of Science Technology and Research (JSTAR)*, 3(1), 568–583.
36 K. Jothimani, S. Thangamani, K. Ushmansherif, P. Manojkumar, S. Gowrishankar. 2022. Artificial Intelligent Based Computational Model For Detecting Chronic-Kidney Disease. *Journal of Science Technology and Research (JSTAR)*, 3(1), 723–735.
37 Yawei Yue, Shancang Li, Phil Legg, Fuzhong Li. 2021. Deep Learning-Based Security Behaviour Analysis in IoT Environments: A Survey. *Hindawi Security and Communication Networks*, 2021, Article ID 8873195, 13 pages,. DOI: 10.1155/2021/8873195

Chapter 15

Blockchain-energized smart healthcare monitoring system

P. Vanitha, G.K. Kamalam, and S. Subashini
Kongu Engineering College, Erode, India

15.1 INTRODUCTION

Blockchain technology (BCT) was introduced in the year 2008 by Satoshi Nakamoto [5] and defined through various dimensions like a term for data structure, a suite of technologies, an algorithm, and peer to peer systems. Blockchain, a technology that was created from bitcoin and has since revolutionised every sector, is one of the groundbreaking innovations of our time. Various businesses and researchers have attempted to integrate the technology into a variety of sectors, including marketing, banking, human resources, operations, and supply chain [1–3]. By offering a secure and distributed infrastructure, BCT provides a wide range of solutions to address various issues in healthcare.

Blockchain is a decentralised, open-source ledger of data that keeps track of activities on numerous computers in a way that prevents any record from being changed retroactively without also changing any subsequent blocks. Blockchain is a long series of verified blocks connected to one another. Blockchain offers a lot of accountability because every transaction is recorded and verified in public. No one has the ability to change any of the data that has been put into the blockchain. It acts as proof that the information is accurate and unaltered. Blockchain improves reliability and demonstrates its vulnerability to hacking by maintaining data on networks rather than a central database. Blockchain provides an excellent platform for development and competition with conventional businesses.

Blockchain was initially created as a money substitute, a safe form of currency that can be used as a medium of trade everywhere. Without the involvement of a third-party vendor, online payments could be sent straight from one party to another through a peer-to-peer network [6]. The network timestamps would give an additional layer of security by preventing any data eavesdropping on the active transaction. Blockchain technology is far superior to other technologies because of this characteristic and various benefits, which increases its application in diverse areas. It is able to record the transactions that occur in the network over a period of time. This property increases the usage of BCT for data transfer, currency exchange, or any valid information.

The most recent advancements in blockchain technology are focused on non-financial uses of the technology. As a result, research is ongoing to apply

DOI: 10.1201/9781003474524-15

this technology to other sectors, including human resources, authentication and identification, logistics management, and so on [7, 8]. Healthcare is one of the most popular fields where blockchain technology is being applied and where numerous applications and prototypes have recently been built. Maintaining the records of patients and dealing with postponed and fragmentary data are difficult for healthcare professionals. A blockchain-based architecture is also provided to collect insightful data using embedded predictive approaches [9]. Several healthcare stakeholders may be united under one roof with the use of blockchain technology for better information sharing.

The most apparent benefit of BCT is the fact that it allows two individuals to trade with one another in a distributed environment, eliminating the need for an intermediary. The removal of the third party increases transaction efficiency while lowering transaction costs. In essence, blockchain derives the majority of its characteristics from cryptography. Every user on the network is classified as a node and is equipped with a set of keys that are both private and public. The authentication is achieved through private key, while the public key serves as the participant's public address. The public keys of the sender, receiver, and transaction message are all included when a transaction is made. It is then signed by cryptography using the private key and the same is transmitted over the blockchain network.

The lack of secure links connecting all the independent healthcare systems together to form an end-to-end system ensuring data protection and privacy is a significant issue with today's healthcare systems [14]. There is a maintenance fee associated with maintaining data standards across platforms for data exchange. Therefore, a more open approach that respects privacy and security would be better for the system. Implementing blockchain, which enables transactions via decentralisation, provides a more adequate answer to these issues. The healthcare industry's complex system makes it more difficult to adopt blockchain technology.

Numerous blockchain implementations are accessible for different healthcare functionalities. The main advantage of blockchain is that it makes it possible for healthcare users to share data effectively while maintaining anonymity and security [15]. The distinct blockchain uses for various healthcare system features would complicate the current system, though. In order to create a smart healthcare system that is specifically suited according to the concept of smart cities, this chapter suggests an integrated framework of the healthcare system. Any difficulties encountered during the application of blockchain technology will be mitigated by other technologies like Internet of Things (IoT), analytics, and artificial intelligence (AI).

15.2 HEALTHCARE SYSTEM

The healthcare business is made up of a number of sectors that are focused on offering health-related services and goods. According to the

United Nations International Standard Industry Classification, it is a sector of the economy made up of a hospital, healthcare, and dentistry operations managed by various primary healthcare professionals and other healthcare workers. The primary focus of healthcare involves maintaining and improving health through the avoidance, diagnosis, and treatment of diseases, illnesses, injuries, and mental issues. The public's access to healthcare may differ between nations and societies, and it is primarily determined by economic and social circumstances as well as by governmental policies [16–18]. However, the medical requirements of the general public were addressed through various healthcare organisations. Despite having the same basic objective, there are three distinct categories of healthcare professionals.

Primary healthcare:
This is provided by those specialists who serve as patients' initial points of interaction and sources of consultation. This group offers the most options for people from all age groups, socioeconomic groups, and chronic illness types. As a result, it is anticipated of the main provider of healthcare to possess extensive knowledge in numerous fields. Primary healthcare frequently has an impact on the neighborhood.
Secondary healthcare:
This area involves intensive treatment for a severe but transient illness. It could be used as a synonym for a medical emergency room. Patients may occasionally be required to see a primary care physician for a recommendation before seeking secondary treatment, depending on the organisation's healthcare policies.
Tertiary healthcare:
This is a particular field that is typically consulted after receiving a recommendation from either primary care or secondary care. Those in this healthcare sector need more attention and time to recover from the illness, and patients are transferred only when it is anticipated that they will have chronic conditions.

Due to the participation and contributions of the various professional groups working in the healthcare industry, it is treated as functional where the experts are termed as stakeholders. They are the organisations that are either fully or partially active in the sector and have a significant impact on how the system works. Patients, doctors, drug companies, insurance firms, and governments are the company's main stakeholders. The relationships between the many industrial stakeholders are, nevertheless, very complicated. A few of them are privately owned, while others are run by individual employees. As a result, the laws and policies are different for each relationship between the stakeholders. The relationship between various stakeholders is given in Figure 15.1.

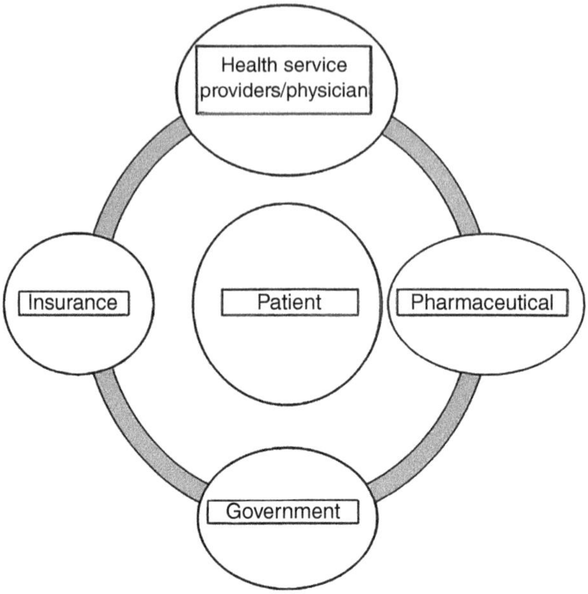

Figure 15.1 Relationship among modules.

15.2.1 Insurance industry

Health insurance is a type of coverage that pays for all or a portion of medical expenses incurred by an individual or group of individuals. Depending on the number of people who will be covered under the plan, the insurance firm creates a standard financial structure. A central body, such as a government, a business, or a nongovernmental group, administers such benefits. Since medical treatment has become more expensive and chronic illnesses are becoming more common nowadays, it is now essential to carry insurance for everyone.

15.2.2 Pharmaceutical industry

As patients rely on the medications they produce, pharmaceutical businesses are essential to the healthcare sector. The sector finds, creates, manufactures, and sells the pharmaceuticals used to treat patients. The businesses must abide by the rules and regulations set forth by public or private authorities. The expenses incurred due to marketing and supply chain, corporate investments in clinical research, and many other factors all contribute to the increase in price. Before they reach the end user, drugs pass through the entire supply chain and work their way through a number of vendors. The established blockchain system can really monitor and trace the medical products and also track the drug's origin.

15.2.3 Doctors

To make sure that patients get the right care, doctors are essential. Only doctors are capable of making judgements regarding patients due to their clinical complexities and inconsistencies. They can do this by using their clinical expertise and knowledge to grasp the patient's circumstance. The responsibility of the doctors is transforming in healthcare due to medical science progress, experience-based mistakes, and diagnosis according to symptom-based illness. The adoption of BCT in healthcare sector has made doctors' jobs considerably easier. Owing to the rapidly growing population, doctors wouldn't have sufficient time to review each patient's history and provide treatment in accordance. With the use of blockchain, all medical records are made accessible on a unified platform, which enables doctors to interpret it in minimal time.

15.2.4 Patients

The patient is one of the most significant stakeholders and the final consumer of healthcare services. Unfortunately, the focus of healthcare institutions and policymakers has switched from patient-driven to profit-driven. Although patients have more knowledge than ever before, they are nonetheless given very little or no opportunity to participate in the formulation of policies and decision-making processes. The patients will gain through BCT in a broader sense with the ability to access their health records from anywhere [12]. This framework not only enables patients to gain access to their data, but also to secure their data and control the accessibility of their data.

15.2.5 Government

Over time, the government's responsibility in the healthcare sector has grown, and it now significantly affects the political discussion and balance of power in the sector [19]. The government improves the efficiency of the healthcare sector by adopting various preventative interventions for chronic diseases and is also involved in cost savings. It also takes part in creating budgets and carrying out other planning tasks connected to expenditures in the healthcare sector [20]. Nonetheless, it has a lot of space to impact the marketplace and change policies in the healthcare industry to get better results.

Since its inception, healthcare has evolved and becomes more complex over years [21]. Growing demand for healthcare services is being driven by the ageing population and the prevalence of chronic diseases. This system is particularly vulnerable to mistakes and frauds that endanger the lives of patients because of the industry's complicated communication between its various stakeholders. Industry should think about safeguarding data security and privacy in addition to patient care [11]. It is motivated by record-keeping, regulatory compliance, and ethical principles [22]. It is essential to manage the healthcare system effectively given the significance of this sector for both economic development and for individuals across the globe.

15.3 MAJOR ISSUES FACING HEALTHCARE AND CHALLENGES IN THE HEALTHCARE SYSTEM

Today's healthcare industry faces ongoing challenges in trying to develop ways to raise revenue, lower expenses, and enhance quality [23]. Despite the fact that a variety of solutions have been developed to address the issues facing the sector, privacy and security remain major obstacles. Regardless of numerous advancements, inefficiency still persists and can occasionally pose major risks to patients' lives [24]. Although the adoption of wireless sensor networks (WSNs) in the healthcare sector is becoming more prevalent [25], BCT–based applications will revolutionise the healthcare sector.

Healthcare businesses collect extremely sensitive patient data, making the sector a prime target for cybercrime [26]. As healthcare organisations respond to such problems proactively, this issue will persist. Data is more susceptible to hackers when it is maintained by centralised systems. Healthcare businesses should adopt a system that is more reliable and secure as a permanent solution in addition to combating cybercrime.

Electronic medical records and other heterogeneous information systems and software technologies can exchange data and communicate with one another if they are interoperable [27]. Effective care and safety for persons and communities depend on allowing the information to collaborate within and across organisational boundaries [28]. For instance, interoperability enables healthcare providers to securely communicate patient data with one another, regardless of their locations and the level of trust between the parties [29].

To deliver patients with effective collaborative treatment and care, secure data sharing is essential. Data sharing helps to increase the diagnostic precision [30] by gathering confirmations and recommendations from several experts. Current healthcare organisations require patients to gather and exchange their medical records with doctors via paper copies or electronic copies, notwithstanding the value of data exchange in the medical systems. Since there is currently no interoperability across health systems and applications, there is a lack of confidence among providers as a result of the inefficient data-sharing process [31, 32].

The present mechanisms for managing patient billing are intricate and vulnerable to manipulation by the service providers. Recently, 50% of healthcare billings have been fraudulent, which has resulted in excessive billing or charging patients for services that have not been rendered [35]. For instance, Medicare fraud cost the United States $30 million in 2016 [36]. It is anticipated that healthcare systems powered by blockchain would offer practical answers to such frauds and reduce frauds connected to medical billing.

15.4 CONCEPT AND APPLICATION AREAS OF BLOCK CHAIN TECHNOLOGY

It is essential to know the advantages and applications of blockchain in real life. Analysing the related papers or works in this field can help you fully comprehend this idea. The findings of related studies shows that not all of the ideas of healthcare applications of blockchain have turned to functional prototypes. Understanding how the real-time blockchain prototype works in relation to other existing blockchain use cases in the healthcare industry is therefore important. This helps in determining the possible research gaps and the direction of blockchain applications in healthcare as a guiding concept.

Electronic health records are the most well-liked healthcare areas and applications of blockchain. Patient information is particularly sensitive, both in terms of preserving the confidentiality of records for future use and in terms of sharing among stakeholders with the appropriate access rights. As discussed in the preceding sections, the transparency, trust, and traceability that are intrinsic to blockchain technology are achieved through the synthesis of encryption and hashing in time-stamped settings for transactions [6]. Applications of blockchain for sharing healthcare data offer excellent implementation guidance for legal frameworks which includes the General Data Protection Regulation for securely sharing healthcare data among stakeholders while upholding integrity of data.

This technology is regarded as the reliable platform that guarantees suitable and secured data transfers among stakeholders [11]. Liu [37] provided the framework on blockchain and big data, outlining many healthcare obstacles and opportunities. A novel use of blockchain technology called Medrec was introduced by the MIT Digital Media Lab for correctly managing health data sensitively. Dhillon and Smith [4] developed a value-focused thinking-based method for examining blockchain elements that supports upholding the integrity of health data.

Another crucial sector where blockchain solutions are sought as a significant breakthrough is the drug supply chain. A traceability system driven by blockchain can be a very excellent foundation to assure authentic drug products and prevent fraudulent pharmaceuticals from reaching the end user [7, 10, 24, 25]. More sophisticated prototypes have been built over IBM Hyperledger [7] with methods that uphold data quality control and the integrity for the medication chain.

Another use for blockchain in the healthcare industry is in biomedical research, particularly when it comes to clinical trials and ensuring the privacy of individual participants' data [13, 26, 30]. By the insights and data produced by the platforms of blockchain, healthcare training and inference is another aligned use. A prototype blockchain platform was built on Etherium [27] to show how the process of protecting patient data in clinical trials works.

Health insurance may benefit greatly from the auditability, traceability, and transparency characteristics of blockchain technology. Both the insurance firms and the insurers have a history of open and trusting settings to the processing of policy claims, proper accountability, and data sanctity [11, 34]. A prototype of several applications built on the Etherium platform has been created. MI Store, a biomedical insurance system, serves as an example of how blockchain is being used in the health sector.

15.5 CHALLENGES TO BLOCK CHAIN APPLICATIONS IN HEALTHCARE

We identified several issues of blockchain applications in healthcare based on the examination of research papers and practice-based advancements see Table 15.1 for more information. These difficulties may fall under the theoretical, conceptual, or application categories. We recognised the following as a few notable obstacles:

- Many apps have a bias toward the healthcare stakeholders they are aiming to serve [14].
- In order to effectively utilise the platform, the blockchain must be strengthened so that it can operate at the intersection of other technical breakthroughs like IoT, analytics, and so on [35, 37].
- Stakeholder participation and interactions should become more and more straightforward and thorough in order to include behavioral data that is as accurate to reality as feasible [33, 35].
- The majority of apps are still conceptual in nature, necessitating real-time testing and usage. This will enable us to evaluate the blockchain technology's actual practical use [36].
- Working toward the inter-operability and integration of various blockchain applications in healthcare and usage of an end-to-end architecture in blockchain is the most crucial and vital component.
- Examining the blockchain applications scalability in healthcare is another significant problem [26].
- Ensuring stakeholder participation on a blockchain platform for the healthcare sector is one of the biggest challenges, considering the involvement of young people and the elderly in blockchain applications [30].

Selected data preservation on the blockchain platform is suggested as a solution to some issues regarding blockchain applications in the healthcare. To improve the effectiveness and its scalability in data storage, it is stated that only the encrypted health data is retained on the platform and can be computationally decrypted when needed [12]. Additionally, permissioned blockchain infrastructure may be used to manage access to the blockchain platform, and an appropriate reversal mechanism must be built in order to undo fraudulent transactions.

Table 15.1 Applications development using blockchain on healthcare platform

Platform of blockchain	*Blockchain applications*
Etherium	Medical Chain [12], GHN, Medre, MIStor, Healthchain, Ancile, Medium.io

The advantages of blockchain technology have attracted many research- and practice-based groups, improving the negative effects in technology to create a good and more reliable version. The quick rate it is progressing suggests that these technological constraints will soon be lessened, and the healthcare sector is not an issue in those situations.

15.6 INTEGRATED BLOCKCHAIN-ENERGISED SMART HEALTHCARE SYSTEM FRAMEWORK

The blockchain technology's importance in the field of healthcare is known by examination of several research publications and practitioner documents. But the technology is still too new for widespread usage. It becomes extremely crucial and vital to examine the problems with the current scope of blockchain applications in healthcare, put them into context, and create a strategy to address them. A few notable issues are discovered when analysing the previously offered relevant research, as follows:

- *Feasibility issue*: Blockchain technology is used in various applications to address problems in the healthcare industry. It becomes crucial to examine their viability and the applicability of technology in certain sectors. For instance, despite what many conceptual models imply, blockchain's cognitive intelligence is still a distant hope.
- *Scalability analysis*: Scoping the adoption of blockchain technology in terms of scalability is another crucial problem. But blockchain is a suitable technology and scalability is not workable in terms of adoption, according to several applications. The field of healthcare data analytics is an example.
- *Technology convergence*: If blockchain integrates effectively with forthcoming technologies like ML, AI, IoT, cloud, and so forth, it can become incredibly valuable. To make this element mature, a lot of study is still required. Business intelligence and analytical techniques in the healthcare industry serve as an example.
- *Interoperability dimension*: The most difficult aspect of blockchain applications in healthcare systems is their lack of interoperability because they are often built on separate platforms and operate in silos. A new method is required to make block chains compatible. Examples are Drug Chain on Hyperledger and MedRec on Etherium.

- *Stakeholder unbiased*: Although most apps are centred on the interests of a single stakeholder, they should contain smart contracts to enable impartial conduct toward all stakeholders.
- *Technology adoption*: In order to increase usage and awareness, technology adoption needs a strong push. Healthcare stakeholders should benefit from blockchain.
- *Cognitive capability integration*: The applications are created in such a way that they provide mature use and behavioral intelligence starts driving them, incorporating stakeholder perspectives to understand healthcare ideas by effective deep learning concepts.
- *Real time applications*: From the period of ideation to implementation, there is a need to move swiftly and effectively. Prototypes should be used to evaluate conceptual models, and after enough testing, prototypes should be transformed into real-time applications.

After learning about some significant problems with the application of blockchain technology in healthcare, it is crucial to consider potential solutions. The area of potential resolution revolves around creating and putting into use an integrated framework that may assist various healthcare stakeholders in integrating blockchain applications. Designing this framework should take into account the compatibility of multiple blockchain apps in their contracts and models allowing the transfer of data, rules, and communication in a certain fashion. To develop and set the stage for potential future applications, we provide a complete blockchain platform for the healthcare industry.

The following blockchain application prototypes make up the majority of the proposed framework:

- Electronic records on health
- Monitoring remote patient
- The drug supply chain
- Biomedical research
- Health insurance
- Healthcare analytics

This framework aims to combine the applications around the following factors:

- Interoperability
- Functionality in real time
- Cognitive capability

These are used as individual layers of various applications of blockchain, enabling a connected ecosystem of applications in the healthcare industry. It becomes crucial to plan how these layers will operate in terms of the various smart contracts that will make it easier for various levels to be executed. By using the principle of many layers, any underlying logic in a blockchain context may be implemented. Therefore, these layers should be designed in a way that establishes a unique contract for deciding the layer should be invoked. The types of smart contracts required to implement this architecture may be classified into the following groups:

- Layer Functionality Smart Contracts (LFSC)
- Layer Coordination/Cognitive Intelligence Smart Contracts (LCSC)
- Layer Interoperability Smart Contracts (LISC)

According to the needs, any additional significant feature of smart contracts that may be considered an essential part of the integration process can be included. A framework of blockchain applications in healthcare is presented in Figure 15.2 for a synergistic and effective use of blockchain technology.

The aforementioned integrated framework addressed issues with real-time functioning, cognitive intelligence use in blockchain in healthcare, and its interoperability. Let's quickly review the functions of each form of smart contract that makes this integrated framework for blockchain applications in the healthcare industry functional:

- *LFSC*
  LFSCs are fundamentally designed to make it easier for each layer to function. The interoperability, or cognitive intelligence layer, undergoes a check of applicable restrictions on the layer construction in aiding every layer's successful operation in its designated field.
- *LCSC*
  LCSC handles information access and the proper use of cognitive intelligence tools like AI, ML, and nNatural language processing, among

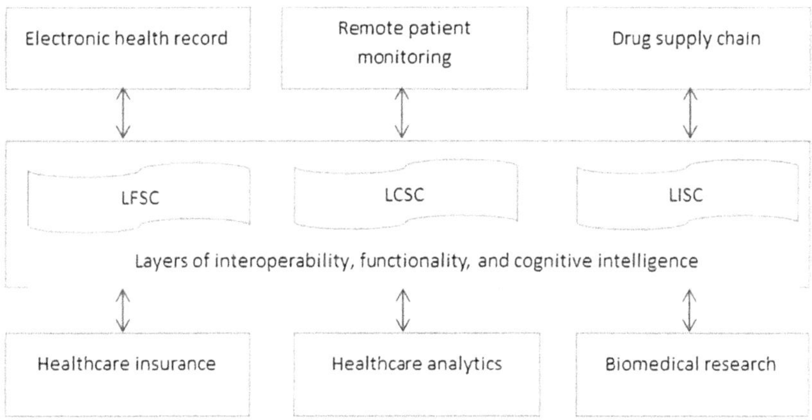

Figure 15.2 A framework for smart healthcare enabled by blockchain integration.

others. This increases the robustness of blockchain applications in data sanctity and its analytical findings that serve as parameters for healthcare.

- *LISC*
  One of the most notable and difficult situations in the current paradigm of blockchain applications is handled by LISC. When blockchain apps are properly connected with one another, this creates the foundation for a smart healthcare system that is synergistic and integrated and has end-to-end compliance.

Scalability, prototype development, raising technology awareness, and developing technological usefulness are some additional crucial blockchain application ideas needed for both technical and non-technical components of research and development. With more focus on blockchain technology, its advancements particularly in creating applications from an interoperability concept will have a significant influence on the acceptance and use of this technology. Applications created on several platforms like Coda, Hyperledger, Etherium, and others are to the point where they will be merged and have widespread value. By employing the smart contract characteristics that enable the majority of the logic in blockchain technology, the suggested framework can serve as a guiding principle in this area.

15.7 CONCLUSION AND INFERENCE

Since its creation, blockchain technology has changed over time. Blockchain has evolved from being the underlying technology for the ground-breaking cryptocurrency bitcoin to being a facilitator of mechanisms like trust, transparency, traceability, authentication, auditability, and more, enabling applications in several commercial and societal sectors. The result of this technology is given in the context under healthcare in this chapter. Understanding the current state of healthcare systems, as well as the problems and difficulties they confront, served as the foundation for our debate. We introduced the idea of blockchain before highlighting its most important uses in the healthcare industry. We also discussed the obstacles to these blockchain applications in healthcare and how to overcome them. We also offered a concept for a blockchain system that is integrated and interoperable from end to end.

15.8 FUTURE RESEARCH DIRECTIONS

The concept of blockchain is a dynamic and rapid technology, with applications on healthcare in the development phase see Table 15.2 for more information. As we've already stated, researchers must act quickly to transition from the ideation phase to the execution or implementation phase if they are

Table 15.2 Recent research and its analysis implications

Content of discussion	*Research and analysis implications*
Analysing the viability	• Transition from concept generation to creation of prototype.
Analysis of scalability	• Blockchain doesn't provide a solution for all problems.
Advancing technologies	• Assess the necessity for the adoption of blockchain technology from the feasibility point. • Develop and test models for scaling up blockchain technology. • When data is kept in blockchain and can be acquired, saved, maintained, and analysed with the integration of some other technologies, blockchain can be utilised most effectively.
The interoperability factor	• Create and evaluate models for advancing blockchain technology.
Stakeholder unbiased	• Blockchain may be used most effectively when data is kept there and can be collected, accessed, maintained, and analysed through the use of various technologies.
Acceptance of technology	• Healthcare systems whose data that are closely sensitive can use these features more aggressively.
Cognitive intelligence integration	• The next major thing for blockchain technology to aim towards is the intersection using AI, ML (machine learning), IoT, and cloud. It is important to consider integrating these silo blockchain and making them interoperable because there are many applications for healthcare systems that have been created on various platforms.
Real-time software	• The majority of apps are patient focused, but as data is a valuable resource for all parties involved in healthcare, there is a way to gain the data without requiring permission in both permissioned and non-permissioned scenarios. • Stakeholders should be informed about the advantages of using technology and encouraged to do so.
	When compared to the rate of technological evolution, blockchain adoption is still moving along at a very slow pace.

to make blockchain applications for healthcare systems successful. It is quite challenging to estimate the actual usage of blockchain apps with no physical models and end-to-end operation is not yet visualised. Blockchain's catering interoperability issue is another crucial and significant problem. The likelihood that blockchain will support the healthcare system will significantly increase if these problems are adequately addressed. Many obstacles to the current structure of healthcare systems would be reduced with the appropriate use of blockchain technology, leading to a more powerful and stable healthcare system.

REFERENCES

[1] S. Olnes, J. Ubacht, M. Janssen, Blockchain in government: Benefits and implications of distributed ledger technology for information sharing, Elsevier, 2017.

[2] R. Beck, J. Stenum Czepluch, N. Lollike, S. Malone, Blockchain—The gateway to trust-free cryptographic transactions. *ECIS 2016 Proc.*, 2016, pp. 1–15.

[3] Statistica, Global Market for Blockchain Technology 2018 2023|Statistic [WWW Document], 2019. Available from: https://www.statista.com/statistics/647231/worldwide-blockchain-technology-market-size

[4] K. Smith, G. Dhillon, Blockchain for digital crime prevention: The case of health informatics, *Inf. Syst. Secur. Priv.* 1 (2017) 1 10.H.D. Zubaydi, Y.-W. Chong, K. Ko, S.M. Hanshi, S. Karuppayah, A review on the role of blockchain technology in the healthcare domain, *Electronics* 8 (2019) 679.

[5] S. Nakamoto, Bitcoin: A peer-to-peer electronic cash system. (2008). Available from: http://bitcoin.org/bit-coin.pdf

[6] E. Androulaki, A. Barger, V. Bortnikov, C. Cachin, K. Christidis, A. De Caro, et al., Hyperledger fabric: A distributed operating system for permissioned blockchains. *Proceedings of the Thirteenth Euro Sys Conference.* ACM, 2018, p. 30.

[7] C. Wood, B. Winton, K. Carter, S. Benkert, L. Dodd, J. Bradley, How blockchain technology can enhance EHR operability, *ARK. Invest. Res.* 1 (2016) 1–13.

[8] T.-T. Kuo, L. Ohno-Machado, Modelchain: Decentralized privacy-preserving healthcare predictive modeling framework on private blockchain networks, ArXiv Prepr 1 (2018) 1–13.

[9] J.-H. Tseng, Y.-C. Liao, B. Chong, S. Liao, Governance on the drug supply chain via gcoin blockchain, *Int. J. Environ. Res. Public. Health* 15 (2018) 1055.

[10] M.A. Engelhardt, Hitching healthcare to the chain: An introduction to blockchain technology in the healthcare sector, *Technol. Innov. Manag. Rev.* 7 (2017), pp. 22–34.

[11] S. Jiang, J. Cao, H. Wu, Y. Yang, M. Ma, J. He, Blochie: A blockchain-based platform for healthcare information exchange. *2018 IEEE International Conference on Smart Computing (SMARTCOMP)*, IEEE, 2018, pp. 49–56.

[12] A. Dubovitskaya, Z. Xu, S. Ryu, M. Schumacher, F. Wang, Secure and trustable electronic medical records sharing using blockchain, AMIA. *Annual Symposium Proceedings*, 2017 (2018) 650–659.

[13] I.C. Lin, T.C. Liao, A survey of blockchain security issues and challenges, *J. Netw. Secur.* 19 (2017) 653–659. https://doi.org/10.6633/IJNS.201709.19(5).01

[14] D. Mingxiao, M. Xiaofeng, Z. Zhe, W. Xiangwei, C. Qijun, A review on consensus algorithm of blockchain. *2017 IEEE International Conference on Systems, Man, and Cybernetics (SMC)*. Presented at the 2017 IEEE International Conference on Systems, Man, and Cybernetics (SMC), 2017, pp. 2567–2572. https://doi.org/10.1109/SMC.2017.8123011

[15] J. Park, J. Park, Blockchain security in cloud computing: Use cases, challenges, and solutions, *Symmetry* 9 (2017) 164.

[16] M. Vukolić, Rethinking permissioned blockchains. *Proceedings of the ACM Workshop on Blockchain, Cryptocurrencies and Contracts*, ACM, 2017, pp. 3–7.

[17] Z. Zheng, S. Xie, H. Dai, X. Chen, H. Wang, An overview of blockchain technology: Architecture, con- sensus, and future trends. Presented at the *2017 IEEE International Congress on Big Data (BigData Congress)*, 2017, pp. 557–564. https://doi.org/10.1109/BigDataCongress.2017.85

[18] Price Water House Coopers, Blockchain is here. What's your next move? [WWW Document]. PwC, 2019. Available from: https://www.pwc.com/gx/en/issues/blockchain/blockchain-in-business.html
[19] R. Zambrano, A. Young, S. Verhulst, Connecting Refugees to Aid Through Blockchain-Enabled ID Management: World Food Programme's Building Blocks, GOVLAB, 2018.
[20] D. Galvin, IBM and Walmart: Blockchain for food safety, PowerPoint Present, 2017.
[21] M. Mettler, Blockchain technology in healthcare: The revolution starts here. *2016 IEEE 18th International Conference on E-Health Networking, Applications and Services (Healthcom)*, IEEE, 2016, pp. 1–3.
[22] K. Peterson, R. Deeduvanu, P. Kanjamala, K. Boles, A blockchain-based approach to health information exchange networks. *Proc. NIST Workshop Blockchain Healthcare*, 2016, pp. 1–10.
[23] T.K. Mackey, G. Nayyar, A review of existing and emerging digital technologies to combat the global trade in fake medicines, *Expert Opin. Drug. Saf.* 16 (2017) 587–602.
[24] T. Bocek, B.B. Rodrigues, T. Strasser, B. Stiller, Blockchains everywhere-a use-case of blockchains in the pharma supply-chain. *2017 IFIP/IEEE Symposium on Integrated Network and Service Management (IM)*, IEEE, 2017, pp. 772–777.
[25] M.N.K. Boulos, J.T. Wilson, K.A. Clauson, Geospatial blockchain: Promises, challenges, and scenarios in health and healthcare, *BioMed. Cent.* 17 (2018), pp. 1–10.
[26] T. Nugent, D. Upton, M. Cimpoesu, Improving data transparency in clinical trials using blockchain smart contracts, *F1000Research* 5 (2016), pp. 1–7.
[27] P. Mytis-Gkometh, G. Drosatos, P.S. Efraimidis, E. Kaldoudi, Notarization of knowledge retrieval from biomedical repositories using blockchain technology. *Precision Medicine Powered by PHealth and Connected Health*, Springer, 2018, pp. 69–73.
[28] Z. Shae, J.J. Tsai, On the design of a blockchain platform for clinical trial and precision medicine. *2017 IEEE 37th International Conference on Distributed Computing Systems (ICDCS)*, IEEE, 2017, 1980, pp. 1972.
[29] I. Radanović, R. Likić, Opportunities for use of blockchain technology in medicine, *Appl. Health Econ. Health Policy*. 16 (2018) 583–590.
[30] D. Ichikawa, M. Kashiyama, T. Ueno, Tamper-resistant mobile health using blockchain technology, *JMIR MHealth UHealth* 5 (2017) e111.
[31] J. Zhang, N. Xue, X. Huang, A secure system for pervasive social network-based healthcare. *IEEE Access* 4 (2016) 9239–9250.
[32] M. Weiss, A. Botha, M. Herselman, G. Loots, Blockchain as an enabler for public mHealth solutions in South Africa. *2017 IST-Africa Week Conference (IST-Africa)*, IEEE, 2017, pp. 1–8.
[33] V. Gatteschi, F. Lamberti, C. Demartini, C. Pranteda, V. Santamaría, Blockchain and smart contracts for insurance: Is the technology mature enough? *Future Internet* 10 (2018) 20.
[34] P. Mamoshina, L. Ojomoko, Y. Yanovich, A. Ostrovski, A. Botezatu, P. Prikhodko, et al., Converging blockchain and next-generation artificial intelligence technologies to decentralize and accelerate biomed- ical research and healthcare, *Oncotarget* 9 (2018) 5665.

[35] M. Saxena, M. Sanchez, R. Knuszka, Method for Providing Healthcare-Related, Blockchain-Associated Cognitive Insights Using Blockchains, 2018a.
[36] M. Saxena, M. Sanchez, R. Knuszka, Providing healthcare-related, Blockchain-Associated Cognitive Insights Using Blockchains, 2018b.
[37] P.T.S. Liu, Medical record system using blockchain, big data and tokenization. *International Conference on Information and Communications Security*, Springer, 2016, pp. 254–261.

Index

Pages in **bold** refer to tables.